80

Agassant/Avenas/Sergent/Carreau

Polymer Processing

J.-F. Agassant
P. Avenas
J.-Ph. Sergent
P. J. Carreau

Polymer Processing

Principles and Modeling

Hanser Publishers, Munich Vienna New York

Distributed in the United States of America and Canada
by Oxford University Press, New York

05145387

CHEMISTRY

Distributed in USA and in Canada by
Oxford University Press
200 Madison Avenue, New York, N.Y. 10016

Distributed in all other countries by
Carl Hanser Verlag
Kolbergerstrasse 22
D-8000 München 80

The use of general descriptive names, trademarks, etc. in this publication, even if the former are not especially identified, is not to be taken as a sign that such names, as understood by the Trade Marks and Merchandise Marks Act, may accordingly be used freely by anyone.

While the advice and information in this book are believed to be true and accurate at the date of going to press, neither the authors nor the editors nor the publisher can accept any legal responsibility for any errors or omissions that may be made. The publisher makes no warranty, express or implied, with respect to the material contained herein.

Library of Congress Cataloging-in-Publication Data

Polymer processing : principles & modeling / J.-F. Agassant . . . [et al.].
     p. cm.
   Includes bibliographical references and index.
   ISBN 0-19-520864-1
   1. Polymers.   I. Agassant, J.-F.
  TP1087.P65 1991
  668.9 - - dc20   90-85975
                                   CIP

CIP-Titelaufnahme der Deutschen Bibliothek

Polymer processing : principles and modeling / J.-F. Agassant
. . . -Munich ; Vienna ; New York : Hanser ; New York:
Oxford Univ. Pr., 1991
 Einheitssacht.: La mise en forme des matières plastiques <engl.>
 ISBN 3-446-14584-2
NE: Agassant, Jean-François; EST

ISBN 3-446-14584-2 Carl Hanser Verlag, Munich, Vienna, New York
ISBN 0-19-520864-1 Oxford University Press

Preface

In this modern age man uses increasingly high molecular weight polymers or plastics for his clothing, shoes and boots, furniture, etc. The processing of metals has been investigated by scientists for many decades; in contrast, the processing of polymers has not been studied as closely. The book of AVENAS, AGASSANT and SERGENT intends to cover the existing gap.

This book is addressed to engineers and other industrialists involved in polymer processing as well as researchers working in the field of rheology and continuum mechanics. Its objective is to foster exchanges between the two groups and hence make a significant contribution to the understanding of polymer processing.

The authors first review basic principles of continuum mechanics and heat transfer. These notions are essential for the analysis of the flow of viscous and viscoelastic materials. To facilitate reading the text, the more difficult notions such as molecular theories related to viscosity and viscoelasticity have been placed in appendices. These remain, however, of particular interest.

Following the review chapters, the modelling of the screw extrusion is extensively covered with a multiple facet study of the screw-barrel system. If conveying solids using an Archimede's screw appears to be a simple problem, the notion is deceptive! This chapter brings in an original contribution to the development of a model which is mathematically correct.

The following chapter discusses the problem of pressure build-up for molten polymers using the analogous concepts of hydrodynamic lubrication. The analogy of the flow of a lubricant in a bearing and that of a polymer in the gap of a calendering unit is striking. The analysis of the coupled equations for the flow and the heat transfer is original and practical results are presented.

The models developed to simulate the stretching and biaxial drawing of polymeric liquids are original contributions. In the case of blown film, the authors have compared the predictions with the data obtained from an industrial line.

In these chapters, the molten polymer is taken to be a non-Newtonian viscous, but inelastic material in simple shear. In fact, it has viscoelastic behaviour, evidenced by various phenomena such as the swelling of extrudates at die exit, Weissenberg effects, extrusion instabilities. The last chapter of the book deals with viscoelasticity of liquids. The viscoelastic models are not yet capable of predicting quantitative effects of practical importance. However, the authors have succeeded in presenting a simple treatment of the basic constitutive equations for viscoelastic liquids. The more complex mathematical developments have been relegated to the appendix.

This book is at the crossroad of research and industrial practice. Engineers and technicians involved in plastics processing will find here the basic notions of continuum mechanics and heat transfer, and simple methods to solve flow problems associated with processing. Researchers in the field of rheology and continuum mechanics will find in this book an original presentation of constitutive equations, themes for research projects and applications for their theories.

Designed as a text for teaching, this book contains numerous problems that require a good grasp of the topics. This is therefore a very useful tool for teachers and students in the field of mechanics and engineering materials, as well as a basic reference for process engineers.

JEAN MANDEL   June 1982
Professor of Mechanics
at École Polytechnique, France,
from 1952 to 1973

Acknowledgements

The Centre for Materials Processing (CEMEF) of École des Mines de Paris was founded in 1974. Since 1976 the Centre has been located at Sophia Antipolis, near Nice in France. It has been associated with the CNRS (Centre national de la recherche scientifique) since 1979. More than 100 individuals are employed by the Centre. That total includes researchers, graduate students, technicians and administrators. The CEMEF has two main objectives.

- The training of researchers in the field of engineering materials. Since 1979, more than sixty engineering doctorate and Ph.D. theses have been completed. Currently, the output of Ph.D. graduates is over ten per year. These research engineers generally are to be found today in the numerous industrial sectors which are served by the CEMEF.
- The conduct of research on industrial problems in the field of processing of metals and polymers. The Centre works under contract with most of the major French and several European and Nord American industrial Companies involved in materials processing.

The first three authors of this book are or have been associated with the CEMEF:

- J.-F. AGASSANT is a graduate of the École des Mines de Paris. He obtained his Ph.D. (thèse d'état) in 1980 on the calendering of polymers. In 1976 he became a group leader at the CEMEF on the thermo-mechanical aspects of polymer processing. He is now Professor at the École des Mines de Paris and Assistant Director of the CEMEF.
- P. AVENAS is a graduate of the École Polytechnique of France and the École des Mines de Paris. He was involved in the foundation of the CEMEF in 1972 to 1974, became director of it in 1975 and left in 1979 to work at the Ministry of Industry. He is currently with ATOCHEM (France).
- J.-PH. SERGENT graduated from the École des Mines de Paris. He was a research associate at the CEMEF from 1974 to 1976 and obtained his Ph.D. in 1977 on film processing. In 1979 he joined ETEX (France).

The fourth author, P. J. CARREAU, was responsible for the translation and adaptation of the original French text. He graduated in Chemical Engineering from the Ecole Polytechnique of Montreal. He obtained his Ph.D. from the University of Wisconsin in 1968 in the field of polymer rheology. He is Professor of Chemical Engineering at the École Polytechnique of Montreal and Director of the Centre on Applied Polymer Research, CRASP. This centre consisting of eleven professors, four research associates, five research engineers and some fifty graduate students conducts multidisciplinary activities in the polymer field.

Both the CEMEF and the CRASP have been associated for many years. Recently, under the France-Quebec collaboration program, quite a few joint research projects have been initiated. This book is obviously one of the major outcomes of this collaboration.

This book is a translation and adaptation of the second edition of the French text, "La mise en forme des matières plastiques, approche thermomécanique", Lavoisier, Paris (1986). In this English version, some corrections and extensions have been made, but the most significant change is in the presentation, now more in line with American and English textbooks. Whenever possible, widely accepted notation in practice is used along with S.I. units. All the references have been put together at the end of the book with the authors placed in alphabetical order for easy reference.

This book bridges a gap between research monographs and textbooks presently available. Obviously, it is a compromise between completeness and simplicity. It is not exhaustive and several topics of polymer processing have not been covered. This text, however, is a good example of the engineering approach to problem solving. The complex problems associated with polymer processing are first simplified and order-of-magnitude calculations are presented. When appropriate, analogies with classical problems of fluid mechanics or heat transfer are discussed. The extensions to more rigorous and more complex solutions are briefly discussed in light of current research activities on those topics. Overall, the reader will gain considerable benefits from the simplified analysis of complex problems. He will have a better understanding of the phenomena encountered in polymer processing, a grasp for shear-thinning effects and some understanding of viscoelastic effects.

Many colleagues and former staff members of the CEMEF have made significant contributions to this book by their useful suggestions. We note particularly: Y. DEMAY, J. M. HAUDIN, B. VERGNES, M. VINCENT, J. P. VILLEMAIRE. Also, many segments of the book are based on research theses done at the CEMEF by: H. ALLES, P. BEAUFILS, M. COEVOET, D. COTTO, M. ESPY, H. MADERS, B. NEYRET, S. PHILIPON, A. PHILIPPE, A. PIANA, P. SAILLARD, G. SORNBERGER, E. WEY. Many colleagues at the ÉcolePolytechnique of Montreal have contributed to the translation and adaptation of the book. We note specifically: S. GAGNON, M. GRMELA, P. G. LAFLEUR and H. P. SCHREIBER. We sincerely thank all of them, as well as the others, in university or industry, who have been working in collaboration with us on rheological and polymer processing problems.

Finally, we are grateful to the late Professor J. MANDEL who wrote the preface of the French text, a translation of which introduces the reader to this book.

Spring 1991

J.-F. AGASSANT
P. AVENAS
J.-PH. SERGENT
P. J. CARREAU

Contents

Introduction

This book is written to help engineers, scientists, and students to understand the basic principles and the major phenomena encountered in polymer flow and processing. The problems found in the plastics industry are largely related to the design, operation and control of equipment in order to obtain satisfactory products at a competitive price. The requirements for the products are:

- the finished products must meet the design specifications with respect to surface quality and overall dimensions in spite of many possible alterations due to swelling, stretching, shrinkage, ...;
- they must have satisfactory properties. This implies in part either a specific molecular orientation or a minimum of residual stresses or a specific crystalline structure at the level of the crystallites and spherolites.

A competitive price means that:

- the process attains rapidly steady-state conditions with a minimum of wastes or rejects;
- a maximum output in the extrusion or calendering process, or a minimum cycle time in injection or blow molding.

Considering these different objectives, there is in all cases an optimum in the design parameters of machines and in the operating conditions: temperature, speed, cooling conditions, etc. When considering polymer production (with or without additives), the challenge is to adapt a product to a given polymer processing operation. This can be done by:

- altering the molecular structure (molecular mass and distribution) or using copolymers, branched, grafted polymers, etc.;
- changing the formulation by using internal or external lubricants, fillers, etc.

Globally, one has to master the relations between the formulation, the rheological behaviour and the transformation properties. All these subjects are parts of the mechanics, thermodynamics and rheology which is the science of flow and deformation. These fields are quite vast and only the notions related to polymer processing will be presented in this text, i.e. basic equations and models directly applicable to the processing of thermoplastics which are transformed essentially through thermo-mechanical processes. When dealing with thermoset plastics, chemical phenomena and in particular interactions between chemistry and rheology must be taken into account.

Processing stages

The processing of thermoplastics may involve three distinct thermo-mechanical stages:

- The plastication stage in which the polymer goes from a solid state (powder, granules, ...) to a homogeneous liquid state. The term plastication covers many different physical phenomena: one can speak of fusion and molten state for crystalline polymers; when dealing with amorphous polymers, one should refer to plastication recalling that these polymers show a smooth transition from the glassy to rubbery state as the temperature is increased to values above the glass transition temperature.

In order to simplify the nomenclature in the text, the terms plastication and molten state will be used for all polymers.

- A second stage, where a specified shape is imposed to the molten polymer by bringing it to flow under pressure into molds or dies.
- A third stage, where the final shape or conformation of the finish product is achieved with cooling and eventually drawing, biaxial stretching, blowing...

Logic for polymer process design

The logic used for the design of polymer transformation processes, the choice of the notions introduced in this text and the way they are presented come from the observation of the physical characteristics which are common to most polymers.

a) Low thermal diffusivity

The thermal diffusivity of most polymers is about $10^{-7}\,\mathrm{m^2/s}$ that is a thousand times less than the value for copper. Because of this low thermal diffusivity, it would require:

- 17 min. for any significant heating or cooling effects to attain a depth of 10 mm;
- 10 s for a depth of 1 mm;
- 0.1 s for a depth of 0.1 mm.

This shows that plastication of a polymer using pure conduction from walls of containers would lead to inadmissibly long residence times and low flow rates. This is why in plastication equipment, most designs are based on thin layers of polymer.

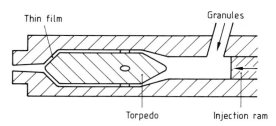

Figure 1   Torpedo plastication system

A simple idea, industrially used on certain low capacity injection machines, is the torpedo, as illustrated in Figure 1. A thin layer of polymer (about 1 mm) is pushed in the gap between the barrel and the torpedo, heated by electricity or by a circulating liquid. But to attain continuous and large plastication rates, for extrusion and injection processes, the screw-barrel system (plasticating extruder) is now universally used. The basic principle comes from the Archimede's screw as shown in Figure 2. The solid polymer, fed through a hopper, progressively melts as it flows along the barrel under the drag action of the rotating screw. The very high efficiency of the plasticating extruder comes from the forced convection between

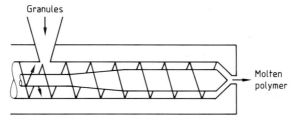

Figure 2   Plasticating screw

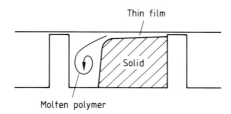

Figure 3   Side view of a screw channel in the plastication zone

the liquid and the solid bed which permits plastication through a very thin layer of molten polymer (about 0.1 mm) as illustrated in Figure 3. This is again an application of the thin layer principle.

b) High viscosity

The viscosity of molten polymers at usual operating temperatures is in the range of 10^2 to 10^4 Pa · s, that is 10^5 to 10^7 times greater than that of water. This very high viscosity has two practical implications:

–   Heating of polymers by viscous dissipation is easy to achieve and is therefore exploited in plastication systems. Under standard operating conditions in a plasticating extruder, the required energy comes from both the viscous dissipation and the heat conducted from the barrel wall; the relative contribution of the two sources is about equal. For certain types of extruders, called adiabatic extruders, the total energy input comes from viscous dissipation.

–   Because of the high viscosity, high pressures are built up by the screw to ensure large enough flow rates in dies or molds. Typical values are 100 MPa (1000 atmospheres) or more for injection molding, and 5 to 50 MPa (50 to 500 atmospheres) for extrusion dies.

The pressure can be built up in two ways:

–   In injection molding, at the beginning of the filling stage, the screw stops rotating and the entire screw moves as a plunger driven by a hydraulic system, as shown in Figure 4.

–   In extrusion, the screw-barrel system plays the role of a pump; the pressure built up at the extruder head results from the equilibrium of pressure and drag flows in the helical screw channel and the pressure flow in the die.

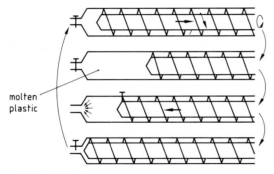

Plastication:

The plastic accumulates
in front of the screw which
backs up while rotating.

Injection:

The screw acts as a piston.

molten
plastic

Figure 4   The injection system

c) Combined effects of low conductivity and high viscosity

On one hand, a low thermal conductivity favors the existence of high thermal gradients. The risk of thermal degradation thus limits the quantity of energy which may be provided by viscous dissipation. On the other hand, the high viscosity associated to the low thermal diffusivity makes it possible to extrude molten polymers in air before the ultimate shaping or stretching. This has resulted into well established post-extrusion techniques, some of them being inspired from glass making techniques:

–   fiber spinning, that is the stretching of polymeric liquid filaments to produce textile fibers;
–   blow molding of bottles and other hollow items (in discontinuous processes);
–   continuous extrusion of tube;
–   stretching and biaxial stretching of films;
–   wire coating techniques.

d) Viscoelasticity

Most polymers are viscoelastic and the flow phenomena encountered because of viscoelasticity are often difficult to understand and the mathematical modelling leads to very difficult problems.

Overall, viscoelasticity favors stretching of material, but is unfavorable in shear flow situations. Indeed, viscoelasticity stabilizes the uniaxial and biaxial stretching of molten polymers thus making spinning possible at high strain rates. In blowing and thermoforming, viscoelasticity has a regulating effect on the wall thickness of the molded objects.

On the contrary, flow rates in shear flows are restricted due the occurrence of viscoelastic instabilities, responsible for various extrudate distortions: melt fracture, sharkskin in extrusion, surface distortions in calendering...

Finally viscoelasticity is responsible for the swelling of the polymer extrudate at the die exit. This complicates remarkably the design of dies used to manufacture profiles.

Plastics sales

Most of the important industrial processes for the transformation of thermoplastic materials have been mentioned in this introduction. The relative importance of the various thermoplastics compared to the thermosets is shown in Table 1 that reports the plastics sales (in ktons) in the U.S.A. for 1985 to 1989. It is interesting to note that the net increase for the period of four years is over 24%, that is considerably larger than the growth of more conventional materials for the same period.

Table 1 Sales of plastics (ktons) in the U.S.A. for 1985 – 1989
(adapted from Modern Plastics, January 1987, 1988, 1989 and 1990)

Material	ktons				
	1985	1986	1987	1988	1989
ABS	460	490	548	580	564
Acrylic	253	287	302	316	335
Alkyd	127	132	138	145	147
Cellulosics	50	40	40	41	41
Epoxy	167	169	190	212	223
Nylon	183	201	226	260	270
Phenolic	1199	1209	1364	1385	1434
Polyacetal	51	53	55	63	64
Polycarbonate (PC)	144	163	178	267	282
Polyester, thermoplastic (PBT, PET)	654	713	802	927	953
Polyester, unsaturated	558	560	596	623	601
Polyethylene, high density (HDPE)	3024	3190	3699	3667	3681
Polyethylene, low density[a] (LDPE)	3991	3994	4360	4966	4824
Polyphenylene-based alloys[b]	71	73	77	78	80
Polypropylene (PP) and copolymers	2395	2644	3048	3215	3287
Polystyrene (PS)	1865	2026	2204	2280	2351
Other styrenics	458	518	569	499	531
Polyurethane	1052	1199	1283	1467	1472
Polyvinyl chloride (PVC) and copolymers	3051	3365	3666	3759	3768
Other vinyls	455	463	414	431	435
Styrene acrylonitrile (SAN)	38	40	53	66	62
Thermoplastic elastomers	170	183	200	223	246
Urea and melamine	634	662	710	687	420
Others[c]	89	98	118	131	139
Total	21140	22470	24840	26288	26410

[a]   Includes LLDPE
[b]   Includes modified phenylene oxide and modified phenylene ether
[c]   Fluoroplastics, polybutylene, polyphenylene sulfide, polysulfone, etc.

Table 2 summarizes the sales in the U.S.A. in 1989 per process for the major commodity thermoplastics. This shows the importance of the extrusion, blow film, injection molding, blow molding..., processes analyzed in this book.

Table 2   Sales of the major commodity thermoplastics in U.S.A. by process in 1989 (adapted from Modern Plastics, January 1990)

Material	ktons				
	Extrusion[a]	Injection molding	Film	Blow Molding	Other uses[b]
ABS	202	273	–	–	89
Nylon	42	–	26	–	201
PET	–	–	273	426	165
HDPE	467	700	347	1331	846
LDPE & LLDPE	708	395	2608	41	1073
PP & copol.	1015	839	278	57	1098
PS	870[c]	58	108	4	1312[d]
PVC & copol.	2224	182	166	94	1103[e]

[a]   All extrusion processing except film;
[b]   includes export and sales for blending;
[c]   includes foam PS;
[d]   mostly thermoforming;
[e]   includes 442 ktons for calendering.

As indicated above, these processes are designed from considering three essential properties of molten polymers: low thermal conductivity, high viscosity and viscoelasticity. All of these considerations have lead to the following logical planning of this work:

– Chapter 1:   Continuum mechanics – review of principles
– Chapter 2:   Energy and heat transfer processes
– Chapter 3:   Flow of molten polymers in various geometries
– Chapter 4:   Extrusion of polymer
– Chapter 5:   Dynamics of film and fiber stretching
– Chapter 6:   Viscoelasticity of polymeric liquids.

Chapter 1
Continuum Mechanics –
Review of Principles

1.1   Strain and rate-of-strain tensor

1.1-1   Strain tensor

a) Phenomenological definitions

Phenomenological definitions of strain are first presented in the following examples.

Extension (or compression)

In extension, a volume element of length l is elongated by Δl in the x-direction as illustrated by Figure 1.1-1. The strain can be defined, from a phenomenological point of view, as $\varepsilon = \Delta l / l$.

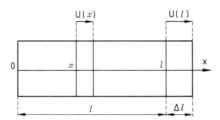

Figure 1.1-1   Strain in extension

For an homogeneous deformation of the volume element, the displacement U on the x-axis is $U(x) = \Delta l \cdot x / l$, and $dU/dx = \Delta l / l$. Hence another definition of the strain is $\varepsilon = dU/dx$.

Pure shear

A volume element of square section h by h in the x-y plane is sheared by a value a in the x and y directions as shown in Figure 1.1-2. Intuitively the strain may be defined as $\gamma = a/h$. For an homogeneous deformation of the volume element, the displacement (U, V) of point $M(x, y)$ is:

$$U(y) = a\left(\frac{y}{h}\right)$$

$$V(x) = a\left(\frac{x}{h}\right)$$

(1.1-1)

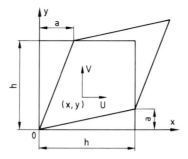

Figure 1.1-2   Strain in pure shear

Hence, another possible definition of the strain is $\gamma = dU/dy = dV/dx$.

b) Displacement gradient

More generally, any strain in a continuous medium is defined through a field of the displacement vector $\boldsymbol{U}(x, y, z)$ with coordinates:

$$U(x, y, z), \quad V(x, y, z), \quad W(x, y, z)$$

The intuitive definitions of strain make use of the derivatives of U, V, W, with respect to x, y, z, i.e. of their gradients. For a three-dimensional flow, one feels that a priori the material can be deformed in 9 different ways: 3 in extension (or compression) and 6 in shear. Therefore, it is natural to introduce the 9 components of the displacement gradient tensor $\nabla \boldsymbol{U}$:

$$[\nabla \boldsymbol{U}] = \begin{bmatrix} \dfrac{\partial U}{\partial x} & \dfrac{\partial U}{\partial y} & \dfrac{\partial U}{\partial z} \\[2mm] \dfrac{\partial V}{\partial x} & \dfrac{\partial V}{\partial y} & \dfrac{\partial V}{\partial z} \\[2mm] \dfrac{\partial W}{\partial x} & \dfrac{\partial W}{\partial y} & \dfrac{\partial W}{\partial z} \end{bmatrix} \qquad (1.1\text{-}2)$$

c) Rigid rotation

If this notion is applied to a volume element which has rotated θ degrees without being deformed, as shown in Figure 1.1-3, the displacement vector can be written as:

$$U(x, y) = x(\cos\theta - 1) - y\sin\theta$$
$$V(x, y) = x\sin\theta + y(\cos\theta - 1) \qquad (1.1\text{-}3)$$

For a very small value of θ:

$$U(x, y) \approx -y\theta$$
$$V(x, y) \approx x\theta \qquad (1.1\text{-}4)$$

hence:

$$[\nabla \boldsymbol{U}] = \begin{bmatrix} 0 & -\theta & 0 \\ \theta & 0 & 0 \\ 0 & 0 & 0 \end{bmatrix} \qquad (1.1\text{-}5)$$

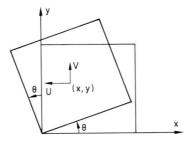

Figure 1.1-3   Rigid rotation

It is obvious from this result that ∇U cannot physically describe the strain of the material since it is not equal to zero when the material is under rigid rotation without being deformed.

d) Deformation or strain tensor ε

To obtain a tensor which physically represents the local deformation, we must make the tensor ∇U symmetrical, i.e.:

– write the transposed tensor (symmetry with respect to the principal diagonal); the transposed deformation tensor is:

$$[\nabla U]^{\dagger} = \begin{bmatrix} \frac{\partial U}{\partial x} & \frac{\partial V}{\partial x} & \frac{\partial W}{\partial x} \\ \frac{\partial U}{\partial y} & \frac{\partial V}{\partial y} & \frac{\partial W}{\partial y} \\ \frac{\partial U}{\partial z} & \frac{\partial V}{\partial z} & \frac{\partial W}{\partial z} \end{bmatrix} \qquad (1.1\text{-}6)$$

– write the half sum of the two tensors, each transposed with respect to the other:

$$\varepsilon = \frac{1}{2}(\nabla U + [\nabla U]^{\dagger})$$

or

$$\varepsilon_{ij} = \frac{1}{2}\left(\frac{\partial U_i}{\partial x_j} + \frac{\partial U_j}{\partial x_i}\right) \qquad (1.1\text{-}7)$$

where $U_i = (U, V, W)$ and $x_i = (x, y, z)$.

Let us now reexamine the three previous cases:

– in *extension* (or compression):

$$\left. \begin{array}{l} U(x) = \frac{\Delta l}{l}x \\ V \ = W = 0 \end{array} \right| \implies [\varepsilon] = \begin{bmatrix} \frac{\Delta l}{l} & 0 & 0 \\ 0 & 0 & 0 \\ 0 & 0 & 0 \end{bmatrix}, \quad \varepsilon_{xx} = \frac{\Delta l}{l} \qquad (1.1\text{-}8)$$

— in *pure shear:*

$$\left.\begin{array}{ll} U(y) & = \dfrac{a}{h}y \\[2mm] V(x) & = \dfrac{a}{h}x \\[2mm] W & = 0 \end{array}\right| \implies [\varepsilon] = \begin{bmatrix} 0 & \frac{a}{h} & 0 \\ \frac{a}{h} & 0 & 0 \\ 0 & 0 & 0 \end{bmatrix}, \quad \varepsilon_{xy} = \varepsilon_{yx} = \gamma \qquad (1.1\text{-}9)$$

— in *rigid rotation:*

$$\left.\begin{array}{ll} U(y) & = -\theta y \\ V(x) & = +\theta x \\ W & = 0 \end{array}\right| \implies [\varepsilon] = \begin{bmatrix} 0 & 0 & 0 \\ 0 & 0 & 0 \\ 0 & 0 & 0 \end{bmatrix}, \qquad (1.1\text{-}10)$$

The definition of ε is such that the deformation is nil in rigid rotation; it is physically satisfactory whereas the use of ∇U for the deformation is not correct.

Important remark

The definition of the tensor ε used here is a simplified one. One can show rigorously that the strain tensor in a material is mathematically described by:

$$\Delta_{ij} = \frac{1}{2}\left\{\left(\frac{\partial U_i}{\partial x_j}\right) + \left(\frac{\partial U_j}{\partial x_i}\right) + \sum_k \left(\frac{\partial U_k}{\partial x_i}\right)\left(\frac{\partial U_k}{\partial x_j}\right)\right\}$$

$$= \varepsilon_{ij} + \frac{1}{2}\sum_k \left(\frac{\partial U_k}{\partial x_i}\right)\left(\frac{\partial U_k}{\partial x_j}\right). \qquad (1.1\text{-}11)$$

This definition of the tensor ε is approximately valid only if the terms $\partial U_i/\partial x_j$ are small. So the expressions for the tensor written above are usable only if ε, γ, θ, ... are small (typically less than 5%). This condition is not generally satisfied for the flow of polymer melts. As will be shown, in those cases, we will then use the rate-of-strain tensor $\dot{\varepsilon}$.

e) Simple shear and symmetry of the tensor ε

This type of deformation is more easy to obtain experimentally than pure shear. Let us consider a volume element of a square section h by h, sheared by the quantity $2a$ in the direction x as illustrated by Figure 1.1-4.

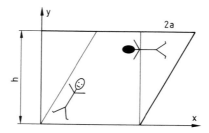

Figure 1.1-4   Symmetry in simple shear

We have:

$$U(y) = \frac{2ay}{h} \atop V = W = 0 \quad \Longrightarrow \quad [\varepsilon] = \begin{bmatrix} 0 & \frac{a}{h} & 0 \\ \frac{a}{h} & 0 & 0 \\ 0 & 0 & 0 \end{bmatrix} \qquad (1.1\text{-}12)$$

and if $a \ll h$, then:

$$\gamma = \frac{a}{h} = \frac{1}{2} \frac{dU}{dy} = \varepsilon_{xy} = \varepsilon_{yx} \qquad (1.1\text{-}13)$$

We see that pure shear is physically imposed in a non symmetrical manner with respect to x and y; however the strain experienced by the material is symmetrical. Indeed, two observers, initially located along the Ox and Oy directions, will feel the deformation of the square material element on their left (see Figure 1.1-4) in an identical manner.

As a general result, the tensor ε is always symmetrical, i.e. it contains only 6 independent components.

- 3 in extension or compression: $\varepsilon_{xx}, \quad \varepsilon_{yy}, \quad \varepsilon_{zz}$.
- 3 in shear: $\varepsilon_{xy} = \varepsilon_{yx}, \quad \varepsilon_{yz} = \varepsilon_{zy}, \quad \varepsilon_{zx} = \varepsilon_{xz}$.

f) Volume variation during deformation

Only the strain components in extension or in compression may result in a variation of the volume. If l_x, l_y, l_z are the dimensions along the three axes and V is the volume, then:

$$V = l_x l_y l_z ,$$

$$\boxed{ \begin{aligned} \frac{dV}{V} &= \frac{dl_x}{l_x} + \frac{dl_y}{l_y} + \frac{dl_z}{l_z} \\[2mm] &= \varepsilon_{xx} + \varepsilon_{yy} + \varepsilon_{zz} \end{aligned} } \qquad (1.1\text{-}14)$$

1.1-2 Rate-of-strain tensor

For a velocity field $u(x, y, z)$, the rate-of-strain tensor is defined as the limit:

$$\dot{\varepsilon} = \lim_{dt \to 0} \frac{[\varepsilon]_t^{t+dt}}{dt} \qquad (1.1\text{-}15)$$

But in the interval t and $t + dt$, the displacement vector is $d\boldsymbol{U} = \boldsymbol{u}\,dt$. Hence:

$$[\varepsilon_{ij}]_t^{t+dt} = \frac{1}{2}\left(\frac{\partial u_i}{\partial x_j} + \frac{\partial u_j}{\partial x_i}\right)dt \tag{1.1-16}$$

where $u_i = (u, v, w)$. The rate-of-strain tensor becomes:

$$\dot{\varepsilon}_{ij} = \frac{1}{2}\left(\frac{\partial u_i}{\partial x_j} + \frac{\partial u_j}{\partial x_i}\right) \tag{1.1-17}$$

As in the case of ε, this tensor is symmetrical:

$$\dot{\varepsilon} = (\nabla\boldsymbol{u} + [\nabla\boldsymbol{u}]^\dagger)$$

Comment

Equation (1.1-17) is the general expression for the rate-of-strain tensor, but its derivation from the expression (1.1-16) for the strain tensor is correct only if the deformations and the displacements are infinitely small (as in the case of a high modulus elastic body). For a liquid material, it is not possible, in general, to make use of expression (1.1-16). Indeed, a liquid experiences very large deformations for which the tensor ε has no physical meaning.

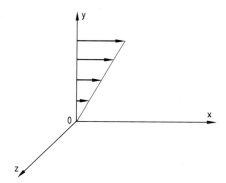

Figure 1.1-5   Simple shear in the $x - y$ plane

Let us consider for example the simple shear experiment for which the velocity profile is illustrated in Figure 1.1-5. We have then:

$$\boldsymbol{U}(t) = \begin{vmatrix} U(t) = \dot{\gamma}yt \\ V(t) \equiv 0 \\ W(t) \equiv 0 \end{vmatrix} \qquad \boldsymbol{u} = \begin{vmatrix} u = \dot{\gamma}y \\ v \equiv 0 \\ w \equiv 0 \end{vmatrix} \tag{1.1-18}$$

$$[\dot{\varepsilon}] = \lim_{dt \to 0}\frac{[\varepsilon]_t^{t+dt}}{dt} = \begin{bmatrix} 0 & \dot{\gamma}/2 & 0 \\ \dot{\gamma}/2 & 0 & 0 \\ 0 & 0 & 0 \end{bmatrix}$$

From the origin on the time scale and for a small time interval compared to $1/\dot{\gamma}$:

$$\dot{\gamma}t \ll 1, \qquad [\varepsilon] = \begin{bmatrix} 0 & \dot{\gamma}t/2 & 0 \\ \dot{\gamma}t/2 & 0 & 0 \\ 0 & 0 & 0 \end{bmatrix}$$

This expression has a physical meaning and $\dot{\varepsilon} = d\varepsilon/dt$. For a longer interval, $\dot{\gamma}t$ is no longer small and ε is no more representative of the strain in the material.

1.1-3 The equation of continuity

a) Mass balance

Let us consider a volume element of fluid $dx\,dy\,dz$. The fluid density is $\varrho(x,y,z,t)$. The variation of mass in the volume element with respect to time is:

$$\frac{\partial \varrho}{\partial t} dx\,dy\,dz$$

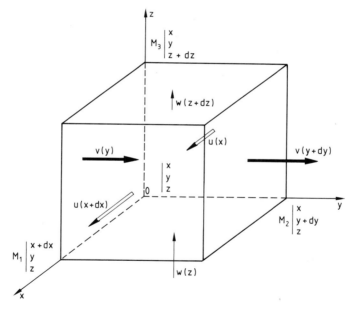

Figure 1.1-6 Mass balance on a cubic volume element

This variation is due to a balance of mass fluxes across the faces of the volume element:

- in the x-direction: $\{\varrho(x+dx)\cdot u(x+dx) - \varrho(x)\cdot u(x)\}\,dy\,dz$
- in the y-direction: $\{\varrho(y+dy)\cdot v(y+dy) - \varrho(y)\cdot v(y)\}\,dz\,dx$
- in the z-direction: $\{\varrho(z+dz)\cdot w(z+dz) - \varrho(z)\cdot w(z)\}\,dx\,dy$

Hence, dividing by $dx\,dy\,dz$ and taking the limits, we get:

$$\frac{\partial\varrho}{\partial t} + \frac{\partial}{\partial x}(\varrho u) + \frac{\partial}{\partial y}(\varrho v) + \frac{\partial}{\partial z}(\varrho w) = 0 \qquad (1.1\text{-}19)$$

which can be written through the definition of the divergence as:

$$\boxed{\frac{\partial\varrho}{\partial t} + \nabla\cdot(\varrho\boldsymbol{u}) = 0} \qquad (1.1\text{-}20)$$

This is the equation of continuity.

b) Incompressible materials

For incompressible materials ϱ is a constant and the equation of continuity then reduces to:

$$\boxed{\nabla\cdot\boldsymbol{u} = 0} \qquad (1.1\text{-}21)$$

This result can be obtained from the expression for the volume variation.

$$\frac{dV}{V} = tr[\varepsilon] = \varepsilon_{xx} + \varepsilon_{yy} + \varepsilon_{zz} \qquad (1.1\text{-}22)$$

also:

$$\frac{1}{V}\frac{dV}{dt} = tr[\dot{\varepsilon}] = \dot{\varepsilon}_{xx} + \dot{\varepsilon}_{yy} + \dot{\varepsilon}_{zz}$$

$$= \frac{\partial u}{\partial x} + \frac{\partial v}{\partial y} + \frac{\partial w}{\partial z} = \nabla\cdot\boldsymbol{u} \qquad (1.1\text{-}23)$$

It follows that:

$$\frac{dV}{dt} = 0 \;\Rightarrow\; tr[\dot{\varepsilon}] = 0 \;\Rightarrow\; \nabla\cdot\boldsymbol{u} = 0.$$

Problems 1.1

1.1-A   Analysis of simple shear flow

Simple shear flow is representative of the rate of deformation experienced in many practical situations. Homogeneous, simple planar shear flow is defined by the following velocity field:

$$u(y) = \dot{\gamma}y$$
$$v \equiv 0$$
$$w \equiv 0$$

where Ox is the direction of the velocity, Oxy is the plane perpendicular to the flow and $\dot{\gamma}$ is the shear rate. Planes parallel to Oxz are sheared surfaces.

a)   Write down the expression for the tensor $\dot{\varepsilon}$ for simple planar shear flow. Justify with physical arguments the symmetry of the tensor.

b)   Assuming that any flow situation is locally simple shear if, at that given point, the rate-of-strain tensor is given by the expression obtained in a), then show that all the following flows, encountered in practical situations, are locally simple shear flows. Obtain in each case the directions 1, 2, 3 (i.e. equivalent to x, y, z for planar shear) and the expression of the shear rate $\dot{\gamma}$ (use the expressions of $\dot{\varepsilon}$ in cylindrical and spherical coordinates given in Appendix A).

–   Flow between parallel plates:

$$u = 0$$
$$v = 0$$
$$w = w(y)$$

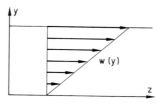

Figure 1.1-7   Flow between parallel plates

–   Flow in a circular tube:

$$u = 0$$
$$v = 0$$
$$w = w(r)$$

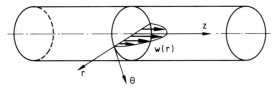

Figure 1.1-8   Flow in a circular tube

— Flow between two parallel disks:
The sheared planes are assumed to be parallel to the disks and rotate at an angular velocity $\Omega(z)$, radiants/s.

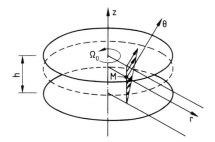

Figure 1.1-9   Flow between parallel disks

— Flow between a cone and a plate:
The sheared planes are assumed to be cones with the same axis and apex as the cone-and-plate system; they rotate at an angular velocity $\Omega(\theta)$, radiants/s.

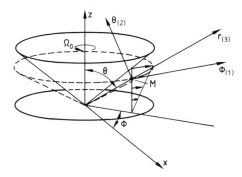

Figure 1.1-10   Flow in a cone-and-plate system

1.1-B   Pure elongational flow

Write down the expression for the tensor $\dot{\varepsilon}$ for the pure extension (or compression) of an incompressible material.

1.2   Stresses and force balances

1.2-1   The stress tensor

a)   Phenomenological definitions

As in the previous section we introduce first phenomenological definitions:

Extension (or compression)

An extension force applied on a cylinder of surface S induces a normal stress: $\sigma_n = F/S$.

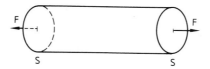

Figure 1.2-1   Stress in extension

Simple shear

A force tangentially applied to a surface S yields a shear stress: $\sigma = F/S$.

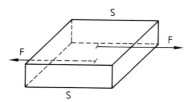

Figure 1.2-2   Stress in simple shear

The units of the stresses are those of pressure, i.e. Pa.

b)   The stress vector

Let us consider, in a more general situation, a surface element dS of a continuum, and a force element $d\boldsymbol{F}$ exerted on the surface. The stress vector $\boldsymbol{T}$ at a point P on this surface is defined as the limit:

$$\boldsymbol{T} = \lim_{dS \to 0} \frac{d\boldsymbol{F}}{dS} \qquad (1.2\text{-}1)$$

At point P, the normal to the surface is defined by the unit vector, $\boldsymbol{n}$, in the outward direction, as illustrated in Figure 1.2-3.

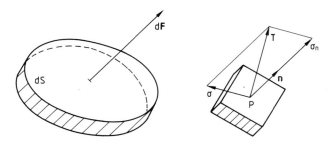

Figure 1.2-3 Stress applied to a surface element

The stress components can be obtained from projections of the stress vector:

– Projection on $\boldsymbol{n}$: $\sigma_n = \boldsymbol{T} \cdot \boldsymbol{n}$

σ_n is the normal stress (in extension, $\sigma_n > 0$; in compression, $\sigma_n < 0$).

– Projection on the surface: σ is the shear stress.

c) The stress tensor

The stress vector cannot characterize the state of stresses at a given point since it is a function of the orientation of the surface element, i.e. of $\boldsymbol{n}$. Thus a tensile force induces a stress on a surface element perpendicular to the orientation of the force, but induces no stress on a parallel surface.

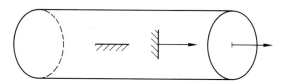

Figure 1.2-4 Stress vector and surface orientation

The state of stresses is in fact characterized by the relation between $\boldsymbol{T}$ and $\boldsymbol{n}$, and as we will see, this relation is tensorial. Let us consider an elementary tetrahedron $OABC$ along the axes $Oxyz$: the x, y, z-components of the unit normal vector to the ABC plane are the ratios of the surfaces OAB, OBC, OCA to ABC, i.e.:

$$n_x = \frac{OBC}{ABC}, \quad n_y = \frac{OCA}{ABC}, \quad n_z = \frac{OAB}{ABC}$$

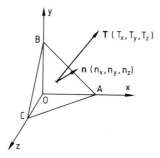

Figure 1.2-5 Stresses exerted on an elementary tetrahedron

Let us define the components of the stress tensor in the following table:

Projection on $\downarrow$	of the stress vector exerted on the face normal to $\downarrow$		
	Ox	Oy	Oz
Ox	σ_{xx}	σ_{xy}	σ_{xz}
Oy	σ_{yx}	σ_{yy}	σ_{yz}
Oz	σ_{zx}	σ_{zy}	σ_{zz}

The net surface forces acting along the three directions of the axes are:

$$T_x(ABC) - \sigma_{xx}(OBC) - \sigma_{xy}(OAC) - \sigma_{xz}(OAB)$$
$$T_y(ABC) - \sigma_{yx}(OBC) - \sigma_{yy}(OAC) - \sigma_{yz}(OAB)$$
$$T_z(ABC) - \sigma_{zx}(OBC) - \sigma_{zy}(OAC) - \sigma_{zz}(OAB)$$

OA, OB, OC being of the order of d, the surfaces OAB, OBC, and OCA are of the order of d^2 and the volume $OABC$ is of the order of d^3. The surface forces are of the order of Td^2 and the volume forces of the order of Td^3 (e.g. $F = \varrho g$ for the gravitational force per unit volume).

When the dimension d of the tetrahedron tends to zero, the volume forces become negligible compared with the surface forces and the net forces, as expressed above, are equal to zero. Hence, in terms of the components of $\boldsymbol{n}$:

$$T_x = \sigma_{xx}n_x + \sigma_{xy}n_y + \sigma_{xz}n_z$$
$$T_y = \sigma_{yx}n_x + \sigma_{yy}n_y + \sigma_{yz}n_z \qquad (1.2\text{-}2)$$
$$T_z = \sigma_{zx}n_x + \sigma_{zy}n_y + \sigma_{zz}n_z$$

This result can be written in tensorial notation as:

$$\boldsymbol{T} = \boldsymbol{\sigma} \cdot \boldsymbol{n} \qquad (1.2\text{-}3)$$

where $\boldsymbol{\sigma}$ is the stress tensor which, a priori, contains three normal components and six shear components defined for the three axes. As in the case of the strain, the state of stresses is described by a tensor.

d) Isotropic stress or hydrostatic pressure

The hydrostatic pressure translates into a stress vector which is in the direction of $\boldsymbol{n}$ for any orientation of the surface:

$$\text{Pressure } p \Rightarrow \boldsymbol{T} = -p\boldsymbol{n}, \quad \forall n.$$

The corresponding tensor is proportional to the unit tensor:

$$[\boldsymbol{\sigma}] = \begin{bmatrix} -p & 0 & 0 \\ 0 & -p & 0 \\ 0 & 0 & -p \end{bmatrix} = -p[\boldsymbol{\delta}] \tag{1.2-4}$$

where $[\boldsymbol{\delta}]$ is the unit tensor.

e) The deviatoric stress tensor

For any general state of stresses, the definition of pressure can be generalized by the following:

$$p = -\frac{tr[\boldsymbol{\sigma}]}{3} = -\frac{\sigma_{xx} + \sigma_{yy} + \sigma_{zz}}{3} \tag{1.2-5}$$

The pressure is independent of the axes since the trace of the stress tensor is an invariant (see Appendix B). It could be positive (compressive state) or negative (extensive state, possibly leading to cavitation problems in the liquid).

The stress tensor can be written as a sum of two terms, the pressure term and a traceless stress term, called the deviatoric stress tensor:

$$\boldsymbol{\sigma} = -p\boldsymbol{\delta} + \boldsymbol{\sigma}' \tag{1.2-6}$$

Examples

– *Uniaxial extension* (or compression):

$$[\boldsymbol{\sigma}] = \begin{bmatrix} \sigma_n & 0 & 0 \\ 0 & 0 & 0 \\ 0 & 0 & 0 \end{bmatrix} \Rightarrow p = \frac{\sigma_n}{3}$$

and

$$[\boldsymbol{\sigma}'] = \begin{bmatrix} \frac{2\sigma_n}{3} & 0 & 0 \\ 0 & -\frac{\sigma_n}{3} & 0 \\ 0 & 0 & -\frac{\sigma_n}{3} \end{bmatrix} \tag{1.2-7}$$

– *simple shear* under a hydrostatic pressure p:

$$[\boldsymbol{\sigma}] = \begin{bmatrix} -p & \sigma & 0 \\ \sigma & -p & 0 \\ 0 & 0 & -p \end{bmatrix} \quad \text{and} \quad [\boldsymbol{\sigma}'] = \begin{bmatrix} 0 & \sigma & 0 \\ \sigma & 0 & 0 \\ 0 & 0 & 0 \end{bmatrix} \tag{1.2-8}$$

1.2-2 Force Balances

Considering an elementary volume of material with a characteristic dimension d:

- the surface forces are of the order of d^2, but the definition of the stress tensor is such that their contribution to a force balance is nil.
- the volume forces (gravity, inertia, ...) are of the order of d^3 and they must balance the derivatives of the surface forces which are also of the order of d^3.

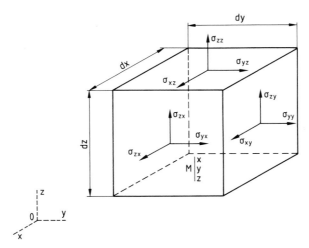

Figure 1.2-6 Balance of forces exerted on a volume element

We will write that the resultant force is nil. The forces acting on a volume element $dx\, dy\, dz$ are then the following ones:

- the mass forces (generally gravity): $\boldsymbol{F}\, dx\, dy\, dz$
- the inertial forces. $\varrho\boldsymbol{\gamma}\, dx\, dy\, dz = \varrho(D\boldsymbol{u}/Dt)\, dx\, dy\, dz$
- the net surface forces exerted by the surroundings, i.e. in the x-direction:

$$\begin{aligned} &\{\sigma_{xx}(x+dx)-\sigma_{xx}(x)\}\, dy\, dz \\ +\ &\{\sigma_{xy}(y+dy)-\sigma_{xy}(y)\}\, dz\, dx \\ +\ &\{\sigma_{xz}(z+dz)-\sigma_{xz}(z)\}\, dx\, dy \end{aligned}$$

and similar terms for the y and z-directions.

Dividing by $dx\, dy\, dz$ and taking the limits, we obtain for the x, y, and z components:

$$\begin{aligned} (F_x - \varrho\gamma_x) + \frac{\partial\sigma_{xx}}{\partial x} + \frac{\partial\sigma_{xy}}{\partial y} + \frac{\partial\sigma_{xz}}{\partial z} &= 0 \\ (F_y - \varrho\gamma_y) + \frac{\partial\sigma_{yx}}{\partial x} + \frac{\partial\sigma_{yy}}{\partial y} + \frac{\partial\sigma_{yz}}{\partial z} &= 0 \\ (F_z - \varrho\gamma_z) + \frac{\partial\sigma_{zx}}{\partial x} + \frac{\partial\sigma_{zy}}{\partial y} + \frac{\partial\sigma_{zz}}{\partial z} &= 0 \end{aligned}$$

$$(1.2\text{-}9)$$

The derivatives of σ_{ij} are the components of a vector which is the divergence of the tensor $\boldsymbol{\sigma}$; Equation (1.2-9) may be written as:

$$\boxed{\nabla \cdot \boldsymbol{\sigma} + (\boldsymbol{F} - \varrho\boldsymbol{\gamma}) = 0} \qquad (1.2\text{-}10)$$

This is the equation of motion. It is often convenient to express the stress tensor as the sum of the pressure and the deviatoric stress:

$$\nabla \cdot \boldsymbol{\sigma} = -\nabla p + \nabla \cdot \boldsymbol{\sigma}' \qquad (1.2\text{-}11)$$

1.2-3   Torque Balances

Let us consider a small volume element of linear dimension d; the mass forces of the order of d^3 induce torques of the order of d^4. There is no mass torque which would result into torques of the order of d^3 (as in the case of a magnetic medium). Finally, the surface forces of the order of d^2 induce torques of the order of d^3, so that only the net torque resulting from these forces must be equal to zero.

If we consider an element $dx\,dy\,dz$, only the shear stresses exert torques: the moments about the z-axis due to the components σ_{xy} and σ_{yx} are obtained by taking the following vector products:

$$\sigma_{xy} : \quad \begin{vmatrix} 0 \\ dy \\ 0 \end{vmatrix} \times \begin{vmatrix} \sigma_{xy}\,dx\,dz \\ 0 \\ 0 \end{vmatrix} = \begin{vmatrix} 0 \\ 0 \\ -\sigma_{xy}\,dx\,dy\,dz \end{vmatrix}$$

$$\sigma_{yx} : \quad \begin{vmatrix} dx \\ 0 \\ 0 \end{vmatrix} \times \begin{vmatrix} 0 \\ \sigma_{yx}\,dy\,dz \\ 0 \end{vmatrix} = \begin{vmatrix} 0 \\ 0 \\ \sigma_{yx}\,dx\,dy\,dz \end{vmatrix}$$

A torque balance, in the absence of a mass torque, yields: $\sigma_{xy} = \sigma_{yx}$. In a similar way: $\sigma_{yz} = \sigma_{zy}$, $\sigma_{zx} = \sigma_{xz}$. The absence of volume torque then implies the symmetry of the stress tensor. Therefore, as for the strain tensor ε, the stress tensor has only 6 independent components (3 normal and 3 shear components).

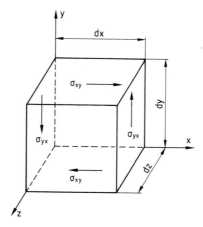

Figure 1.2-7   Torque balance on a volume element

Problems 1.2

1.2-A

A cylinder of length L and radius R is filled with an homogeneous material. A pressure difference Δp is applied between the end sections. The material necessarily exerts a shear stress on the wall of the cylinder. If the shear stress σ is uniform, show that:

$$\sigma = \Delta p \frac{R}{2L}$$

1.2-B

Consider a thin spherical shell of radius R and thickness e, very small compared to R. The internal pressure inside the sphere is p.

– Obtain the expression of the stress tensor in the wall of the sphere.
– What is the corresponding expression for a cylinder?

1.2-C

Consider a pyramid of density ϱ and of a section S which decreases with the elevation z. What is the variation of $S(z)$ that, as a first approximation, would result in a constant normal stress σ_n (due to the weight) in the vertical direction?

1.3   General equations of mechanics

1.3-1   General case

A problem in mechanics involves 10 unknown functions of x, y, z, t:
- the density ϱ
- the vector $\boldsymbol{u}$ with three components u, v, w
  (from which are obtained the strain and the rate of strain)
- the stress tensor $\boldsymbol{\sigma}$ with 6 components.

Four equations are available to solve the problem:
- the equation of continuity [Equation (1.1-20)]:

$$\frac{\partial \varrho}{\partial t} + \nabla \cdot (\varrho \boldsymbol{u}) = 0$$

- the equation of motion [Equation (1.2-10)]:

$$\nabla \cdot \boldsymbol{\sigma} + (\boldsymbol{F} - \varrho \boldsymbol{\gamma}) = 0$$

We need six other equations. These are the constitutive relations between the stresses and the strain in a material element. The elastic behaviour is well known; the constitutive relation in terms of the Lamé and Clapeyron coefficients λ and μ is:

$$\sigma_{ij} = \lambda(\varepsilon_{11} + \varepsilon_{22} + \varepsilon_{33})\,\delta_{ij} + 2\mu\varepsilon_{ij} \qquad (1.3\text{-}1)$$

where δ_{ij} is the Kronecker delta. This relation contains six equations. Other constitutive relations are introduced in subsequent sections.

The constitutive relation is the mathematical expression of the rheological behaviour of a material, determined for experiments conducted in simple flows with a viscometer, a rheometer, a tensional or torsional device, ...

1.3-2   Incompressibility

For incompressible materials, ϱ is no longer unknown and the equation of continuity is simplified, but it is still useful [Equation (1.1-21)]:

$$\boxed{\nabla \cdot \boldsymbol{u} = 0}$$

On the other hand, as shown by MANDEL (1974, page 72), the constitutive relation is no longer written in terms of the stress tensor $\boldsymbol{\sigma}$, but in terms of a tensor that characterizes the state of stresses within an isotropic term (hydrostatic pressure). This stress tensor is frequently taken as the deviatoric stress tensor $\boldsymbol{\sigma}'$. Under these conditions, the constitutive relation contains only five equations. A given problem now represents nine unknowns and may be solved theoretically with the help of nine equations:

– $\boldsymbol{u}$ (3)		the equation of continuity (1)
	to satisfy	the constitutive relation (5)
– $\boldsymbol{\sigma}$ (6)		the equation of motion (3)

1.3-3 Planar flow

We are frequently interested in planar flow problems, of which the most important is planar deformation flow, defined by the following. The z-component of the velocity vector is zero and u and v are independent of z.

For any rheological behaviour (see problem 1.3A below), $\sigma_{xz} \equiv \sigma_{yz} \equiv 0$ and the equation of motion (or a force balance) for steady-state conditions implies that σ_{zz} is constant. The problem is then reduced to two dimensions with only five unknowns:

$$u, v \quad \text{and} \quad \sigma_{xx}, \sigma_{xy}, \sigma_{yy} \,.$$

Example

The flow in the gap of two rollers in the calendering process is not strictly speaking a planar flow, except in the plane of symmetry, since the thickness of the bank in the material decreases with the distance from the center. However, over a large portion of the width, the approximation of planar flow is quite acceptable.

Problem 1.3

1.3-A Stress tensor in simple shear flow

A material is submitted to a field of homogeneous simple shear flow in steady-state conditions with the velocity in the x-direction and varying with the y-axis. The stresses in this material are then steady and uniform. For steady and uniform conditions, there is a unique relation between the tensor $\dot{\varepsilon}$ and the tensor $\boldsymbol{\sigma}$.

a) Using these considerations and making an appropriate change of coordinates, show that the stresses developed in simple shear for any rheological behaviour are such that:

$$\sigma_{xz} = \sigma_{yz} = 0$$

b) Show that it is also the case for any type of planar flow.

1.4   Viscosity – the equation of motion for Newtonian liquids

Viscous flow behaviour is first discussed since it represents in a first approximation the rheological behaviour of molten polymers.

1.4-1   The Newtonian behaviour

a) Newton's law (1687)

Let us consider a viscous liquid under simple shear between two parallel plates, separated by a thickness h as illustrated in Figure 1.4.1; one of the plate is displaced with respect to the other at a velocity U:
-   the shear rate is $\dot{\gamma} = U/h$
-   the force F induces a shear stress: $\sigma = F/S$.

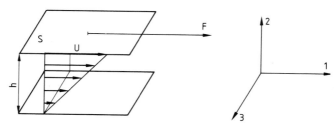

Figure 1.4-1   Simple shear experiment

It has been experimentally observed that most liquids obey Newton's law of viscosity, i.e. at a given temperature and pressure the ratio between σ and $\dot{\gamma}$ is independent of $\dot{\gamma}$ (at least over a restricted range of $\dot{\gamma}$).

Newton defined the viscosity, η, by the following relation:

$$\sigma = \eta\dot{\gamma} \qquad (1.4\text{-}1)$$

A physical interpretation of this law is presented in Appendix C, where the dependence of viscosity on temperature and pressure is also discussed. Molten polymers exhibit a particular behaviour, due to their macromolecular structure:
-   on the one hand, their viscosity changes drastically with temperature;
-   on the other hand, their viscosity is non-Newtonian, i.e. it decreases with $\dot{\gamma}$; this is known as the shear-thinning effect or pseudoplastic behaviour. A physical interpretation is presented in Appendix C and mathematical expressions of the shear-thinning viscosity are given in Chapter 1.5.

b) Generalization to three-dimension flows

Newton's law, as defined by Equation (1.4-1), is restricted to one-dimension flows and does not describe the dynamics of a three-dimension flow problem; hence it does not allow a complete solution of complex flow problem. To do that, one must write the viscosity law in a tensorial form as a relation between the stress and the rate of deformation tensor. The result, first obtained by NAVIER in 1823 for incompressible liquids, is given in terms of the deviatoric stress tensor as:

$$\boxed{\sigma' = 2\eta\dot{\varepsilon}} \qquad (1.4\text{-}2)$$

The factor 2 is used to insure accord with the definition of Newton, as shown below.

c) Simple shear flow

In a simple shear experiment where the flow is in the direction of the x-axis (direction 1) and the shear force is acting on the y-plane (direction 2):

$$
\begin{aligned}
u(y) &= \dot\gamma y \\
v &\equiv 0 \\
w &\equiv 0
\end{aligned}
\quad\Rightarrow\quad
\dot{\boldsymbol\varepsilon} =
\begin{bmatrix}
0 & \frac{\dot\gamma}{2} & 0 \\
\frac{\dot\gamma}{2} & 0 & 0 \\
0 & 0 & 0
\end{bmatrix}
\quad\Rightarrow\quad
\boldsymbol\sigma' =
\begin{bmatrix}
0 & \eta\dot\gamma & 0 \\
\eta\dot\gamma & 0 & 0 \\
0 & 0 & 0
\end{bmatrix}
\quad\Rightarrow\quad \sigma = \eta\dot\gamma
$$

Remarks

- The experiment and the definition of Newton are non symmetrical with respect to the directions 1 and 2; the shear stress that physically describes the force per unit area is σ_{12}; the constitutive relation introduces the other shear stress:

$$
\sigma_{21} = \sigma_{12} ;
$$

- on the other hand, the tensor $\boldsymbol\sigma$ is in simple shear:

$$
\begin{bmatrix}
-p & \eta\dot\gamma & 0 \\
\eta\dot\gamma & -p & 0 \\
0 & 0 & -p
\end{bmatrix}
$$

i.e. it has non-zero shear components, but no differences between the normal stresses $\sigma_{11}, \sigma_{22}, \sigma_{33}$. It will be seen later that this indicates the absence of elasticity in the liquid.

1.4-2 Order of magnitude of the various quantities

a) Units of viscosity and characteristic values

The units of the dynamic viscosity η (dimensions $ML^{-1}\,T^{-1}$) are:

$(N/m^2) \cdot s$ or $Pa \cdot s$ in SI units;

the Poise in c.g.s. units; $1\ Pa \cdot s = 10$ Poises.

The kinematic viscosity is defined by:

$$
\nu = \frac{\eta}{\varrho} \quad \text{(dimensions } L^2 T^{-1}) \tag{1.4-3}
$$

with the units of m^2/s in SI, cm^2/s or Stokes in c.g.s.

The magnitude of the viscosity for various materials is given in Table 1.4-1.

Table 1.4-1 Magnitude of the viscosity

	η Pa·s	ν m^2/s
Air (0 °C)	1.7×10^{-5}	1.33×10^{-5}
Water (0 °C)	1.8×10^{-3}	1.79×10^{-6}
(20 °C)	10^{-3}	10^{-6}
Mercury (20 °C)	1.6×10^{-3}	1.2×10^{-7}
Oils	10^{-2} to 1	10^{-5} to 10^{-3}
Molten polymers	10^2 to 10^4	10^{-1} to 10
Molten glass	10^2 to 10^4	10^{-1} to 10

A characteristic of molten polymers shown in Table 1.4-1 is their extremely high viscosity, comparable to that of molten glass.

b) Reynolds number

The Reynolds number is commonly defined in terms of the average velocity U in a tube of diameter D:

$$\mathrm{Re} = \frac{UD}{\nu} = \varrho\frac{UD}{\eta}$$

$$= \frac{\varrho U^2}{\eta U/D} = \frac{\text{inertial forces}}{\text{viscous forces}} . \tag{1.4-4}$$

It is well known in hydrodynamics that the flow in a tube becomes turbulent when Re is larger than some characteristic number in the range between 2,000 and 10,000.
In other flow geometries such as the flow in a slit of thickness h, and for an average velocity U, we write:

$$\mathrm{Re} = \frac{Uh}{\nu} \tag{1.4-5}$$

In the practical situation of polymer processing, we have:

$$U < 1\,\mathrm{m/s}$$
$$h < 10\ \mathrm{mm} \quad \Rightarrow \quad \mathrm{Re} < 0.1$$
$$\nu > 0.1\ \mathrm{m}^2/\,\mathrm{s}$$

Therefore, for most flow situations encountered in polymer processing, the Reynolds number is smaller than 1. This means that the inertial forces may be neglected with respect to the viscous forces. A fortiori, flow rates encountered in polymer processing are much smaller than those corresponding to the transition from laminar to turbulent flow.

c) Effect of gravity

If L is the difference between the highest and lowest points of a given flow, the difference of hydrostatic pressure is $\varrho g L$. This term may be important or not, depending on the magnitude of the shear force due to the viscosity, i.e. $\eta U/h$. Hence in dimensionless form:

$$\frac{\varrho g L h}{\eta U} = \frac{g L h}{\nu U} \tag{1.4-6}$$

Example

For

$$
\left.
\begin{aligned}
U &= 1 \text{ m/s} \\
h &= 0.01 \text{ m} \\
\nu &= 0.1 \text{ m}^2/\text{s} \\
g &= 10 \text{ m/s}^2
\end{aligned}
\right|
\quad \Rightarrow \quad
\begin{aligned}
&\text{the gravitational force} \\
&\text{is non-negligible if } L > 1\,\text{m}.
\end{aligned}
$$

Generally for flows confined to small gap geometries (pumping zone of an extruder, mold, ...), the gravitational forces are negligible. This is not the case for fibre spinning, where large level differences are normally involved.

1.4-3 Navier-Stokes equations

A force balance per unit volume of the liquid yields:

$$\nabla \cdot \boldsymbol{\sigma} + \boldsymbol{F} - \varrho \boldsymbol{\gamma} = 0 \tag{1.4-7}$$

or

$$-\nabla p + \nabla \cdot \boldsymbol{\sigma}' + \boldsymbol{F} - \varrho \boldsymbol{\gamma} = 0 \tag{1.4-8}$$

For an incompressible Newtonian liquid:

$$\boldsymbol{\sigma}' = 2\eta \dot{\boldsymbol{\varepsilon}}$$

and since $\nabla \cdot \boldsymbol{u} = 0$.

$$\nabla \cdot \dot{\boldsymbol{\varepsilon}} = \frac{1}{2}\nabla^2 \boldsymbol{u} \tag{1.4-9}$$

Hence, the Navier-Stokes equation may be written as:

$$\boxed{-\nabla p + \eta \nabla^2 \boldsymbol{u} + \boldsymbol{F} - \varrho \boldsymbol{\gamma} = 0} \tag{1.4-10}$$

We have shown that the inertial and gravitational forces are generally negligible; hence, the Navier-Stokes and the continuity equations form the following set of 4 equations with 4 unknowns:

$$-\frac{\partial p}{\partial x} + \eta\left(\frac{\partial^2 u}{\partial x^2} + \frac{\partial^2 u}{\partial y^2} + \frac{\partial^2 u}{\partial z^2}\right) = 0$$

$$-\frac{\partial p}{\partial y} + \eta\left(\frac{\partial^2 v}{\partial x^2} + \frac{\partial^2 v}{\partial y^2} + \frac{\partial^2 v}{\partial z^2}\right) = 0$$

$$-\frac{\partial p}{\partial z} + \eta\left(\frac{\partial^2 w}{\partial x^2} + \frac{\partial^2 w}{\partial y^2} + \frac{\partial^2 w}{\partial z^2}\right) = 0$$ (1.4-11)

$$\frac{\partial u}{\partial x} + \frac{\partial v}{\partial y} + \frac{\partial w}{\partial z} = 0$$

This is a set of linear equations (non-linearity would come only from inertial terms such as $u\partial u/\partial x \ldots$).

1.4-4   Solution of the Navier-Stokes equations

a)   Analytical solutions

Analytical solutions of the Navier-Stokes equations exist in a small number of favorable cases. The most important of these are given as problems at the end of this section and solutions to these problems are given in Table 1.5-1. In subsequent chapters we will cover approximate methods that are valid in two important practical situations:
–   the hydrodynamic lubrication approximations for shear flows (Chapter 3);
–   the approximation of thin shell flows in the case of drawing (Chapter 5).

b)   Numerical methods

There exist also numerical methods that can be used to solve the Navier-Stokes equations for two-dimension flows (and with more difficulty for three-dimension flows). These are known as the finite-differences and the finite-elements methods. It is impractical to present here the details of these numerical methods, but we note that they are of two types:
–   Methods based on the primary variables, which consist of solving a set of equations with respect to unknowns u, v, p

$$\eta\nabla^2 u = \frac{\partial p}{\partial y}$$

$$\eta\nabla^2 v = \frac{\partial p}{\partial y}$$ (1.4-12)

$$\frac{\partial u}{\partial x} + \frac{\partial v}{\partial y} = 0$$

One of the two first equations may be replaced by:

$$\nabla^2 p = 0$$ (1.4-13)

– methods which consist of solving the stream and vorticity functions defined below.

• The stream function is defined by the following relations:

$$u = \frac{\partial \psi}{\partial y}$$
$$v = -\frac{\partial \psi}{\partial x}$$

(1.4-14)

Any flow field derived from a stream function as defined by Equation (1.4-14) automatically satisfies the incompressibility condition since:

$$\frac{\partial u}{\partial x} = \frac{\partial^2 \psi}{\partial x\, \partial y} = -\frac{\partial v}{\partial y}$$

The difference between values of the stream function at two points represents the flow rate across an arbitrary line connecting these two points, as shown in Figure 1.4-2:

$$Q = \int_{AB} \boldsymbol{u} \cdot \boldsymbol{n}\, dl = \int_{AB} \frac{\partial \psi}{\partial x}\, dx + \frac{\partial \psi}{\partial y}\, dy = \psi_B - \psi_A .$$

(1.4-15)

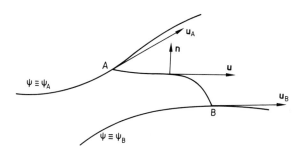

Figure 1.4-2   The stream function

At any time t, the stream lines of constant ψ are the tangents to the velocity vector; in steady-state flows, the streamlines coincide with the pathlines.

• The vorticity for three-dimension flows is taken to be the curl of the velocity vector:

$$\boldsymbol{\Omega} = \frac{1}{2} \nabla \times \boldsymbol{u}$$

(1.4-16)

For two dimension flows, the vorticity may be taken as a scalar function:

$$\Omega = \frac{1}{2}\left(\frac{\partial v}{\partial x} - \frac{\partial u}{\partial y} \right) = -\frac{1}{2}\nabla^2 \psi$$

(1.4-17)

Under these conditions the two-dimension Navier-Stokes equations may be written as two independent equations with two unknowns, p and ψ:

$$\nabla^2 p = 0$$
$$\nabla^2(\nabla^2 \psi) = 0 \tag{1.4-18}$$

where ψ is a biharmonic function.

The methods based on the stream and the vorticity functions consist of solving these equations rewritten as follows:

$$\Omega = -\frac{1}{2}\nabla^2 \psi$$
$$\nabla^2 \Omega = 0 \tag{1.4-19}$$

Examples of numerical results

The following results have been obtained for flow geometries related to polymer processing:

– Calendering flow (MITSOULIS et al., 1985; AGASSANT and ESPY, 1985), where the liquid in the gap is set in motion by the two cylinders rotating at velocity V. Figure 1.4-3 shows the formation of recirculation zones.

Figure 1.4-3 Numerical solution of Newtonian calendering flow

– Sudden contraction flow (AGASSANT et al., 1984). The situation is encountered in extrusion. As illustrated in Figure 1.4-4, secondary flows or vortices are developed in the entry reservoir of extrusion dies. These vortices lead to hydrodynamic instabilities (see Chapter 6) and to chemical degradation of material that stays in the stagnant zones. Through such numerical calculations, these formations can be predicted and eliminated.

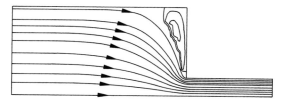

Figure 1.4-4 Numerical solution of Newtonian flow in a sudden planar contraction

Problems 1.4

1.4-A Navier-Stokes equations for two-dimension flow

Consider a Newtonian liquid of viscosity η, and assume that the inertial and gravitational forces are negligible and that the flow is bi-directional under steady-state conditions (these hypotheses apply in the following problems).

Obtain the simplified components of the Navier-Stokes equations.

1.4-B Simple shear flow between two parallel plates

A plate is displaced at constant velocity U in the z-direction with respect to a fixed plate separated by a distance h. The flow field is assumed to be $u = 0$, $v = 0$, $w = w(y)$.

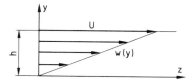

Figure 1.4-5 Simple shear flow between two parallel plates

Show that the pressure is uniform and that $w(y) = Uy/h$.

1.4-C Flow in a thin slit

Consider the flow in a thin slit made of two parallel walls a distance h apart, of width W and of length L, as illustrated in Figure 1.4-6. The pressure difference between the entrance and exit sections is Δp; it is assumed that the velocity is in the z-direction and that w is only a function of y (the edge effects are hence assumed to be negligible, which is valid for $h \ll W$). The flow is therefore uniform with respect to x and z:

$$u = 0$$

$$v = 0$$

$$w = w(y)$$

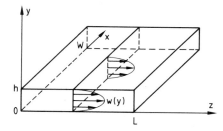

Figure 1.4-6 Flow in a thin slit

a) Obtain the distribution function for the pressure $p(x, y, z)$.
b) Obtain the expression for the velocity profile $w(y)$ and the relation between Δp and the volumetric flow rate Q.

1.4-D   Flow through a circular tube (Poiseuille flow)

A liquid is flowing through a circular tube of radius R and of length L, as illustrated in Figure 1.4-7. Cylindrical coordinates are used with z being the axis of the tube. A pressure difference Δp is exerted between the entrance and exit sections and the velocity is assumed to be in the z-direction only:

$$u = 0$$
$$v = 0$$
$$w = w(r)$$

The flow is symmetric with respect to the radial position r and uniform with respect to z.

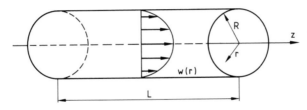

Figure 1.4-7   Flow through a circular tube

a) Obtain the pressure distribution $p(r, \theta, z)$.
b) Derive the expression for the velocity profile $w(r)$ and obtain the relation between Δp and the volumetric flow rate Q.

1.4-E   Simple shear flow between parallel disks

A disk is rotating at angular velocity Ω_o (radiants/s) with respect to another parallel disk at a distance h. The flow hypotheses of Problem 1.1-A are considered valid here.

a) Show that the pressure is uniform and that:

$$\Omega(z) = \Omega_0 \frac{z}{h}$$

b) What is the expression of the shear rate at any point?
c) Derive a relation between the torque and the angular velocity.

1.4-F Flow in a cone-and-plate viscometer

A cone of small angle $\Delta\theta$ is rotating at angular velocity Ω_o above a fixed disk as illustrated in Figure 1.4-8.

a) Show that the shear rate is practically uniform, i.e. constant across the gap.
b) Obtain the expression for the torque.

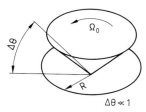

Figure 1.4-8 Flow in a cone-and-plate viscometer

1.4-G Couette flow

Using the same approach as for the previous problems, solve for the flow between two concentric cylinders, one of the cylinder rotating at the angular velocity Ω_o.

1.4-H Converging flow in a dihedron or in a cone

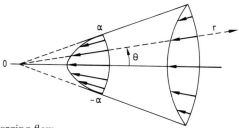

Figure 1.4-9 Converging flow

I) *dihedron:*
   We assume that, in cylindrical coordinates,

$$u = u(r, \theta), \quad v = w = 0.$$

a) Show that:

$$u = \frac{f(\theta)}{r}$$

b) Using the boundary conditions

$$f(\alpha) = f(-\alpha) = 0$$

   obtain a function f that satisfies the equations of Navier-Stokes.

c) Draw the lines of constant pressure (isobars) and the stream lines.

II) *cone:*
This is similar to case I) but in spherical coordinates. Show that:

$$u = \frac{f(\theta)}{r^2}$$

$$v = w = 0$$

1.4-I   Expansion of a cylindrical element

A cylindrical shell of material is submitted to a radial pressure difference $p_1 - p_2$ (without stresses along the axis of the cylinder) as shown in Figure 1.4-10. The velocity field is assumed to be of the following form:

$$u = \frac{A}{r}$$

$$v = w = 0$$

a) Obtain the expression for the velocity in the material.
b) Determine the pressure and the state of stresses in the material.
c) Repeat a) and b) for the limiting case of a very thin wall.

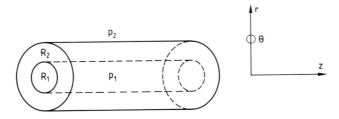

Figure 1.4-10 Expansion of a cylindrical element

1.4-J   Expansion of a spherical element

Solve the case of a spherical shell for the same conditions as in Problem 1.4-I.

1.5   Shear-thinning behaviour

1.5-1 Phenomenological description

It is well known that molten polymers do not behave as Newtonian liquids. Their apparent viscosity is usually a decreasing function of the shear rate $\dot{\gamma}$; this is referred to as shear-thinning or pseudoplastic behaviour. The most commonly used expression for the shear-thinning behaviour is the power-law or Ostwald-de Waele model. In simple shear:

$$\sigma = K|\dot{\gamma}|^{n-1}\dot{\gamma} \tag{1.5-1}$$

which gives the following expression for the viscosity.

$$\eta(\dot{\gamma}) = K|\dot{\gamma}|^{n-1} \tag{1.5-2}$$

where K is generally called the consistency index and n, the power-law index. The two extremes of behaviour are:
- Newtonian behaviour if $n = 1$
- Plastic behaviour   if $n = 0$

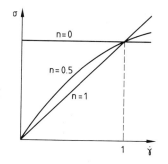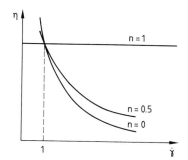

Figure 1.5-1   Shear-thinning behaviour

The relation between η and $\dot{\gamma}$ is usually represented on log-log scales; the power-law behaviour is then given by a straight line of slope $n - 1$.

A major drawback of the power-law model is that it predicts an infinite value for the viscosity as the shear rate tends to zero. Most molten polymers, on the contrary, show a finite and constant (Newtonian) viscosity at very low shear rates. Figure 1.5-3 illustrates a typical viscosity curve $\eta(\dot{\gamma})$ for a polymer melt.

To describe such behaviour, various empirical or semitheoretical models have been proposed:
- The *Spriggs truncated power-law model* (SPRIGGS, 1965)

$$\eta = \eta_0 \qquad\qquad \text{if } \dot{\gamma} < \dot{\gamma}_0$$

$$\eta = \eta_0 \left|\frac{\dot{\gamma}}{\dot{\gamma}_0}\right|^{n-1} \qquad \text{if } \dot{\gamma} > \dot{\gamma}_0 \tag{1.5-3}$$

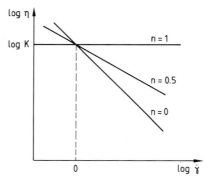

Figure 1.5-2   The viscosity as described by power law

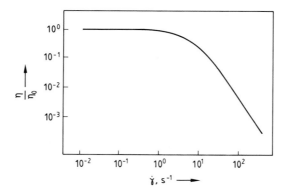

Figure 1.5-3   Typical viscosity curve for a polymer melt

where η_0 is the zero-shear viscosity.

–   The *Carreau model* (CARREAU et al., 1979)

$$\eta - \eta_0 = \frac{\eta_0 - \eta_\infty}{[1 + (\lambda\dot{\gamma})^2]^{(1-n)/2}} \qquad (1.5\text{-}4)$$

where λ is a characteristic time and η_∞ is a constant viscosity at very high shear rates.

–   The *hyperbolic tangent model* (ALSTON and ASTILL, 1973)

$$\eta = A - B \tanh\left(\frac{\dot{\gamma}}{k}\right)^n \qquad (1.5\text{-}5)$$

where the parameters A, B, k, and n are determined from the experimental data. This model verifies the two Newtonian limits:

$$\dot{\gamma} \to 0 \qquad \eta = A = \eta_0$$
$$\dot{\gamma} \to \infty \qquad \eta = A - B = \eta_\infty$$

Such empirical or semi-theoretical expressions have rarely been used for modelling polymer processes, since the power-law expression, in spite of its large departure from reality at low shear rates, generally gives a satisfactory description of the flow.

1.5-2 Three-dimension models

A generalized shear rate can be written as (see Appendix C):

$$\dot{\gamma} = \sqrt{2 \sum_{i,j} \dot{\varepsilon}_{ij}^2} \qquad (1.5\text{-}6)$$

The power-law model can then be generalized for three-dimension flow problems as (see Appendix C):

$$\sigma' = 2K(\dot{\bar{\gamma}})^{n-1}\dot{\varepsilon} \qquad (1.5\text{-}7)$$

In the case of simple shear:

$$
\left.\begin{array}{l}
u(y) = \dot{\gamma}y \\
v \equiv 0 \\
w \equiv 0
\end{array}\right\}
\Rightarrow
[\dot{\varepsilon}] = \begin{bmatrix} 0 & \frac{\dot{\gamma}}{2} & 0 \\ \frac{\dot{\gamma}}{2} & 0 & 0 \\ 0 & 0 & 0 \end{bmatrix}, \quad \dot{\bar{\gamma}} = \dot{\gamma}
$$

and consequently:

$$
[\boldsymbol{\sigma}] = \begin{bmatrix} -p & K|\dot{\gamma}|^{n-1}\dot{\gamma} & 0 \\ K|\dot{\gamma}|^{n-1}\dot{\gamma} & -p & 0 \\ 0 & 0 & -p \end{bmatrix} \qquad (1.5\text{-}8)
$$

Obviously, this last result describes the shear stress behaviour illustrated in Figure 1.5-1. We note, however, that the power-law model, as Newton's law, does not yield any normal stress components in simple shear.

1.5-3 Simple shear flow between parallel plates

Let us now consider a simple shear flow in a split (two parallel plates) where the velocity u varies from zero to U at the position h:

$$
\begin{array}{l}
u = u(y) \\
v = 0 \\
w = 0
\end{array}
$$

The ambient pressure is assumed to be uniform and the stress tensor is given, for $du/dy > 0$, by the following expression:

$$
[\boldsymbol{\sigma}] = \begin{bmatrix} -p & K\left(\frac{du}{dy}\right)^n & 0 \\ K\left(\frac{du}{dy}\right)^n & -p & 0 \\ 0 & 0 & -p \end{bmatrix} \qquad (1.5\text{-}9)
$$

The three components of the equation of motion become:

$$\frac{\partial p}{\partial x} = K \frac{d}{dy} \left| \frac{du}{dy} \right|^n \qquad (1.5\text{-}10)$$

$$\frac{\partial p}{\partial y} = 0 \qquad (1.5\text{-}11)$$

$$\frac{\partial p}{\partial z} = 0 \qquad (1.5\text{-}12)$$

It follows from Equations (1.5-11) and (1.5-12) that the pressure is a function of x alone and consequently, Equation (1.5-10) can be written as:

$$\frac{dp}{dx} = K \frac{d}{dy} \left| \frac{du}{dy} \right|^n \qquad (1.5\text{-}13)$$

Since u is a function of y alone, we get:

$$\frac{dp}{dx} = K \frac{d}{dy} \left| \frac{du}{dy} \right|^n = \text{constant}$$

and the constant is equal to zero since the pressure of the surroundings is uniform.
    The velocity profile established in shear flow is, in the absence of a pressure gradient, linear for any value of the power-law parameter n, except for $n = 0$. In that case the profile can be any monotonous function varying from 0 to U, and the velocity profile may show a discontinuity of the following type:

$$u = 0 \quad \text{if } y < y_D$$
$$u = U \quad \text{if } y > y_D$$

where y_D is not fixed between 0 and h.

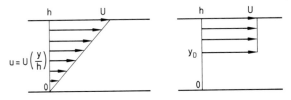

Figure 1.5-4   Velocity profile of a shear-thinning material in simple shear flow

1.5-4   Flow through a circular tube (Poiseuille flow)

We consider the flow of a shear-thinning fluid through a circular tube of radius R and length L, with a pressure difference Δp applied between the entrance and exit sections.

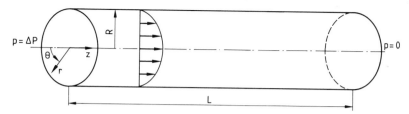

Figure 1.5-5   Flow through a circular tube

In cylindrical coordinates, the velocity profile for steady-state conditions becomes:

$$u = 0, \quad v = 0, \quad w = w(r)$$

which implies that:

$$[\dot{\varepsilon}] = \begin{bmatrix} 0 & 0 & \frac{1}{2}\frac{dw}{dr} \\ 0 & 0 & 0 \\ \frac{1}{2}\frac{dw}{dr} & 0 & 0 \end{bmatrix}$$

Hence the stress tensor is given by:

$$[\sigma] = \begin{bmatrix} -p & 0 & K\left|\frac{dw}{dr}\right|^{n-1}\frac{dw}{dr} \\ 0 & -p & 0 \\ K\left|\frac{dw}{dr}\right|^{n-1}\frac{dw}{dr} & 0 & -p \end{bmatrix} \qquad (1.5\text{-}14)$$

and the equation of motion (see Appendix A) reduces:

$$\frac{\partial \sigma_{rr}}{\partial r} = 0 \;\Rightarrow\; \frac{\partial p}{\partial r} = 0 \qquad (1.5\text{-}15)$$

$$\frac{1}{r}\frac{\partial \sigma_{\theta\theta}}{\partial \theta} = 0 \;\Rightarrow\; \frac{\partial p}{\partial \theta} = 0 \qquad (1.5\text{-}16)$$

$$\frac{\partial \sigma_{rz}}{\partial r} + \frac{\partial \sigma_{zz}}{\partial z} + \frac{\sigma_{rz}}{r} = 0 \;\Rightarrow\; \frac{\partial p}{\partial z} = K\frac{d}{dr}\left[\left|\frac{dw}{dr}\right|^{n-1}\frac{dw}{dr}\right] + \frac{K}{r}\left|\frac{dw}{dr}\right|^{n-1}\frac{dw}{dr}$$

$$\qquad (1.5\text{-}17)$$

It follows from Equations (1.5-15) and (1.5-16) that the pressure is a unique function of z and Equation (1.5-17) is reduced to:

$$\frac{dp}{dz} = K\frac{d}{dr}\left[\left|\frac{dw}{dr}\right|^{n-1}\frac{dw}{dr}\right] + \frac{K}{r}\left|\frac{dw}{dr}\right|^{n-1}\frac{dw}{dr} = \text{constant}.$$

The constant is evaluated from the pressure drop Δp, i.e.:

$$\frac{dp}{dz} = -\frac{\Delta p}{L} \qquad (1.5\text{-}18)$$

We can now write the expression for the velocity profile:

$$w(r) = \frac{n}{n+1}\left[\frac{1}{2K}\frac{\Delta P}{L}\right]^{1/n} R^{1+1/n}\left[1 - \left(\frac{r}{R}\right)^{1+1/n}\right] \qquad (1.5\text{-}19)$$

and the expression for the volumetric flow rate:

$$Q = \int_0^R w(r)\,2\pi r\,dr = \pi\frac{n}{3n+1}\left[\frac{1}{2K}\frac{\Delta P}{L}\right]^{1/n} R^{3+1/n} \qquad (1.5\text{-}20)$$

The velocity profile can be expressed in terms of the average velocity $V = Q/\pi R^2$:

$$w(r) = \frac{3n+1}{n+1}V\left[1 - \left(\frac{r}{R}\right)^{1+1/n}\right] \qquad (1.5\text{-}21)$$

Figure 1.5-6 shows that the velocity profile becomes flatter as the power-law index decreases from 1 to 0 (n for polymer melts is usually in the range of 0.2 to 0.5). For low values of n, it is difficult to distinguish between the flow of a shear-thinning liquid with no slip conditions at the wall, and the flow of an homogeneous liquid with slip.

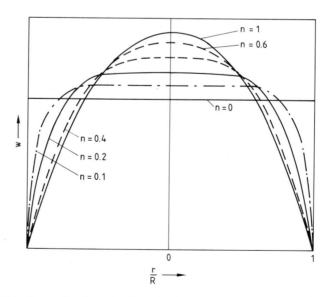

Figure 1.5-6   Velocity profiles in a circular tube for various values of the power-law index

1.5-5 Viscometric flows

Table 1.5-1 summarizes the results for the main flow geometries used to measure the viscosity of Newtonian and non-Newtonian liquids. The Newtonian viscosity is obtained either through a pressure drop – flow rate or a torque – angular rotation relationship. Also the power-law index n can be determined from expressions of the following types:

$$\text{Log}\frac{\Delta P}{L} = n\,\text{Log}Q + \text{const.} \qquad (1.5\text{-}22)$$

$$\text{Log}C = n\,\text{Log}\Omega + \text{const.} \qquad (1.5\text{-}23)$$

However, in a number of geometries, such as circular tubes, the shear rate varies considerably across the flow section from the center to the periphery, while expressions such as Equation (1.5-22) are only representative of an average value of the viscosity. Capillary flow data can be used to obtain the viscosity corresponding to the shear rate evaluated at the wall. The method is the following:

– for any rheological behaviour, the shear stress at the wall under steady-state conditions is given by the expression:

$$\Delta P \pi R^2 = \sigma_w 2\pi RL, \quad \text{hence } \sigma_w = \frac{\Delta PR}{2L} \qquad (1.5\text{-}24)$$

– the shear rate evaluated at the wall is obtained from the expressions of the velocity profile given in Table 1.5-1:

a) *for Newtonian fluids:*

$$\dot{\Gamma}_w = \frac{4Q}{\pi R^3} \qquad (1.5\text{-}25)$$

which is a unique function of the flow rate Q;

b) *for shear-thinning fluids:*

$$\dot{\gamma}_w = \frac{3n+1}{n}\frac{Q}{\pi R^3} \qquad (1.5\text{-}26)$$

which depends on the power-law index as well as on the flow rate. Obviously $\dot{\gamma}_w$ can be written as:

$$\dot{\gamma}_w = \frac{3n+1}{4n}\dot{\Gamma}_w \qquad (1.5\text{-}27)$$

– the relationship between the shear stress and the shear rate at the wall is obtained for a power-law fluid from Equation (1.5-1):

$$\sigma_w = K\dot{\gamma}_w{}^n$$

and combining with Equation (1.5-27) we get:

$$\sigma_w = K\left(\frac{3n+1}{4n}\right)^n \dot{\Gamma}_w{}^n \qquad (1.5\text{-}28)$$

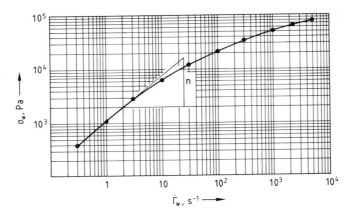

Figure 1.5-7   Determination of the power-law index (polypropylene)

The power-law index n can be obtained from a plot on log-log scales of σ_w (which is a unique function of the pressure drop $\Delta P/L$) as a function of $\dot{\Gamma}_w$ which depends only on the flow rate Q. Such a plot is presented in Figures 1.5-7.

The shear rate at the wall can now be evaluated from Equation (1.5-27); hence the non-Newtonian viscosity evaluated at the wall is given by:

$$\eta = \frac{\sigma_w}{\dot{\gamma}_w} \tag{1.5-29}$$

Figure 1.5-8 reports viscosity data obtained from capillary flow for a polypropylene.

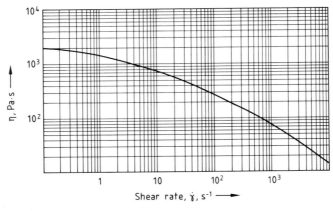

Figure 1.5-8   Viscosity data from capillary flow (polypropylene)

1.5-6   Computation methods in more complex flows

The power-law behaviour allows analytical solutions of viscometric flows (see Table 1.5-1) or of more complex flows when hydrodynamic approximations or narrow shell approximations are valid. Numerical solutions are possible as in the Newtonian case with methods using primary variables (AGASSANT et al., 1984) as well as with methods based on the stream and vorticity functions (AVENAS, 1980).

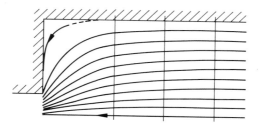

Figure 1.5-9   Computed stream lines for a power-law fluid ($n = 0.4$) in an abrupt contraction (from AVENAS, 1980). The corner vortices are no longer visible.

Table 1.5-1 Viscometric flows

	Newtonian	Shear – thinning (power laws)
Simple shear flow	$u(y) = U\dfrac{y}{h}$ $F = WL\eta\dfrac{U}{h}$	$u(y) = U\dfrac{y}{h}$ $F = WLK\left(\dfrac{U}{h}\right)^n$
Flow in a slit	$u(y) = \dfrac{3}{2}V\left[1 - \left(\dfrac{2y}{h}\right)^2\right]$ $Q = \dfrac{1}{12\eta}\dfrac{\Delta P}{L}h^3 W$	$u(y) = \dfrac{2n+1}{n+1}V\left[1 - \left(\dfrac{2|y|}{h}\right)^{(1/n)+1}\right]$ $Q = \dfrac{n}{2(2n+1)}\left(\dfrac{1}{2K}\dfrac{\Delta P}{L}\right)^{1/n}h^{(1/n)+2}W$
Flow in a tube	$w(r) = 2V\left[1 - \left(\dfrac{r}{R}\right)^2\right]$ $Q = \dfrac{\pi}{8\eta}\dfrac{\Delta P}{L}R^4$	$w(r) = \dfrac{3n+1}{n+1}V\left[1 - \left(\dfrac{r}{R}\right)^{(1/n)+1}\right]$ $Q = \pi\dfrac{n}{3n+1}\left[\dfrac{1}{2K}\dfrac{\Delta P}{L}\right]^{1/n}R^{(1/n)+3}$
Couette flow	$v(r) = \dfrac{\Omega_0}{(R_2/R_1)^2 - 1}\cdot$ $\cdot\left(\dfrac{R_2^2 - r^2}{r}\right)$ $C = 4\pi\eta\dfrac{R_2^2\Omega_0}{(R_2/R_1)^2 - 1}L$	$v(r) = \dfrac{\Omega_0}{(R_2/R_1)^{2/n} - 1}\left[\dfrac{R_2^{2/n} - r^{2/n}}{r^{2/n-1}}\right]$ $C = 2^{n+1}\pi K(R_1^2 R_2^2)\left[\dfrac{\Omega_0}{n(R_2^{2/n} - R_1^{2/n})}\right]^n L$
Flow between parallel disks	$v(r,z) = r\Omega_0\dfrac{z}{h}$ $C = \dfrac{\pi}{2}\eta\dfrac{\Omega_0}{h}R^4$	$v(r,z) = r\Omega_0\dfrac{z}{h}$ $C = \dfrac{2\pi}{n+3}K\left(\dfrac{\Omega_0}{h}\right)^n R^{n+3}$
Cone-and-plane flow	$w(r,\theta) = r\dfrac{\Omega_0}{\Delta\theta}\theta$ $C = \dfrac{2\pi}{3}\eta\dfrac{\Omega_0}{\Delta\theta}R^3$	$w(r,\theta) = r\dfrac{\Omega_0}{\Delta\theta}\theta$ $C = \dfrac{2\pi}{3}K\left(\dfrac{\Omega_0}{\Delta\theta}\right)^n R^3$

Chapter 2
Energy and Heat Transfer Processes

2.1 Basic notions on energy

2.1-1 Work done by deformation

a) Simple flow geometries

Let us consider first simple cases of work done by deformation.

Extension or compression

The work done to stretch a cylindrical element of material by Δl (in Figure 2.1-1) under a force F is:

$$\Delta W = F \cdot \Delta l = \left(\frac{F}{S} \cdot \frac{\Delta l}{l} \right) (S \cdot l) = \sigma_{xx} \varepsilon_{xx} V \tag{2.1-1}$$

Here the work done per unit volume is:

$$W = \frac{\Delta W}{V} = \sigma_{xx} \varepsilon_{xx} \tag{2.1-2}$$

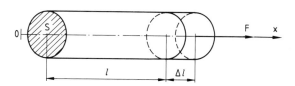

Figure 2.1-1   Work done in extension

Simple shear

The work done to shear an element of material by the quantity a (see Figure 2.1-2) under a force F is:

$$W = F \cdot a = \frac{F}{S} \cdot \frac{a}{h} \cdot S \cdot h = \sigma_{xy} \varepsilon_{xy} V \tag{2.1-3}$$

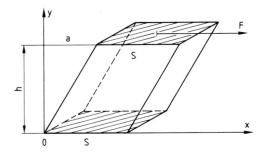

Figure 2.1-2   Work done in simple shear

Hence, the work done per unit volume is:

$$\Delta W = \frac{\Delta W}{V} = \sigma_{xy}\varepsilon_{xy} \tag{2.1-4}$$

b) Generalization

One can show that the general expression, which includes the two previous examples as special cases, is:

$$W = \sum_{i,j} \sigma_{ij}\varepsilon_{ij} = [\boldsymbol{\sigma}] : [\boldsymbol{\varepsilon}] \tag{2.1-5}$$

c) Dissipated power

The work done to deform an element of material during the interval dt is:

$$dW = [\boldsymbol{\sigma}] : [\boldsymbol{\varepsilon}]_t^{t+dt} = [\boldsymbol{\sigma}] : [\dot{\boldsymbol{\varepsilon}}]dt \tag{2.1-6}$$

provided that the deformations are small.
The expression for the dissipated power is then:

$$\boxed{\dot{W} = [\boldsymbol{\sigma}] : [\dot{\boldsymbol{\varepsilon}}] = \sum_{i,j} \sigma_{ij}\dot{\varepsilon}_{ij}} \tag{2.1-7}$$

Using the deviatoric stress tensor and the hydrostatic pressure, Equation (2.1-7) can be written as:

$$\dot{W} = -p\sum_{i,j} \dot{\varepsilon}_{ij}\delta_{ij} + \sum_{i,j} \sigma'_{ij}\dot{\varepsilon}_{ij} \tag{2.1-8}$$

where δ_{ij} is the Kronecker delta. It follows that:

$$\dot{W} = -p\sum_{i} \dot{\varepsilon}_{ii} + \sum_{i,j} \sigma'_{ij}\dot{\varepsilon}_{ij}$$

For incompressible materials ϱ is constant and from Equation (1.1-23):

$$tr[\dot{\varepsilon}] = \sum_i \dot{\varepsilon}_{ii} = 0$$

Therefore

$$\dot{W} = \sum_{i,j} \sigma'_{ij} \dot{\varepsilon}_{ij} \tag{2.1-9}$$

and no work is done by pressure.

d) Newtonian and shear-thinning liquids

For Newtonian liquids, the deviatoric stress tensor is given by Equation (1.4-2):

$$\boldsymbol{\sigma}' = 2\eta\dot{\boldsymbol{\varepsilon}}$$

and the rate of work is (see Appendix B)

$$\dot{W} = 2\eta \sum_{i,j} \dot{\varepsilon}_{ij}{}^2 = \eta\dot{\bar{\gamma}}^2 \tag{2.1-10}$$

For shear-thinning liquids obeying the power-law expression (1.5-7)

$$\boldsymbol{\sigma}' = 2K\left(\dot{\bar{\gamma}}\right)^{n-1}\dot{\boldsymbol{\varepsilon}}$$

the rate of work is

$$\dot{W} = 2K\left(\dot{\bar{\gamma}}\right)^{n-1} \sum_{i,j} \dot{\varepsilon}_{ij}{}^2 = K\left(\dot{\bar{\gamma}}\right)^{n+1} \tag{2.1-11}$$

For $n = 1$ and $K = \eta$, we recover the expression for Newtonian liquids.

2.1-2 The equation of energy

The equation of energy is obtained for a closed system from a balance of the rate of accumulation of internal energy, the rate of heat generation in the system, and the rate of heat exchanged by conduction with the surroundings.

a) Internal energy

If one neglects the effects of possible chemical or physical transformations in the material, the internal energy is related to the material temperature through the definition of the heat capacity c_p. The rate of accumulation of internal energy per unit volume for incompressible materials is then:

$$\frac{DU}{Dt} = \varrho c_p \frac{DT}{Dt} \tag{2.1-12}$$

Here D/DT is a total or substantial derivative, as it represents the variation with time following an element of material in its displacement. For example, in rectangular coordinates:

$$\frac{DT(x,y,z,t)}{Dt} = \frac{\partial T}{\partial t} + u\frac{\partial T}{\partial x} + v\frac{\partial T}{\partial y} + w\frac{\partial T}{\partial z}$$

$$= \frac{\partial T}{\partial t} + \boldsymbol{u} \cdot \nabla T \tag{2.1-13}$$

The term $\varrho c_p \partial T/\partial t$ expresses the rate of variation of internal energy at a fixed position in space (this term is equal to zero for steady-state conditions). The term $\varrho c_p \boldsymbol{u} \cdot \nabla T$ is the rate of heat transferred by motion of the material. It is the convection term and it is equal to zero if the material is at rest.

b) Conduction

The heat exchanged by conduction with the surroundings is related to a transfer flux $\boldsymbol{q}$. From a balance of these exchanges, the rate of heat transferred per unit volume is:

$$Q = -\nabla \cdot \boldsymbol{q} \tag{2.1-14}$$

A constitutive relationship for the thermal behaviour of the material is now required. The simplest one is Fourier's law of heat conduction which states that the heat flux is proportional to the temperature gradient and the direction of heat flux is from high to low temperature:

$$\boldsymbol{q} = -k\nabla T \tag{2.1-15}$$

For a constant heat conductivity k, Equation (2.1-14) becomes:

$$Q = k\nabla \cdot (\nabla T) = k\nabla^2 T = k\left[\frac{\partial^2 T}{\partial x^2} + \frac{\partial^2 T}{\partial y^2} + \frac{\partial^2 T}{\partial z^2}\right] \tag{2.1-16}$$

in rectangular coordinates.

c) Equation of energy

The overall energy balance is then:

$$\underbrace{\varrho c_p \frac{DT}{Dt}}_{\substack{\text{rate of accumulation} \\ \text{of the internal energy} \\ \text{of a material element}}} = \underbrace{\varrho c_p \frac{\partial T}{\partial t}}_{\substack{\text{rate of variation} \\ \text{of the internal} \\ \text{energy at a fixed} \\ \text{point in space}}} + \underbrace{\varrho c_p \boldsymbol{u} \cdot \nabla T}_{\substack{\text{convection of} \\ \text{energy}}} = \underbrace{k\nabla^2 T}_{\substack{\text{conduction} \\ \text{(or diffusion)} \\ \text{of energy}}} + \underbrace{\dot{W}}_{\substack{\text{rate of} \\ \text{generation} \\ \text{of energy}}}$$

$$\tag{2.1-17}$$

The analogy between the equation of energy and the Navier-Stokes equation (Eq. 1.4-10) is interesting; energy is analogous to momentum, which appears in Equation (1.4-10):

$$\underbrace{\varrho\frac{D\boldsymbol{u}}{Dt}}_{\substack{\text{rate of accumulation}\\\text{of the momentum of a}\\\text{material element}}} = \underbrace{\varrho\frac{\partial\boldsymbol{u}}{\partial t}}_{\substack{\text{rate of variation}\\\text{of the momentum}\\\text{at a fixed point}\\\text{in space}}} + \underbrace{\varrho\boldsymbol{u}\cdot\nabla\boldsymbol{u}}_{\substack{\text{convection of}\\\text{momentum}}} = \underbrace{\eta\nabla^2\boldsymbol{u}}_{\substack{\text{diffusion of}\\\text{momentum}}} + \underbrace{\varrho\boldsymbol{F} - \nabla p}_{\substack{\text{rate of}\\\text{generation}\\\text{of momentum}}}$$

Remark

The equation of energy describes a local heat balance. If we integrate each of its terms or make a heat balance over a finite volume V of surface S, we obtain the macroscopic equation of energy:

$$\int_V \varrho c_p \frac{DT}{Dt}dV = \int_V \dot{W}\,dV - \int_S (\boldsymbol{q}\cdot\boldsymbol{n})\,dS \tag{2.1-18}$$

where $\boldsymbol{n}$ is the unit vector normal to the surface. As an example, for steady-state flow in the x-direction, we obtain:

$$\varrho c_p \bar{u}\frac{D\bar{T}}{Dx}V = \dot{W}_V - Q_S \tag{2.1-19}$$

where $\bar{u}$ and $\bar{T}$ are respectively the average velocity and average temperature, $\dot{W}_V$ is the total energy dissipated in the given volume and Q_S is the rate of energy transferred by conduction across the surface.

2.1-3 Boundary conditions

The temperature profile in a given material and its variation with time can be obtained in principle by solving the equation of energy. This, however, requires knowing the initial temperature profile and the boundary conditions for the system being considered. The initial temperature profile is usually known, but this is not always the case for the boundary conditions, which may be more or less complex depending on the interfaces.

The usual boundary conditions are:

a) Constant wall temperature

This is the case when the temperature control is very efficient so that the temperature of the wall is approximately constant and equal to T_0. This is known as an isothermal boundary condition:

$$T(x,0) = T_0 \tag{2.1-20}$$

The case is illustrated in Figure 2.1-3:

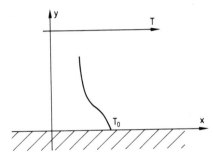

Figure 2.1-3   Isothermal boundary condition

b) Constant heat flux

This is the case when the wall can draw from or transmit heat to the material at a constant rate or flux. At the boundary:

$$k\frac{\partial T}{\partial y}(x, o) = q \qquad (2.1\text{-}21)$$

The situation is described in Figure 2.1-4:

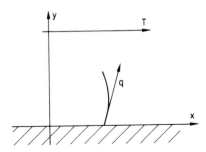

Figure 2.1-4   Constant heat flux boundary condition

The adiabatic condition is a special case for which the heat flux is equal to zero (perfectly isolated wall):

$$k\frac{\partial T}{\partial y}(x, o) = 0 \qquad (2.1\text{-}22)$$

c) Solid-fluid interface

When a solid (or a highly viscous liquid) is in contact with a low viscosity fluid (gas or liquid), heat is transferred through the interface by convection. The heat flux at the interface is then proportional to the temperature difference between the interface temperature and a characteristic temperature of the fluid, T_f (frequently taken as the temperature of the fluid at a distance far enough from the interface):

$$k\frac{\partial T}{\partial y}(x,0) = h[T(x,0) - T_f] \tag{2.1-23}$$

This relation is known as *Newton's law* of cooling and the corresponding boundary condition is illustrated in Figure 2.1-5.

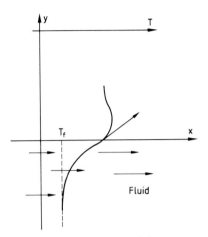

Figure 2.1-5   Convective heat transfer boundary condition

The heat transfer coefficient, h, cannot in general be accurately determined. Appendix D shows how h can be calculated using semi-empirical expressions for various flow geometries. The adiabatic boundary condition is obtained as a special case when h is equal to zero (no heat transferred through the interface).

Finally, in some applications (mainly for heat transfer in air), a large portion of the heat flux through the interface may be due to radiation. It is shown in Appendix D.3 that the radiation heat flux is proportional to the difference of the fourth power of the absolute temperature of the two surfaces exposed to each other (for example, a wall of an oven and walls of a room). This boundary condition, illustrated in Figure 2.1-6, is:

$$k\frac{\partial T}{\partial y}(x,0) = \sigma\varepsilon[T^4(x,0) - T_2{}^4] \tag{2.1-24}$$

where σ is the Stefan-Boltzman constant and ε is the surface emissivity.

Relation (2.1-24) is seldom used as such. By analogy with the convection heat transfer, we write:

$$k\frac{\partial T}{\partial y}(x,o) = h_r[T(x,o) - T_2] \tag{2.1-25}$$

where

$$h_r = \sigma\varepsilon\frac{T^4 - T_2{}^4}{T - T_2} \tag{2.1-26}$$

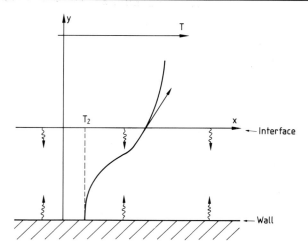

Figure 2.1-6   Radiation between two surfaces

The radiative heat transfer coefficient h_r varies with temperature, but for a small variation in T, it can be considered as a constant.

In summary, the isothermal, adiabatic or convective boundary conditions are used in most cases.

2.1-4 Solution of the heat transfer equation

The originality and the particularity of heat transfer problems in polymer processing are due to the following:

–   The heat dissipated by viscous forces of high viscosity polymer melts play an important role in many processes (see Section 2.3).
–   The viscosity is strongly dependent on temperature (see Appendix C); the importance of heat dissipation decreases as the melt processing temperature increases.
–   The low heat conductivity of polymers reduces the heat transferred to the surroundings (see Table 2.1-1 and Sections 2.2 and 2.3).

Various numerical techniques (finite differences, finite elements) can be used to solve the equation of energy which is usually coupled with the equations of motion. These techniques lead to complete solutions of flows in complex geometries under large temperature variations. However, approximate methods can be used to solve simple cases and also more complex cases so as to readily obtain a good approximation of temperature profiles. The approximate solutions are also useful for a rational development of more sophisticated numerical techniques.

The thermal properties of various materials are reported in Table 2.1-1.

Table 2.1-1   Thermal properties of various materials

	Conductivity, k		Heat capacity, ϱc_p	
	W/°C · m	cal/°C · cm	J/°C · m^3	cal/°C · m^3
Air	2.63 10^{-2}	0.63 10^{-4}	1.3 10^3	0.312 10^{-3}
Water	0.63	1.50 10^{-3}	4.18 10^6	1.0
Glass	0.84	2.0 10^3	2.17 10^6	0.52
Copper	3.89 10^2	0.93	3.43 10^6	0.82
Carbon steel	46	0.11	3.59 10^6	0.86
Stainless steel	14.6	3.5 10^{-2}	3.59 10^6	0.86
Polyethylene	0.33	8.0 10^{-4}	2.51 10^6	0.60
Polystyrene	0.125	3.0 10^{-4}	1.92 10^6	0.46

	Thermal diffusivity, α		Thermal effusivity, b	
	m^2/s	cm^2/s	J/°C · m^2 · s$^{1/2}$	cal/°C · cm^2 · s$^{1/2}$
Air	2.0 10^{-5}	2.0 10^{-1}	5.85	1.4 10^{-4}
Water	1.5 10^{-7}	1.5 10^{-3}	1.63 10^3	3.9 10^{-2}
Glass	3.8 10^{-7}	3.8 10^{-3}	1.34 10^3	3.2 10^{-2}
Copper	1.13 10^{-4}	1.13	3.76 10^4	0.90
Carbon steel	1.3 10^{-5}	0.13	1.25 10^4	0.30
Stainless steel	4.0 10^{-6}	4.0 10^{-2}	7.52 10^3	0.18
Polyethylene	1.3 10^{-7}	1.3 10^{-3}	9.2 10^2	2.2 10^{-2}
Polystyrene	6.5 10^{-7}	6.5 10^{-4}	5.02 10^2	1.2 10^{-2}

2.2   Cooling in molds, in air and in water

In this section, we are concerned mostly with cooling problems:
- cooling of parts in molds
- cooling of fibers and films in air or water
- cooling of polymer extrudates in air or in contact with a metallic surface.

The same approach can be applied to heating problems.

We consider that the heat generated or dissipated is zero or negligible in the case of fiber or film drawing. In problems of cooling of parts in molds, there is no motion and we look for solutions of the temperature profile as a function of time. In cases of cooling or heating of films or extrudates in contact with air, water or metallic surfaces, we seek solutions for the temperature profile as a function of the axial distance. We will show that these two classes of problems can be reduced to the same mathematical formulation, which draws this section together.

2.2-1   Heat transfer equations

In a body at rest, the heat transfer equation is reduced to

$$\frac{\partial T}{\partial t} = \alpha \nabla^2 T \qquad (2.2-1)$$

where ∇^2 is the Laplacian operator and α is the thermal diffusivity defined by

$$\alpha = \frac{k}{\varrho c_p} \qquad (2.2\text{-}2)$$

This section will be restricted to unidimensional problems for which the heat fluxes are transferred in the y-direction. The heat transfer equation is simplified to:

$$\boxed{\frac{\partial T}{\partial t} = \alpha \frac{\partial^2 T}{\partial y^2}} \qquad (2.2\text{-}3)$$

For a *body in motion*, as in the case of a film cooling in air or water, illustrated in Figure 2.2-1, the heat transfer equation is:

$$\varrho c_p u \frac{\partial T}{\partial x} = k \frac{\partial^2 T}{\partial y^2} \qquad (2.2\text{-}4)$$

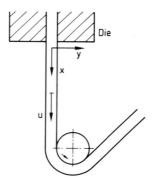

Figure 2.2-1   Cooling of a film

If there is no stretching of the film, the velocity u is constant and Equation (2.2-4) reduces to Equation (2.2-3) with

$$\frac{u \partial T}{\partial x} = \frac{\partial T}{\partial t} \qquad (2.2\text{-}5)$$

This problem is solved in Appendix D.4.

2.2-2 Heat penetration thickness and thermal diffusivity

Let us consider a semi-infinite body occupying the space from $y = 0$ to $y = \infty$. Initially, at time $t \leq 0$, the temperature of the body is T_0 and at $t = 0$, the surface at $y = 0$ is suddenly lowered to temperature T_S. That temperature is maintained constant for $t > 0$ as heat diffuses progressively in the semi-infinite body.

The solution to this problem is (see BIRD et al., 1960, Section 11.1):

$$\frac{T - T_s}{T_0 - T_s} = \text{erf} \frac{y}{2\sqrt{\alpha t}} \tag{2.2-6}$$

where erf( ) is the error function defined by:

$$\text{erf}(x) = \frac{2}{\sqrt{\pi}} \int_0^x e^{-u^2} du \tag{2.2-7}$$

Typical temperature profiles are illustrated in Figure 2.2-2 and Figure 2.2-3 shows how the dimensionless temperature $(T - T_s)/(T_0 - T_s)$ varies as a function of $y/2\sqrt{\alpha t}$.

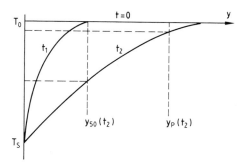

Figure 2.2-2   Temperature profiles in a semi-infinite body

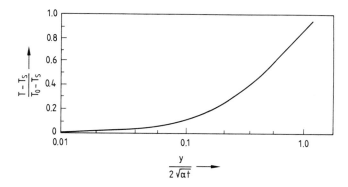

Figure 2.2-3   Dimensionless temperature profile in a semi-infinite body

For $t > 0$, heat penetrates in the body. The penetration thickness is arbitrary however: we may take it at a distance y_p where 1% of the temperature difference is felt:

$$T = T_0 + 0.01(T_s - T_0)$$

This penetration thickness is attained at a time t_p. From Equation (2.2-6), y_p and t_p are related by:

$$0.99 = \operatorname{erf} \frac{y_p}{2\sqrt{\alpha t_p}}$$

and from Table A.3 (Appendix A):

$$\frac{y_p}{2\sqrt{\alpha t_p}} = 1.82 \quad \text{and} \quad t_p = \frac{y_p^2}{13.25\alpha} \tag{2.2-8}$$

Similarly, 50% of the temperature difference is felt at distances and times related by:

$$y_{50} = 0.96\sqrt{\alpha t_{50}} \quad \text{and} \quad t_{50} = \frac{y_{50}^2}{0.92\alpha} \tag{2.2-9}$$

More generally, the penetration thickness increases with the square root of t. The time at which most of the temperature difference is felt at a distance y is of the following order:

$$\boxed{t = \frac{y^2}{\alpha}} \tag{2.2-10}$$

The thermal diffusivity for polymers is about $10^{-7}\,\mathrm{m^2/s}$, which yields the following approximate values for the penetration thickness as a function of time.

Table 2.2-1   Penetration thickness in polymers

y	$t = y^2/\alpha$
10 μ	0.001 s
100 μ = 0.1 mm	0.1 s
1 mm	10 s
10 mm	1000 s ($\approx$ 16.7 min.)

Table 2.2-1 shows that heating or cooling times are very short for thickness values below $100\,\mu$. This is the case for fibers, films and coatings. Heating and cooling times become important for thickness above 1 mm; this is one of the reasons why in classical injection molding thickness above 2 mm is rarely used. Table 2.2-1 also stresses the necessity for plasticating polymers in thin coats (of the order of 0.1 mm).

2.2-3 Interfacial temperature and effusivity

Let us consider two semi-infinite slabs initially at temperature T_1 and T_2 respectively. At $t = 0$, their surfaces ($y = 0$) are brought into contact. The thermal

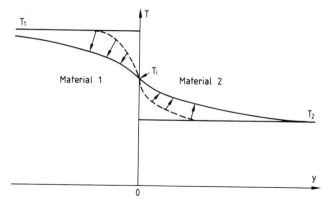

Figure 2.2-4   Evolution of the temperature profile for two semi-infinite slabs in close contact

diffusivities are respectively α_1 and α_2. The evolution of the temperature profile with time in the two slabs is shown in Figure 2.2-4.

The solution is obtained in a way similar to the previous case (Section 2.2-3). For each material, the temperature profile is given by an equation of the type (2.2-6):

$$\text{Material 1}: \quad T_1(y,t) = A_1 + B_1 \operatorname{erf} \frac{y}{2\sqrt{\alpha_1 t}} \tag{2.2-11}$$

$$\text{Material 2}: \quad T_2(y,t) = A_2 + B_2 \operatorname{erf} \frac{y}{2\sqrt{\alpha_2 t}} \tag{2.2-12}$$

If the contact between the two surfaces is perfect (no interfacial resistance, as in the case of a molten polymer which adheres very well to a metallic surface), the boundary conditions yield two relations:
– Temperature continuity of the interface:

$$T_1(0,t) = T_2(0,t)$$

which implies that $A_1 = A_2$.
– Heat flux continuity at the interface (provided no heat is generated at the interface as in the case of phase changes):

$$k_1 \frac{\partial T_1}{\partial y} = k_2 \frac{\partial T_2}{\partial y}$$

Hence:

$$\frac{k_1 B_1}{2\sqrt{\alpha_1 t}} \frac{2}{\sqrt{\pi}} = \frac{k_2 B_2}{2\sqrt{\alpha_2 T}} \frac{2}{\sqrt{\pi}}$$

Finally, from the initial conditions:

$$T_1(y,0) = T_1 \Rightarrow A_1 - B_1 = T_1$$
$$T_2(y,0) = T_2 \Rightarrow A_2 + B_2 = T_2$$

The four constants, A_1, A_2, B_1 and B_2, can be evaluated from the four boundary or initial conditions and the expressions for the temperature profiles are:

$$T_1(y,t) = \frac{b_1 T_1 + b_2 T_2}{b_1 + b_2} + \frac{b_2}{b_1 + b_2}(T_2 - T_1)\,\text{erf}\,\frac{y}{2\sqrt{\alpha_1 t}} \tag{2.2-13}$$

$$T_2(y,t) = \frac{b_1 T_1 + b_2 T_2}{b_1 + b_2} + \frac{b_1}{b_1 + b_2}(T_2 - T_1)\,\text{erf}\,\frac{y}{2\sqrt{\alpha_2 t}} \tag{2.2-14}$$

where $b = \sqrt{k \varrho c_p}$ is called the thermal effusivity. Values of b for various substances are given in Table 2.1-1. The interfacial temperature is readily obtained from Equations (2.2-13, 14):

$$T_i = \frac{b_1 T_1 + b_2 T_2}{b_1 + b_2} \tag{2.2-15}$$

It is interesting to note that the interfacial temperature is constant for any time $(t > 0)$ after contacting the two surfaces.

Numerical applications

–   *Contact polyethylene – stainless steel*

$b_{\text{polyethylene}} = 9.2 \cdot 10^2\,\text{J} \cdot {}^\circ\text{C}^{-1} \cdot \text{m}^{-2} \cdot \text{s}^{-1/2}$
$b_{\text{stainless steel}} = 7.52 \cdot 10^3\,\text{J} \cdot {}^\circ\text{C}^{-1} \cdot \text{m}^{-2} \cdot \text{s}^{-1/2}$

For $T_{\text{polymer}} = 200^\circ\,C$ and $T_{\text{metal}} = 20^\circ\,C$, equation (2.2-15) gives: $T_i = 40^\circ\,C$.

This is only approximate since polyethylene solidifies at a temperature around $110^\circ\,C$ and the result (2.2-15) does not take into account the heat of solidification.

In injection molding, the temperature of the polymer at the surface of the mold is therefore quite close to the wall temperature of the mold.

–   *Contact polyethylene – air* $(b = 5.85\,\text{J} \cdot {}^\circ\text{C}^{-1} \cdot \text{m}^{-2} \cdot \text{s}^{-1/2})$

We assume, as a first approximation, that heat is transferred in air by conduction only.

For $T_{\text{air}} = 20\,^\circ\text{C}$ and $T_{\text{polymer}} = 200\,^\circ\text{C}$, we obtain: $T_i = 199\,^\circ\text{C}$.

The cooling of a polymer in air is slow. The skin of the extrudate remains practically at the temperature of the core for an appreciable lapse of time. It is possible therefore to expose the extrudate to air on a distance of a few centimeters and keep the polymer in the molten state for calendering or other transformation processes.

— *Contact polyethylene – water* $(b = 1.63 \cdot 10^3 \, \text{J} \cdot {}^{\circ}\text{C}^{-1} \cdot \text{m}^{-2} \cdot \text{s}^{-1/2})$

For $T_{\text{water}} = 20\,^{\circ}\text{C}$ and $T_{\text{polymer}} = 200\,^{\circ}\text{C}$, we get: $T_i = 84\,^{\circ}\text{C}$.

Here, the skin of the polyethylene film solidifies readily on contact with water and crystallization is initiated at an intermediate temperature between that of water and the initial temperature of the molten polymer.

For the last two cases, free or forced convection in air or water results in variations of the interfacial temperature with time. This is discussed in the next section.

2.2-4 Interfacial temperature with convection

The contact between a semi-infinite solid (1) and a semi-infinite liquid (2) with a heat transfer coefficient h_2 (as illustrated in Figure 2.2-5) can be treated as follows:

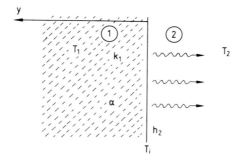

Figure 2.2-5   Interfacial temperature with convection

The temperature profile in the solid (1) is given by:

$$T_1(y, t) = A_1 + B_1 \operatorname{erf} \frac{y}{2\sqrt{\alpha_1 t}}$$

From the continuity of the temperature at the interface:

$$T_1(0, t) = T_i \;\Rightarrow\; A_1 = T_i \,,$$

from the continuity of the heat flux at the interface:

$$k_1 \frac{\partial T_1}{\partial y} = h_2(T_i - T_2) \;\Rightarrow\; \frac{k_1 B_1}{2\sqrt{\alpha_1 t}} \frac{2}{\sqrt{\pi}} = h_2(T_i - T_2)$$

and from the initial condition:

$$T_1(y, 0) = T_1 \;\Rightarrow\; A_1 + B_1 = T_1$$

Hence:

$$\frac{k_1(T_1 - T_i)}{\sqrt{\pi \alpha_1 t}} = h_2(T_i - T_2)$$

and

$$T_i(t) = \frac{k_1\left(T_1/\sqrt{\pi \alpha_1 t}\right) + h_2 T_2}{\left(k_1/\sqrt{\pi \alpha_1 t}\right) + h_2} \qquad (2.2\text{-}16)$$

The interfacial temperature may take the two following limits:

as

$$t \to 0, \quad T_i \to T_1 \quad \text{and} \quad T_i(t) \approx T_1 - \sqrt{\pi}\,\frac{h_2}{b_2}(T_1 - T_2)\,\sqrt{t}$$

and

$$t \to \infty, \quad T_i \to T_2$$

Figure 2.2-6 compares the variation of $T_i(t)$ with time assuming only convection in the fluid and assuming both conduction and convection in the fluid. At time $t = 0$ the interfacial temperature T_{ic} is calculated from Equation (2.2-15):

$$T_{ic} = \frac{b_1 T_1 + b_2 T_2}{b_1 + b_2} = \frac{k_1\left(T_1/\sqrt{\pi \alpha_1 t}\right) + k_2\left(T_2/\sqrt{\pi \alpha_2 t}\right)}{\left(k_1/\sqrt{\pi \alpha_1 t}\right) + \left(k_2/\sqrt{\pi \alpha_2 t}\right)}$$

Comparing this result to Equation (2.2-16), one observes that

$$T_i(t) = T_{ic} \quad \text{for} \quad t = t_c = \frac{k_2 \varrho_2 c_{p2}}{\pi h_2{}^2} = \frac{1}{\pi}\left(\frac{b_2}{h_2}\right)^2 \qquad (2.2\text{-}17)$$

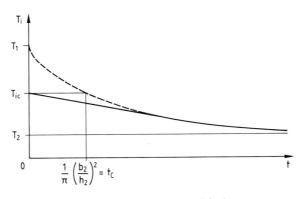

Figure 2.2-6   Variations of the interfacial temperature with time
———— exact solution, - - - - - approximation solution (Equation (2.2-16))

For $t < t_c$, one may consider that the interfacial temperature is controlled by conduction; Equation (2.2-16), based on the assumption of a constant h_2 for $t \geq 0$, is not applicable, at least in the case of free convection. For $t > t_c$, the interfacial temperature is controlled by convection. The dimensionless number $b/h\sqrt{t}$ represents the ratio of the heat transferred by conduction to the heat transferred by convection.

Orders of magnitude

–   In air: $b \approx 5.85 \, \text{J/m}^2 \cdot \text{s}^{0.5} \cdot {}^\circ\text{C}$
                $h \approx 20 \, \text{W/m}^2 \cdot {}^\circ\text{C}$
    hence $t_c \approx 10^{-2} \, \text{s}$

The interfacial temperature is therefore controlled almost instantaneously by convection. However, the polymer effusivity being considerably larger than that of air, the temperature T_{ic} is practically that of the polymer.

–   In water: $b \approx 1.63 \cdot 10^3 \, \text{J/m}^2 \cdot \text{s}^{0.5} \cdot {}^\circ\text{C}$
    for $h = 4 \, \text{W/m}^2 \cdot {}^\circ\text{C}, \quad t_c \approx 1 \, \text{h}$
                   $h = 40 \, \text{W/m}^2 \cdot {}^\circ\text{C}, \quad t_c \approx 30 \, \text{s}$
                   $h = 400 \, \text{W/m}^2 \cdot {}^\circ\text{C}, \quad t_c \approx 0.3 \, \text{s}$

Depending on the value of h, the conduction-controlled regime may be long or short.

In all cases, the temperature T_{ic} is the instantaneous temperature of the skin of a polymer film or sheet when placed in contact with a solid or a fluid. In the case of a fluid, the interfacial temperature changes more or less rapidly whether or not convection is significant. Since the thickness of the polymer film or sheet is not infinite, it is obvious that the interfacial temperature will vary with the internal temperature as the heat penetration reaches the center of the material.

2.2-5   Example: heating (or cooling) of a plate

As real objects are never of infinite dimensions, we present here the extension of the previous cases to the case of bodies of finite thickness. Let us consider a plate of thickness $2e$ and of very large surfaces. The initial temperature of the plate is T_0 and at $t = 0$, the two planar surfaces are brought to specific temperature or heat flux conditions. As far as the heat penetration thickness (for both sides) is small with respect to the half thickness of the plate, the solution of the heat transfer in either side is independent of one another and the semi-infinite solutions presented in the previous sections can be considered as first approximations.

The interfacial temperatures of a plate, initially isothermal and immersed in a fluid also initially isothermal, will remain constant during t_1, time period required for the heat penetration to reach the half-thickness e of the plate. From Equation (2.2-10):

$$t_1 \approx \frac{e^2}{\alpha}$$

For a polyethylene plate of thickness $2e = 1 \, \text{mm}$, $t_1 \approx 2.5 \, \text{s}$ and for a polyethylene film of thickness $2e = 100 \, \mu$, $t_1 \approx 0.025 \, \text{s}$. For a time longer than t_1, the heat fluxes from both sides merge and the heat transfer problem cannot anymore be solved using the semi-infinite body approximation.

a) Isothermal boundary conditions

Let us consider that at instant $t = 0$, the surface temperature of a plate, with thickness $2e$ and initial temperature T_0, is brought to T_s and maintained at that temperature. The exact solution to this problem is given by CARSLAW and JAEGER (1959). The temperature profile expressed by

$$\frac{T_s - T}{T_s - T_0} = 2 \sum_{n=0}^{\infty} \frac{(-1)^n}{(n + 1/2)\pi} \exp\{-(n + 1/2)^2 \pi^2 \alpha t / e^2\} \cos\{(n + 1/2)\pi y / e\}$$

$$(2.2\text{-}18)$$

is illustrated in Figure 2.2-7 for various values of the parameter $\alpha t / e^2$. The values of the dimensionless temperature profile are obtained by summing a large number of terms of the series.

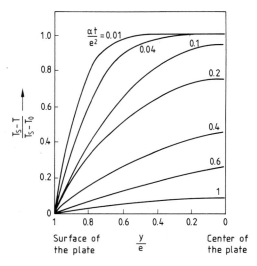

Figure 2.2-7   Temperature profile in a plate of finite thickness

Approximate solution

The series converges very rapidly for large values of $\alpha t / e^2$, and the solution is approximately equal to the first term of the series. The temperature of the center of the plate $(y = 0)$, T_c, is given for the approximate solution by

$$\frac{T_s - T_c}{T_s - T_0} = \frac{4}{\pi} \exp\left\{-\frac{\pi^2 \alpha t}{4e^2}\right\}$$

$$(2.2\text{-}19)$$

whereas the exact solution is from Equation (2.2-18):

$$\frac{T_s - T_c}{T_s - T_0} = 2 \sum_{n=0}^{\infty} \frac{(-1)^n}{(n + 1/2)\pi} \exp\{-(n + 1/2)^2 \pi^2 \alpha t / e^2\}$$

$$(2.2\text{-}20)$$

Figure 2.2-8 compares the exact and approximate solutions as a function of $\alpha t / e^2$. It is clear from the figure that the approximate solution is quite acceptable for values of $\alpha t / e^2$ larger than 0.1. For smaller values, the approximate solution

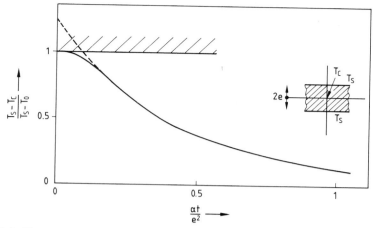

Figure 2.2-8   Temperature variation at the center of a finite plate
——— exact solution, - - - - - approximate solution

makes no sense with a dimensionless temperature $(T_s - T_c)/(T_s - T_0)$ larger than 1. The cooling time of a plate can be calculated as a good approximation by taking $t_1 = e^2/\alpha$.

Numerical applications

Let us consider a polyethylene plate of 2 mm thickness, initially at 200° C, and placed in a mold for which the wall temperature is maintained constant at 50° C. From Figure 2.2-8, the temperature at the center of the plate ($\alpha = 13\,10^{-8}$ m^2/s for polyethylene) shows the following variations:

At
$$
\begin{aligned}
t &= 0.1\,\text{s} & T_c &= 200°\ \text{C} \\
t &= 1\ \ \text{s} & T_c &= 185°\ \text{C} \\
t &= 5\ \ \text{s} & T_c &= \ \ 92°\ \text{C} \\
t &= 10\ \ \text{s} & T_c &= \ \ 58°\ \text{C}
\end{aligned}
$$

As mentioned previously, polyethylene would solidify at a temperature around 110° C. This analysis does not account for the heat of solidification and the cooling time could be considerably longer than estimated from Equation (2.2-19) or (2.2-20).

b) Convective boundary conditions

This type of boundary conditions has been defined in section 2.1:

$$
\text{at}\quad y = +e, \quad -\frac{k\partial T(t)}{\partial y} = h[T(t) - T_f]
$$
$$
\text{at}\quad y = -e, \quad \frac{k\partial T(t)}{\partial y} = h[T(t) - T_f]
$$

where T_f is the fluid temperature far away from the interface and h is the heat transfer coefficient.

With these two boundary conditions and the initial condition that the temperature is constant ($T = T_0$, at $t = 0$) anywhere in the plate, the solution is obtained from CARSLAW and JAEGER (1959):

$$\frac{T - T_f}{T_0 - T_f} = \sum_{n=1}^{\infty} \left[\frac{4 \sin M_n}{2M_n + \sin(2M_n)} \right] \exp\left\{ -\left(M_n^2 \frac{\alpha t}{e^2} \right) \right\} \cos\left(\frac{M_n y}{e} \right) \quad (2.2\text{-}21)$$

where M_n is the nth positive root of the following equation:

$$M_n \tan M_n = \frac{he}{k} = \text{Bi} \quad (2.2\text{-}22)$$

and Bi is the Biot number. We note that its definition is similar to that of the Nusselt number used in convection (see Appendix D). The variations of the surface temperature for various values of the Biot number are presented in Figure 2.2-9, whereas the variations of the center temperature as a function of the dimensionless time are illustrated in Figure 2.2-10. To obtain these curves, one has to obtain for each value of Bi, the roots M_n from Equation (2.2-22) (a graphical method is satisfactory), then sum a large number of terms of the series in Equation (2.2-21).

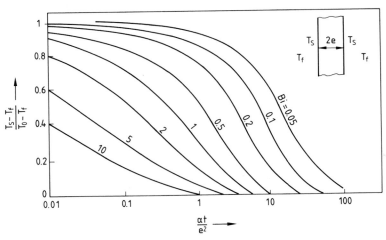

Figure 2.2-9   Dimensionless surface temperature variations in a plate for convective boundary conditions

Approximate solution

In many cases of interest, an approximate solution is obtained by taking only the first term of the series. We will show below that this is acceptable if $\text{Bi} = h\,e/k$ is much smaller than 1. For $\text{Bi} \ll 1$,

$$M_1 \tan M_1 = \text{Bi} \Rightarrow M_1 \approx \sqrt{\text{Bi}}$$

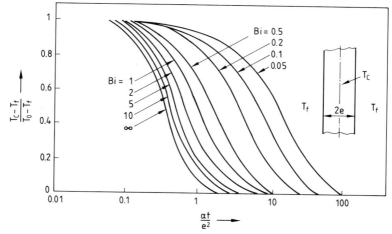

Figure 2.2-10 Dimensionless center temperature variations in a plate for convective boundary conditions

since M_1 must be also very small and $\tan M_1$ is approximately equal to M_1. The approximate solution is readily obtained from Equation (2.2-21):

$$\frac{T - T_f}{T_0 - T_f} \approx \exp\left\{-\text{Bi}\,\frac{\alpha t}{e^2}\right\}\cos\left(\sqrt{\text{Bi}}\,\frac{y}{e}\right) \qquad (2.2\text{-}23)$$

since

$$\frac{4\sin M_1}{2M_1 + \sin 2M_1} \approx 1\,.$$

Replacing Bi by $h\,e/k$ and α by $k/\varrho c_p$, we get:

$$\frac{T - T_f}{T_0 - T_f} = \exp\left\{-\frac{h}{\varrho c_p}\,\frac{t}{e}\right\}\cos\left(\sqrt{\text{Bi}}\,\frac{y}{e}\right) \qquad (2.2\text{-}24)$$

We note, moreover, that for very small values of Bi:

$$\cos\left(\sqrt{\text{Bi}}\,\frac{y}{e}\right) \approx 1 - \frac{1}{2}\,\text{Bi}\left(\frac{y}{e}\right)^2 \approx 1$$

and the approximate solution can be simply written as:

$$\frac{T - T_f}{T_0 - T_f} \approx \exp\left\{-\frac{h}{\varrho c_p}\,\frac{t}{e}\right\} \qquad (2.2\text{-}25)$$

This approximate solution shows no dependence on y, i.e. at any instant t, the temperature is constant everywhere in the plate and the surface and center temperatures are equal. Figure 2.2-11 compares the approximate and exact solution for the dimensionless temperature at the center of a plate. It is clear from

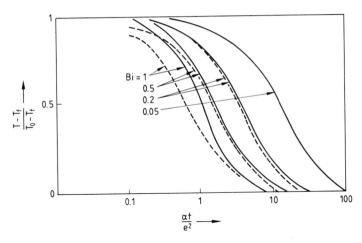

Figure 2.2-11   Temperature variations at the center of a plate
———— exact solution, - - - - - approximate solution

the figure that the approximate solution is valid for Bi less than 0.1. For values of Bi greater than 1, the approximate solution is no longer acceptable.

The limit of applicability of the approximate solution (Bi < 0.1) can be obtained by comparing Figure 2.2-9 and 2.2-10. For approximate solution, the surface and center temperature are equal. The approximate solution is then valid if the curves for the exact solution of $(T_c - T_f)/(T_0 - T_f)$ and $(T_s - T_f)/(T_0 - T_f)$ can be superposed. The reader will verify that the superposition is good for Bi smaller than 0.1.

Remarks

- The limiting case of Bi equal to infinity is equivalent to the case of isothermal boundary conditions. In Figure 2.2-9, the curve for Bi equal to infinity coincides with the abscissa $(T_S = T_f)$ and in Figure 2.2-10, the temperature at the center of a plate for Bi $= \infty$ is the same as given by Figure 2.2-7 for isothermal boundary conditions.
- The limit of Bi very small corresponds to cases where the heat conduction in the solid (k/e) is much more rapid than the convection at the interface (h). The temperature variations with time in the solid can be obtained from a heat balance per unit surface, assuming a uniform temperature in the solid equal to $\overline{T}$:

$$-\varrho c_p e d\overline{T} = 2h(\overline{T} - T_f)\, dt \tag{2.2-26}$$

Hence, after integration:

$$\overline{T} - T_f = (T_0 - T_f) \exp\left\{-\frac{h}{\varrho c_p e} t\right\}$$

This is identical to Equation (2.2-25).

Orders of magnitude

$$h \approx 20 \quad \mathrm{W/m^2 \cdot {}^\circ C}$$
$$k \approx \;\; 0.2\,\mathrm{W/m \cdot {}^\circ C}$$

Hence

$$\mathrm{Bi} = \frac{he}{k} \approx 10^2 e$$

and

$$\text{for} \quad e < \;\; 1\,\mathrm{mm} \Rightarrow \mathrm{Bi} < 0.1 \;\text{(good approximation)}$$
$$\text{for} \quad e > 10\,\mathrm{mm} \Rightarrow \mathrm{Bi} > 1 \quad \text{(non valid approximation)}$$

2.3 Viscous dissipation

Molten polymers are highly viscous but have low thermal conductivity. It is therefore inefficient to try to transfer heat by conduction to molten polymer across a large thickness. On the other hand, the heat generated by viscous forces may be of importance and can possibly represent an efficient mode of heating as heat losses to the surrounding equipment are not rapid.

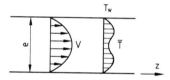

Figure 2.3-1   Characteristic dimensions in planar flow

The relative importance of the heat generated by viscous forces to the heat transferred by conduction is characterized by the Brinkman number. For a planar flow situation, with characteristic dimensions given in Figure 2.3-1, the following order of magnitude analysis allows us to obtain a definition of the Brinkman number.
–   The power dissipated by viscous forces is

$$\dot{W} \approx \eta \frac{V^2}{e^2} \cdot edz$$

where V is the average velocity of the flow;
–   the heat transferred by conduction is

$$Q \approx k \frac{|\bar{T} - T_w|}{e} \cdot dz$$

where $\bar{T}$ and T_w are respectively the average temperature of the flowing material and the wall temperature.

The Brinkman number is the ratio of these two expressions:

$$\mathrm{Br} = \frac{\eta V^2}{k(\bar{T} - T_w)}$$

(2.3-1)

The Brinkman number is usually defined with the absolute value of the temperature difference. Here, a negative Brinkman number will mean that heat is added to the polymer by conduction from the walls.

Taking a viscosity value of 10^3 Pa $\cdot$ s, a conductivity of 0.2 W/m°C, and a typical average velocity equal to 0.2 m/s, we observe that the dissipated power is larger than the conduction term whenever $(\bar{T} - T_w)$ is smaller than 100° C.

2.3-1   Heat generation in Poiseuille flow

Let us consider the flow of a Newtonian polymer in a circular tube with a constant wall temperature equal to T_0, the initial temperature of the polymer.

The evolution of the temperature profile in the polymer is qualitatively described in Figure 2.3-2.

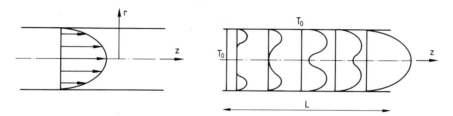

Figure 2.3-2   Velocity (left) and temperature (right) profiles for the flow in a circular tube

Near the entrance, the maximum temperature is obtained close to the wall, since the shear forces are maximum at the wall. But progressively, due to the heat conduction towards the center, the temperature profile evolves towards an equilibrium profile with a maximum at the center of the tube. In general, three distinct zones can be described:

–   Near the entrance of the tube, the temperature rise is small and the heat transferred by conduction through the tube wall is negligible. The power dissipated is used almost entirely to heat the polymer, as if the liquid was thermally isolated from the surroundings. This is called the *adiabatic* regime.
–   Somewhat downstream from the entrance, the temperature rise is such that the conduction cannot be any longer neglected. The dissipated power is distributed into internal heat and heat transferred by conduction to the surroundings. This is known as the *transition* regime.
–   Further downstream, the temperature rise in the liquid attains a limiting value and all the dissipated power is transferred by conduction to the surroundings. This is the *equilibrium* regime.

The rigorous solution of the energy equation is quite complex, even for this simple flow situation. The temperature rise in the flowing polymer results in variations of the polymer viscosity which then affect the velocity profile. A priori, the equations of motion and of energy are coupled and must be solved by an iterative method for each point of the flow system (see for example GEE and LYON (1957), MORETTE and GOGOS (1968), MENNIG (1972), FOREST and WILKINSON (1973), COX and MACOSKO (1974), WINTER (1975), NUNN and FENNER (1977). However, as shown for example by AGUR and VLACHOPOULOS (1977), the uncoupled equations of motion and energy can be solved to yield useful results that are close to the real situation in many instances.

Let us assume that the polymer is a Newtonian liquid and that the variations of its viscosity with temperature are negligible. The velocity profile is from Section 1.5:

$$w(r) = 2V\left[1 - \left(\frac{r}{R}\right)^2\right]$$

and the shear rate at any point is:

$$\dot{\gamma}(r) = -4r\frac{V}{R^2}$$

The energy equation in cylindrical coordinates can be written as:

$$\varrho c_p w(r)\frac{\partial T}{\partial z} = k\left[\frac{1}{r}\frac{\partial}{\partial r}\left(r\frac{\partial T}{\partial r}\right) + \frac{\partial^2 T}{\partial z^2}\right] + \frac{16\eta V^2}{R^4}r^2 \qquad (2.3\text{-}2)$$

In all cases encountered in polymer processing, the heat conduction in the axial direction is negligible with respect to the convection term. Indeed, the order of magnitude of the convection term is:

$$\varrho c_p V\frac{\Delta T}{L}$$

whereas the order of magnitude of the conduction term in the axial direction is:

$$k\frac{\Delta T}{L^2}$$

The ratio of the two terms is the Peclet number, defined by:

$$\boxed{\text{Pe} = \varrho c_p\frac{VL}{k} = V\frac{L}{\alpha}} \qquad (2.3\text{-}3)$$

For polymers, $\alpha \cong 10^{-7}\,\text{m}^2/\text{s}$, and taking:

$$V = 0.1\,\text{m/s}$$
$$L > 0.1\,\text{m}$$

we get:

$$\text{Pe} > 10^5$$

Clearly, the axial conduction can be neglected and the energy equation reduces to:

$$\varrho c_p w(r)\frac{\partial T}{\partial z} = k\left[\frac{1}{r}\frac{\partial}{\partial r}\left(r\frac{\partial T}{\partial r}\right)\right] + \frac{16\eta V^2}{R^4}r^2 \qquad (2.3\text{-}4)$$

This equation can be solved numerically (BRINKMAN, 1951). HULATT and WIL-
KINSON (1978), VI DUONG DANG (1979) and DINH and ARMSTRONG (1982) have
compared the computed temperature profiles to experimental ones. Figure 2.3-3
shows the results for two different boundary conditions.

–   the solid lines are the solutions of the dimensionless temperature profiles for
     the case of a constant wall temperature equal to the initial temperature of
     the liquid, T_0. The equilibrium temperature profile is shown by the curve at
     $Z = \infty$:

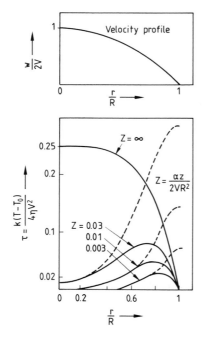

Figure 2.3-3
Temperature profiles for the flow in a circular
tube

———   constant wall temperature,
- - - - - -   adiabatic wall

–   the dashed lines are the solutions for the adiabatic wall conditions. In this case,
     there is no equilibrium temperature profile, as the temperature will increase
     indefinitely.

     In the following paragraphs, we examine the three limiting cases corresponding
to the three regimes presented above.

a) Equilibrium regime

The convection term becomes equal to zero and Equation (2.3-4) reduces to:

$$\frac{d}{dr}\left[r\frac{dT}{dr}\right] = -16\frac{\eta V^2}{k}\frac{r^3}{R^4} \tag{2.3-5}$$

and the temperature profile is readily obtained:

$$T(r) = T_0 + \frac{\eta V^2}{k}\left[1 - \frac{r^4}{R^4}\right] \tag{2.3-6}$$

The difference between the center and the wall temperature, called the shear-heating temperature, is

$$\Delta T = \eta \frac{V^2}{k}$$

(2.3-7)

This result can be expressed differently using the relationship between the average velocity and the pressure drop in the tube:

$$\Delta T = \frac{\Delta P^2}{64\eta k L^2} R^4$$

(2.3-8)

The following order-of-magnitude analysis shows that shear-heating can be very important and incompatible with the thermal stability of polymers: Taking:

$$\eta = 10^3 \, \text{Pa} \cdot \text{s}$$
$$k = 0.2 \, \text{W/m} \cdot {}^\circ\text{C}$$
$$R = 5 \, \text{mm}$$
$$\Delta P = 6.4 \times 10^6 \, \text{Pa}$$

we obtain, for $L = 1\,\text{m}$, and $V = 20\,\text{mm/s}$, $\eta V^2/k = 2^\circ\,\text{C}$, and for $L = 0.10\,\text{m}$ and $V = 0.2\,\text{m/s}$, $\eta V^2/k = 200^\circ\,\text{C}$.

In practical situations, the equilibrium regime is rarely attained.

b) Adiabatic regime

Near the entrance of the tube, the radial conduction is negligible and Equation (2.3-4) reduces to:

$$2\varrho c_p V \left[1 - \left(\frac{r}{R}\right)^2\right] \frac{\partial T}{\partial z} = 16 \frac{\eta V^2}{R^4} r^2$$

(2.3-9)

We look for an expression for the mean or bulk temperature $\overline{T}(z)$ of the liquid. Knowing that V represents the average velocity, we write from an energy balance over a volume $\pi R^2 dz$ as in Section 2.1:

$$\pi R^2 \varrho c_p V \frac{d\overline{T}}{dz} = \int_0^R \dot{W}(r) \, 2\pi r \, dr$$

(2.3-10)

where the power dissipated per unit volume, $\dot{W}(r)$, is given by the right-hand side of Equation (2.3-9).

Hence

$$\frac{d\overline{T}}{dz} = \frac{8\eta V}{\varrho c_p R^2}$$

(2.3-11)

and the integration of this equation gives the expression for the mean temperature:

$$\overline{T}(z) = 8 \frac{\eta V}{\varrho c_p R^2} z + T_0$$

(2.3-12)

This represents an approximation to the mean temperature that could be obtained from the Brinkman solution. This is valid for a certain length of tube, as shown below in the case of isothermal walls. In the case of adiabatic walls, as the liquid temperature increases indefinitely, this expression is exact for any length of tube. Replacing the average velocity by the expression

$$V = \frac{1}{8\eta} \frac{\Delta P}{L} R^2 \qquad (2.3\text{-}13)$$

we get

$$\boxed{\Delta \bar{T} = \bar{T}(z) - T_0 = \frac{\Delta P}{\varrho c_p} \frac{z}{L}} \qquad (2.3\text{-}14)$$

This result is of interest as it shows that the expression for the increase in temperature due to viscous dissipation, written in that form, is independent of the tube radius and of the liquid viscosity.

Using the same conditions as taken in the numerical applications in the case of the equilibrium regime a), we obtain for both tubes at $z = L$:

$$\Delta \bar{T} = \frac{\Delta P}{\varrho c_p} = 3.2\,°\mathrm{C}$$

when taking ϱc_p equal to $2 \cdot 10^6$ J/m$^3 \cdot$ °C. This is typical of temperature rises in such cases.

In general terms, relation (2.3-14) indicates how mechanical energy, $\Delta Pz/L$, per unit volume is converted into thermal energy, $\varrho c_p \Delta T(z)$, per unit volume. In any Poiseuille flow situation, and for any rheological behaviour, the temperature rise under adiabatic conditions is simply given by:

$$\Delta \bar{T}(L) = \frac{\Delta P}{\varrho c_p} \qquad (2.3\text{-}15)$$

c) Transition regime

In the case where the walls are maintained at temperature T_0, a part of the dissipated energy is transferred to the walls by conduction. An energy balance over a tubular differential volume yields by extending Equation (2.3-10):

$$\pi R^2 \varrho c_p V d\overline{T} = 8\pi \eta V^2 dz - q2\pi R\, dz \qquad (2.3\text{-}16)$$

where q is the heat flux by conduction through the walls.

In order to integrate Equation (2.3-16), an expression for q with respect to $\overline{T}$ must be derived. This is done by using the results obtained for the equilibrium regime. From Equation (2.3-6):

$$T(r) - T_0 = \frac{\eta V^2}{k} \left[1 - \left(\frac{r}{R} \right)^4 \right]$$

we get:

$$\bar{T} = \frac{\int_0^R 2V[1-(r/R)^2]\,T2\pi r\,dr}{V\pi R^2} = T_0 + \frac{5}{6}\frac{\eta V^2}{k} \tag{2.3-17}$$

This follows from the definition of the bulk (or mean) temperature in a Poiseuille flow (BIRD, STEWART and LIGHTFOOT, 1960). The heat flux at the walls is

$$q = -k\frac{dT}{dr}\bigg|_R = 4\eta\frac{V^2}{R} \tag{2.3-18}$$

Then for the equilibrium regime:

$$q = \frac{24}{5}k\left(\frac{\bar{T}-T_0}{R}\right) \tag{2.3-19}$$

If we assume now that this result is valid for the transition regime, Equation (2.3-16) becomes:

$$\pi R^2 \varrho c_p V\,d\bar{T} = 8\pi\eta V^2 dz - \frac{48}{5}\pi k(\bar{T}-T_0)\,dz \tag{2.3-20}$$

and integrating with the initial condition $\bar{T}(0) = T_0$, we get

$$\boxed{\bar{T}(z) = T_0 + \frac{5}{6}\frac{\eta V^2}{k}\left[1 - \exp\left\{-\frac{48}{5}\frac{kz}{\varrho c_p V R^2}\right\}\right]} \tag{2.3-21}$$

The following dimensionless number appears in this result:

$$\boxed{\mathrm{Ca} = \frac{kL}{\varrho c_p V R^2} = \frac{\alpha L}{V R^2}} \tag{2.3-22}$$

It was used by CAMERON (1966) to characterize heating effects in lubrication. We will call it the Cameron number. It can be easily seen that it is the inverse of the Graetz number (GRAETZ, 1889; WINTER, 1977).

Comments

In the transition regime the heat flux through the wall is not constant along the axial direction. We could make use of a heat transfer coefficient, h, for the heat exchange between the polymer and the tube wall:

$$q = h(\bar{T}-T_0) \tag{2.3-23}$$

The coefficient h can be related to a Nusselt number, defined here for forced convection to a liquid (see Appendix D.2):

$$\mathrm{Nu} = h\frac{D}{k} \tag{2.3-24}$$

where D is the tube diameter.

The Nusselt number can be obtained from the numerical solution of Equation (2.3-4). As shown in Figure 2.3-4, the Nusselt number decreases with increasing distance or Cameron number to attain a constant value of 12 at equilibrium (SAILLARD, 1982; AGASSANT et al., 1984).The influence of the Brinkman number is discussed below.

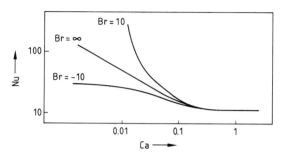

Figure 2.3-4   Variation of the Nusselt number with the Cameron number for three values of the Brinkman number

Equation (2.3-21) becomes:

$$\bar{T}(z) = T_0 + \frac{5}{6}\frac{\eta V^2}{k}\left[1 - \exp\left\{-\frac{48}{5}\,\mathrm{Nu}\,\mathrm{Ca}\,\frac{z}{L}\right\}\right] \qquad (2.3\text{-}25)$$

FOREST and WILKINSON (1973) and PEARSON (1979) have proposed the same kind of correlation for the Nusselt number, but as a function of the Graetz number.

d) Analysis of the different thermal regimes

We can now specify the conditions that control the three thermal regimes: adiabatic, transition and equilibrium regimes, as illustrated in Figure 2.3-5. From Equation (2.3-21):

– If

$$\frac{48}{5}\frac{kz}{\varrho c_p V R^2} \ll 1 \quad \left(\text{e.g.}\;\; \mathrm{Ca}\,\frac{z}{L} < 10^{-2}\right)$$

then

$$\bar{T}(z) \approx T_0 + \frac{8\eta V}{\varrho c_p R^2}z$$

This is the result for the adiabatic regime, always observed in the initial part of the flow. The other two regimes may not necessarily be attained.

– If

$$\frac{48}{5}\frac{kz}{\varrho c_p V R^2} \gg 1 \quad \left(\text{e.g.}\;\; \mathrm{Ca}\,\frac{z}{L} > 1\right)$$

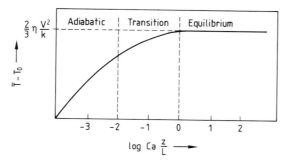

Figure 2.3-5   The three different thermal regimes in Poiseuille flow

then

$$\bar{T}(z) \approx T_0 + \frac{5}{6}\frac{\eta V^2}{k}z$$

This is the result for the equilibrium regime.

– If $10^{-2} < \mathrm{Ca}\, z/L < 1$, the flow is in the transition regime.

e) Application

The following allows us to calculate the bulk axial temperature profile of a polymer flowing in a circular tube of length L and of radius R. The average velocity is V and the polymer viscosity is η, its heat capacity is ϱc_p and its thermal conductivity is k.

First, we calculate the Cameron number from Equation (2.3-22):

$$\mathrm{Ca} = \frac{\alpha L}{V R^2}$$

– If $\mathrm{Ca} < 10^{-2}$, then the flow is in the adiabatic regime and the polymer temperature increases by $\Delta T = \Delta P/\varrho c_p$, where ΔP is the pressure drop across the length of the tube.
– If $\mathrm{Ca} > 1$, the thermal equilibrium regime is attained and the maximum polymer temperature with respect to the wall temperature is: $\Delta T = \eta V^2/k$.
– If $10^{-2} < \mathrm{Ca} < 1$, the flow is in the transition regime and the increase of the polymer bulk temperature is:

$$\Delta T = \frac{5}{6}\eta\frac{V^2}{k}\left(1 - \exp\left\{-\frac{48}{5}\mathrm{Ca}\right\}\right) \tag{2.3-26}$$

Example 2.3-1

Let us reconsider the particular cases presented above in sections a) and b). The properties of the polymer are:

$$\eta = 10^3\,\mathrm{Pa}\cdot\mathrm{s}, \quad k = 0.2\,\mathrm{W/m}\cdot{}^\circ\mathrm{C}, \quad \varrho c_p = 2\times 10^6\,\mathrm{J/m^3}\cdot{}^\circ\mathrm{C}$$

The radius of the tube is $R = 5\,\text{mm}$.

- *For $L = 10\,m$, $V = 2\,\text{mm/s}$, Ca $= 20$. This corresponds to the equilibrium regime and $\Delta T = 0.01°\,C$.*
- *For $L = 1\,m$, $V = 20\,\text{mm/s}$, Ca $= 0.02$. This corresponds to the transition regime and $\Delta T = 1.21°\,C$.*
- *For $L = 0.1\,m$, $V = 0.2\,\text{m/s}$, Ca $= 0.002$. This is the adiabatic regime and $\Delta T = 3°\,C$.*

In each case, one can calculate the bulk temperature profile and the results are shown in Figure 2.3-6.

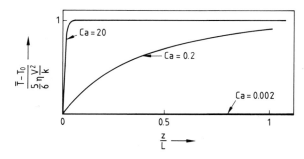

Figure 2.3-6 Bulk temperature profiles in Poiseuille flow for three different values of the Cameron number

The analysis presented in this section is based on hypotheses which are for many applications too restrictive: flow in a tube, Newtonian behaviour, initial temperature of the polymer equal to the wall temperature. In the following sections, these restrictions will be successively removed.

2.3-2 Heat generation in Poiseuille flow – wall temperature different from the initial polymer temperature

This is a more general case and the simplified analysis of Section 2.3-1 is no longer applicable. If the wall temperature is different from the initial polymer temperature, heat conduction plays an important role, even in the initial part of the flow, and the adiabatic regime does not exist any more. However the equilibrium regime is still controlled by the wall temperature and the previous result holds. Hence we will examine particularly the transition regime.

Since the wall temperature is different from the initial polymer temperature, Equation (2.3-20) has to be rewritten as:

$$\pi R^2 \varrho c_p V \, d\bar{T} = 8\pi \eta V^2 dz - \left(\frac{48}{5}\right)\pi k(\bar{T} - T_w)\,dz \qquad (2.3\text{-}27)$$

where T_w is the wall temperature of the tube. This equation can be easily integrated with respect to z to obtain:

$$\bar{T}(z) - T_w = (T_0 - T_w) \exp\left(-\frac{48}{5}\,\mathrm{Ca}\,\frac{z}{L}\right) + \frac{5}{6}\frac{\eta V^2}{k}$$

$$\times \left[1 - \exp\left(-\frac{48}{5}\,\mathrm{Ca}\,\frac{z}{L}\right)\right] \tag{2.3-28}$$

Introducing the Brinkman number evaluated at the entrance of the tube:

$$\mathrm{Br} = \frac{\eta V^2}{k(T_w - T_0)} \tag{2.3-29}$$

we get

$$\boxed{\bar{T}(z) - T_0 = \frac{5}{6}\eta\frac{V^2}{k}\left[1 + \frac{6}{5\,\mathrm{Br}}\right]\left[1 - \exp\left(-\frac{48}{5}\,\mathrm{Ca}\,\frac{z}{L}\right)\right]} \tag{2.3-30}$$

- For $T_w = T_0$, the Brinkman number as defined by Equation (2.3-29) becomes infinite and Equation (2.3-30) reduces to Equation (2.3-21).
- For $T_w < T_0$, the Brinkman number is negative and the bulk temperature of the fluid increases less rapidly than in the previous case. The bulk temperature will decrease if the magnitude of the Brinkman number is small enough.
- For $T_w > T_0$, Equation (2.3-30) may show the following paradox. Referring to Figure 2.3-7, the mean temperature of the polymer is lower than the wall temperature. Nevertheless, because of the high viscous dissipation near the walls, heat may be transferred from the polymer to the walls. This situation is not accounted for in Equation (2.3-30). However, in most flow situations, additional heat is transferred from the walls to the fluid; the Brinkman number is positive and the fluid temperature increases more quickly than in the case discussed in Section 2.3-1.

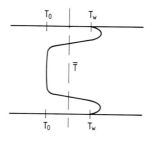

Figure 2.3-7 Paradoxical heat transfer situation encountered when the wall temperature is higher than the initial temperature of the molten polymer.

Comments

As mentioned in Section 2.3-1, the Nusselt number is a function of both the Cameron and the Brinkman numbers (see Fig. 2.3-4). The results were obtained through a solution of Equation (2.3-4) by SAILLARD (1982) and AGASSANT et al. (1984). The curve for Br $= \infty$ represents the case where the wall is maintained at the initial temperature of the polymer. The large differences shown for a positive Brinkman number reflect the heat transfer enhancement when heat conduction from the wall is added to heat generated by viscous forces.

Example 2.3.2

We examine the cases presented in the example of Section 2.3-1. The wall temperature is now set at 180, 200 and 220° C.

- In the first case ($V = 2$ mm/s), the Brinkman number is very small ($\pm 10^{-3}$ for $|T_w - T_0| = 20°$ C) whereas the Cameron number is very large. The bulk temperature is rapidly equal to the wall temperature of the tube. This is illustrated in Figure 2.3-8.

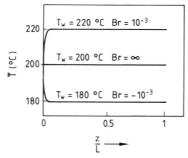

Figure 2.3-8   Influence of the wall temperature on the polymer temperature profile (Ca $= 20$)

- In the second case ($V = 20$ mm/s), the Brinkman number is larger (Br $= \pm 0.1$ for $|T_w - T_0| = 20°$ C), whereas the Cameron number is considerably smaller than that of the first case. The polymer temperature is still controlled by the wall temperature, but the axial distance to obtain the equilibrium temperature is considerably larger, as shown in Figure 2.3-9. This is clearly a transition regime.

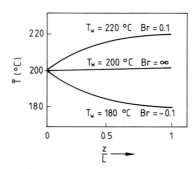

Figure 2.3-9   Influence of the wall temperature on the polymer temperature profile (Ca $= 0.2$)

- In the third case ($V = 0.2$ m/s), the Brinkman number is quite large ($Br = \pm 10$ for $|T_w - T_0| = 20° C$) and the Cameron number very small. The conduction to the wall is relatively not important and the polymer temperature is only slightly affected by the wall temperature. This is illustrated in Figure 2.3-10.

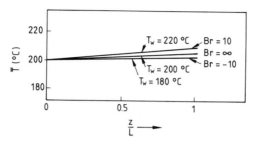

Figure 2.3-10   Influence of the wall temperature on the polymer temperature profile ($Ca = 0.002$)

2.3-3   Heat generation in Poiseuille flow for shear-thinning fluids

For shear-thinning fluids obeying the power-law expression, the velocity profile is given by the following expression (see Table 1.5-1):

$$w(r) = \frac{3n+1}{n+1} V \left[1 - \left(\frac{r}{R} \right)^{1+1/n} \right]$$

The expressions for the shear rate and for the rate of energy dissipated per unit volume are respectively:

$$\dot{\gamma}(r) = -\frac{3n+1}{n} V \frac{r^{1/n}}{R^{1/n+1}} \tag{2.3-31}$$

and

$$\dot{W} = K |\dot{\gamma}|^{n+1} = K \left(\frac{3n+1}{n} V \right)^{n+1} \frac{r^{1+1/n}}{R^{(n+1)^2/n}} \tag{2.3-32}$$

where K and n are the parameters of the power-law expression (1.5-2):

$$\eta = K |\dot{\gamma}|^{n+1}$$

The energy Equation (2.3-4) is then modified to the following form:

$$\frac{3n+1}{n+1} \varrho c_p V \left[1 - \left(\frac{r}{R} \right)^{1+1/n} \right] \frac{\partial T}{\partial z}$$

$$= k \frac{1}{r} \left[\frac{\partial}{\partial r} \left(r \frac{\partial T}{\partial R} \right) \right] + K \left(\frac{3n+1}{n} V \right)^{n+1} \frac{r^{1+1/n}}{R^{(n+1)^2/n}} \tag{2.3-33}$$

We will successively examine the cases for the equilibrium and the transition regimes. The previously obtained solution for the adiabatic case is still valid and given by Equation (2.3-14):

$$\bar{T}(z) = \frac{\Delta P}{\varrho c_p} \frac{z}{L} + T_0$$

The adiabatic regime will not be further discussed in this section.

a) Equilibrium regime

In the equilibrium regime, the convection term becomes equal to zero and Equation (2.3-33) reduces to:

$$\frac{d}{dr}\left(r\frac{dT}{dr}\right) = -\frac{K}{k}\left[\frac{3n+1}{n}V\right]^{n+1}\frac{r^{(2n+1)/n}}{R^{(n+1)^2/n}} \qquad (2.3\text{-}34)$$

which can be readily integrated to obtain the temperature profile:

$$T(r) = T_w + \frac{K}{k}\left(\frac{n}{3n+1}\right)^{1-n}V^{n+1}\frac{R^{3+1/n} - r^{3+1/n}}{R^{(n+1)^2/n}} \qquad (2.3\text{-}35)$$

The radial temperature profile is illustrated in Figure 2.3-11 for different values of the power-law index. As n decreases, the velocity profile becomes flatter, the rate of shear is more important near the tube wall and hence the temperature profile becomes flatter.

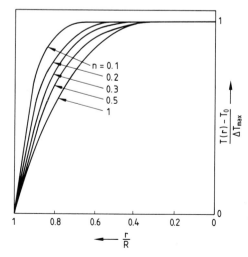

Figure 2.3-11   Radial temperature profiles in Poiseuille flow for different values of the power-law index. Thermal equilibrium regime

The maximum temperature is at the tube center and can be calculated from:

$$\Delta T_{max} = T(0) - T_w = \frac{K}{k}\left(\frac{n}{3n+1}\right)^{1-n}\frac{V^{n+1}}{R^{n-1}} \qquad (2.3\text{-}36)$$

This expression reduces correctly to $\Delta T_{max} = \eta V^2/k$ for a Newtonian fluid ($K = \eta$ and $n = 1$). It can be also expressed in term of the pressure drop (using the relationship of Table 1.5-1 between the flow rate and the pressure drop):

$$\Delta T_{max} = \frac{K}{k}\left(\frac{n}{3n+1}\right)^{2}\left(\frac{1}{2K}\frac{\Delta P}{L}\right)^{1+1/n}R^{3+1/n} \qquad (2.3\text{-}37)$$

Figure 2.3-12 shows the value of ΔT_{max} as a function of n for a polymer of consistency index, K, equal to 10^4 Pa$\cdot$s^n flowing at an average velocity of 0.2 m/s in a capillary of radius equal to 5 mm. The thermal conductivity is equal to 0.2 W/m$\cdot$°C. Obviously, as n decreases the polymer at the same flow rate is much less viscous and viscous heating is reduced considerably.

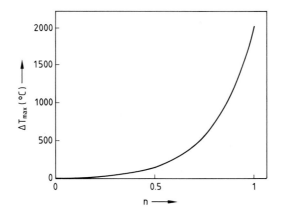

Figure 2.3-12   Viscous heating as a function of the power-law index in the thermal equilibrium regime – Flow conditions and polymer properties reported in the text

b) Transition regime

In the general case, the temperature varies along the flow axis and it is not possible to obtain an analytical solution of $T(r, z)$ from Equation (2.3-33). As previously done, we look for a solution of the average temperature $\overline{T}(z)$. From an energy balance over a differential volume $\pi R^2 dz$, we obtain:

$$\varrho c_p \pi R^2 V d\overline{T}(z) = -2\pi R dz q + \left[\int_0^R \dot{W}(r)\, 2\pi r\, dr\right] dz \qquad (2.3\text{-}38)$$

As done in the Newtonian case, the heat flux to the wall is approximated from the equilibrium temperature profile (Equation (2.3-35)). The mean bulk temperature is at equilibrium:

$$\bar{T} - T_w = \left(\frac{4n+1}{5n+1}\right) \frac{K}{k} \left(\frac{n}{3n+1}\right)^{1-n} \left(\frac{V}{R}\right)^{n+1} R^2 \tag{2.3-39}$$

Hence:

$$q = -k \frac{dT}{dr}\bigg|_{r=R} \approx \left(\frac{3n+1}{n}\right)\left(\frac{5n+1}{4n+1}\right) k \left(\frac{\bar{T}(z) - T_w}{R}\right) \tag{2.3-40}$$

Using the definition (2.3-22) for the Cameron number, i.e.:

$$\mathrm{Ca} = \frac{\alpha L}{V R^2}$$

Equation (2.3-38) can be rewritten as:

$$\frac{d\bar{T}}{dz} = -2\left(\frac{3n+1}{4n+1}\right)\left(\frac{5n+1}{n}\right) \mathrm{Ca} \frac{\bar{T} - T_w}{L} + \frac{1}{\varrho c_p} \frac{\Delta P}{L} \tag{2.3-41}$$

This equation can be easily integrated to obtain the approximate axial temperature profile in the transition regime. For the wall temperature equal to the initial polymer temperature ($T_w = T_0$), the solution is:

$$\boxed{\bar{T}(z) = T_0 + \frac{4n+1}{5n+1} \Delta T_{\max} \left(1 - \exp\left\{-2\left(\frac{5n+1}{n}\right)\left(\frac{3n+1}{4n+1}\right) \mathrm{Ca}\, \frac{z}{L}\right\}\right)}$$

$$\tag{2.3-42}$$

where

$$\Delta T_{\max} = \frac{K}{k}\left(\frac{n}{3n+1}\right)^{1-n}\left(\frac{V}{R}\right)^{n+1} R^2$$

This result reduces to Equation (2.3-21) obtained for a Newtonian fluid, i.e. for $K = \eta$ and $n = 1$. The magnitude of the Cameron number allows us to distinguish between the adiabatic, the transition and the equilibrium regimes.

Example 2.3-3

We examine the flow of a polymer in a tube of radius equal to 5 mm and length equal to 1 m. The average velocity is 0.2 m/s, and the polymer properties are: $K = 10^4$ Pa·s^n, $k = 0.2$ W/m·°C and $\alpha = 10^{-7}$ m^2/s. The Cameron number for these conditions is equal to 0.02; hence the flow is in the transition regime.

The solutions for the temperature increase predicted by Equation (2.3-42) are presented in Figure 2.3-13 for different values of the power-law index, n.

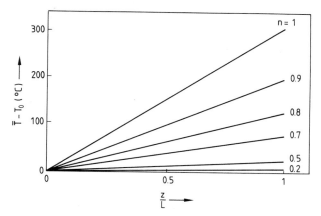

Figure 2.3-13 Influence of the power-law index parameter on the polymer temperature increase. The Cameron number is equal to 0.2.

It is interesting to note that the thermal regime is highly affected by the shear thinning of polymers. For $n \geq 0.5$, we are very near the adiabatic regime even at the tube exit, whereas for $n = 0.2$, we are in the transition regime at $z/L = 1$. Is is clear that equilibrium is attained more rapidly as the factor multiplying the Cameron number in Equation (2.3-42) becomes larger as the shear-thinning properties increase (smaller values of the parameter n).

Remark

It is possible to improve the solution for the transition regime by introducing a Nusselt number which depends on the Cameron number and on the parameter n, as done in Sections 2.3-1 and 2.3-2. SAILLARD (1982) proposed the following correlation:

$$\mathrm{Nu} = b(n) \, \mathrm{Ca}^{-0.6} \tag{2.3-43}$$

where the function $b(n)$ is given in Figure 2.3-14.

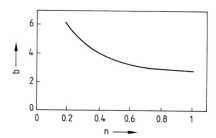

Figure 2.3-14 Coefficient $b(n)$ in Equation (2.3-43) as a function of the shear-thinning parameter n

We will examine in the next sections two flow geometries that are encountered in polymer processing, mainly

– simple shear flow between two parallel plates.
– Poiseuille flow between two parallel plates.

2.3-4 Heat generation in simple shear flow

The velocity profile in simple shear flow between two parallel plates is linear as shown in Figure 2.3-15.

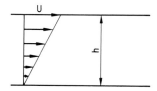

Figure 2.3-15   Simple shear flow

Neglecting the axial conduction, the equation of energy for steady-state conditions is

$$\varrho c_p u \frac{\partial T}{\partial x} = k \frac{\partial^2 T}{\partial y^2} + \eta \left(\frac{U}{h}\right)^2 \tag{2.3-44}$$

The three thermal regimes are now discussed (see GAVIS and LAURENCE, 1968).

a) Equilibrium regime

Equilibrium is reached when the contribution of the convection term is nil, i.e.:

$$k \frac{d^2 T}{dy^2} = -\eta \left(\frac{U}{h}\right)^2 \tag{2.3-45}$$

If both plates are maintained at T_0, integration of Equation (2.3-45) yields

$$T(y) = \frac{\eta}{2k} \left(\frac{U}{h}\right)^2 y(h - y) + T_0 \tag{2.3-46}$$

This is a parabolic temperature profile and a maximum is predicted at the center of the flow geometry, i.e.

$$\Delta T_{\max 1} = \frac{\eta U^2}{8k} = \frac{\eta V^2}{2k} \tag{2.3-47}$$

where the average velocity is $V = U/2$.

If the lower plate is maintained at T_0 whereas the upper plate is isolated so that the heat flux is equal to zero at $y = h$, Equation (2.3-45) leads to:

$$T(y) = \frac{\eta U^2}{2kh^2} y(2h - y) + T_0 \tag{2.3-48}$$

The temperature profile is still parabolic, but the maximum is attained at the upper plate, and

$$\Delta T_{\max 2} = \frac{\eta}{2k} U^2 = 4 \Delta T_{\max 1} \tag{2.3-49}$$

Both relations are compared in Figure 2.3-16.

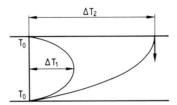

Figure 2.3-16 Temperature profiles in shear flow for two different boundary conditions

Order of magnitudes

- For polymers we take: $\eta = 10^3 \, \mathrm{Pa \cdot s}$
  $k = 0.2 \, \mathrm{W/^\circ C \cdot m}$
  $U = 0.40 \, \mathrm{m/s}$

  we obtain $\Delta T_{\max 1} = 100^\circ \, \mathrm{C}$
  $\Delta T_{\max 2} = 400^\circ \, \mathrm{C}$

These magnitudes of temperature increase due to viscous dissipation are unrealistic. The shear-thinning properties of most polymers will result in much lower effective viscosity and hence in shear-heating effects that could be 100 times smaller.

- For oils we take: $\eta = 0.5 \, \mathrm{Pa \cdot s}$
  $k = 0.14 \, \mathrm{W/^\circ C \cdot m}$
  $U = 5 \, \mathrm{m/s}$

  we obtain $\Delta T_{\max 1} = 10^\circ \, \mathrm{C}$
  $\Delta T_{\max 2} = 40^\circ \, \mathrm{C}$

These values appear to be more reasonable.

b) Adiabatic regime

The adiabatic regime is observed at the beginning of the flow or if both plates are perfectly isolated. A heat balance over a differential volume element yields:

$$\varrho c_p \frac{U}{2} W h \, d\bar{T}(x) = \left[\int_0^h \eta \left(\frac{U}{h} \right)^2 dy \right] W \, dx \tag{2.3-50}$$

where $\bar{T}$ is the mean temperature and W is the width of the plates. Equation (2.3-50) reduces to:

$$\frac{d\bar{T}}{dx} = \frac{2\eta U}{\varrho c_p h^2} \tag{2.3-51}$$

and the integration with the initial temperature T_0 yields:

$$\bar{T}(x) = T_0 + \frac{2\eta U}{\varrho c_p h^2} x \tag{2.3-52}$$

This is a linear temperature profile similar to the expression (2.3-12) obtained for Poiseuille flow.

Numerical applications

- For polymers, we assign the following values:

$$\eta = 10^3 \text{ Pa} \cdot \text{s}$$
$$\varrho c_p = 2 \times 10^6 \text{ J/m}^3 \cdot {}^\circ\text{C}$$
$$U = 0.40 \text{ m/s}$$
$$h = 5 \text{ mm}$$
$$L = 1 \text{ m}$$

and obtain from Equation (2.3-52): $T - T_0 = 15.3^\circ$ C.

- Taking for oils

$$\eta = 0.5 \text{ Pa} \cdot \text{s}$$
$$\varrho c_p = 1.75 \times 10^6 \text{ J/m}^3 \cdot {}^\circ\text{C}$$
$$U = 5 \text{ m/s}$$
$$h = 0.10 \text{ mm}$$
$$L = 0.10 \text{ m}$$

we get: $\bar{T} - T_0 = 28.5^\circ$ C.

c) Transition regime

The conduction term is no longer negligible and Equation (2.3-50) has to be replaced by:

$$\varrho c_p \frac{U}{2} W h \, d\bar{T}(x) = -2W \, dx \, q + \eta \frac{U^2}{h} W \, dx \tag{2.3-53}$$

where q is the heat flux by conduction through the walls. The order of magnitude of q can be obtained from the thermal equilibrium temperature. Hence for the case where both plates are maintained at T_0:

$$\bar{T} = T_0 + \frac{\eta}{12k} U^2 \tag{2.3-54}$$

$$q = -k \left(\frac{dT}{dy}\right)_{y=h} = -6k \frac{\bar{T} - T_0}{h} \tag{2.3-55}$$

We assume, as done in the case of Poiseuille flow, that relation (2.3-55) remains valid in the transition regime. Equation (2.3-53) becomes:

$$\varrho c_p \frac{U}{2} h \, dx \frac{d\bar{T}}{dx} = -12k \frac{\bar{T} - T_0}{h} dx + \eta \frac{U^2}{h} dx \qquad (2.3\text{-}56)$$

which can be readily integrated to get:

$$\bar{T}(x) = T_0 + \frac{\eta U^2}{12k} \left(1 - \exp\left\{ -12 \, \mathrm{Ca} \frac{x}{L} \right\} \right) \qquad (2.3\text{-}57)$$

where

$$\mathrm{Ca} = \frac{kL}{\varrho c_p V h^2} \quad \text{and} \quad V = \frac{U}{2}$$

The Cameron number, Ca, plays the same role as in Poiseuille flow:

if $\mathrm{Ca} < 10^{-2}$,   adiabatic regime
if $10^{-2} < \mathrm{Ca} < 1$, transition regime
if $\mathrm{Ca} > 1$,   equilibrium regime

Example 2.3-4: Heat generation in simple shear

Let us consider the simple shear flow of a polymer ($\eta = 10^3$ Pa·s, $k = 0.2$ W/m·°C, $\varrho c_p = 2 \times 10^6$ J/m^3 · °C). For $U = 0.40$ m/s, $h = 5$ mm and $L = 1$ m, the Cameron number is equal to 0.02. The evolution of the bulk temperature in the adiabatic and in the transition regimes is shown in Figure 2.3-17.

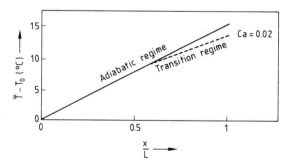

Figure 2.3-17   Evolution of the bulk temperature in simple shear flow (wall temperature maintained at T_0)

Remark

If the wall temperature, T_w, is different from the initial polymer temperature, T_0, the temperature profile is then given by the following expression:

$$\bar{T} - T_0 = \frac{\eta V^2}{3k} \left(1 + \frac{3}{\mathrm{Br}} \right) \left(1 - \exp\left\{ -12 \, \mathrm{Ca} \frac{x}{L} \right\} \right) \qquad (2.3\text{-}58)$$

with $\quad \mathrm{Br} = \dfrac{\eta V^2}{k(T_w - T_0)} \quad$ and $\quad V = \dfrac{U}{2}$

This expression is very similar to result (2.3-30) obtained for Poiseuille flow and similar conclusions can be drawn.

2.3-5   Heat generation in planar Poiseuille flow

The velocity profile in planar Poiseuille flow, illustrated in Figure 2.3-18, is given by

$$u(y) = \frac{3}{2}V\left(1 - \left(\frac{2y}{h}\right)^2\right) \tag{2.3-59}$$

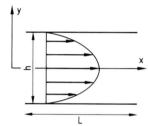

Figure 2.3-18
Planar Poiseuille flow

The shear rate is

$$\dot{\gamma}(y) = -\frac{12V}{h^2}y \tag{2.3-60}$$

and the thermal energy equation becomes:

$$\varrho c_p u(y)\frac{\partial T}{\partial x} = k\frac{\partial^2 T}{\partial y^2} + 144\eta\frac{V^2}{h^4}y^2 \tag{2.3-61}$$

The most important developments on heat generation in planar Poiseuille flow have been proposed by VLACHOPOULOS and KEUNG (1972), COX and MACOSKO (1974), and LIN and HSU (1980).

The reader can easily verify the following results obtained for the three thermal regimes.

a)   Equilibrium regime

$$T(y) = T_w + \frac{3}{4}\frac{\eta V^2}{k}\left[1 - \left(\frac{2y}{h}\right)^4\right] \tag{2.3-62}$$

with a maximum temperature increase given by:

$$\Delta T_{\max} = \frac{3}{4}\eta\frac{V^2}{k} \tag{2.3-63}$$

b) Adiabatic regime

$$\bar{T}(x) = T_0 + \frac{\Delta P}{\varrho c_p} \frac{x}{L} \tag{2.3-64}$$

where T_0 is the initial temperature of the polymer.

c) Transition regime

If the wall temperature is maintained at T_0, the temperature profile is

$$\bar{T}(x) = T_0 + \frac{99}{140} \eta \frac{V^2}{k} \left[1 - \exp\left(- \frac{560}{33} \text{Ca} \frac{x}{L} \right) \right] \tag{2.3-65}$$

where the Cameron number is defined by:

$$\text{Ca} = \frac{\alpha L}{V h^2} \tag{2.3-66}$$

If the wall temperature is controlled at a temperature T_w different from T_0, the temperature profile is a function of the Brinkman number (Equation 2.3-29):

$$\bar{T}(x) = T_0 + \frac{99}{140} \eta \frac{V^2}{k} \left[1 + \frac{140}{99 \, \text{Br}} \right] \left[1 - \exp\left(- \frac{560}{33} \text{Ca} \frac{x}{L} \right) \right] \tag{2.3-67}$$

2.3-6 Applications to polymer processing

The flow situation in most polymer processing applications is locally close to one of the three types described in the previous sections. The mean or bulk polymer temperature can be estimated by calculating two dimensionless numbers.

The Cameron number: $\text{Ca} = \alpha L / V h^2$ (Equation 2.3-22)

where $\alpha =$ thermal diffusivity of the polymer
  $V =$ average velocity of the polymer flow
  $h =$ transverse characteristic dimension of the flow geometry
  $L =$ axial characteristic dimension of the flow geometry.

The magnitude of the Cameron number defines the thermal regime encountered:

– if the Cameron number is very small (typically $\text{Ca} < 10^{-2}$), the polymer temperature will increase linearly with the axial distance (adiabatic regime);
– if the Cameron number is large (typically $\text{Ca} > 1$), the temperature profile is established and its magnitude depends on the wall temperature (equilibrium regime):
– if the Cameron number is included between these two extremes (i.e. $10^{-2} < \text{Ca} < 1$), the bulk temperature will describe an exponential dependence with the axial distance. This is typical of the transition regime.

The Brinkman number: $\mathrm{Br} = \dfrac{\eta V^2}{k(T_w - T_0)}$

where $k =$ thermal conductivity of the polymer
$\quad\quad\quad \eta =$ viscosity of the polymer
$\quad\quad\quad T_w =$ wall temperature
$\quad\quad\quad T_0 =$ initial temperature of the polymer.

The Brinkman number expresses the relative importance of the viscous dissipation with respect to the heat conduction:

- if Br is small, viscous heating is negligible and heat conduction from the wall controls the temperature profile;
- if Br is large, the evolution of the polymer temperature is essentially due to viscous heating.

Example 2.3-5 Polymer temperature in a die

A polymer flows from an extruder to a coat hanger die as illustrated in Figure 2.3-19.

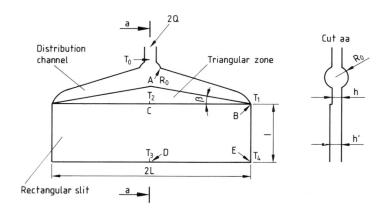

Figure 2.3-19   Sketch of a coat hanger die

The polymer is first fed through a channel to attain equilibrium, then flows in a triangular zone of thickness h and finally through a rectangular slit of thickness h'. The temperature evolution of the polymer may be quite different depending if the polymer flows in the center or along the edges of the die. Large temperature differences will result in serious non uniformities of the flow rate. It is possible to estimate the polymer temperature at any position in the die from the following approximate and simple calculations.

- The polymer is fed at the temperature T_0 and we assume that the flow in the initial channel can be approximated by Poiseuille flow. The polymer temperature at the end of the channel (point B) is T_1, given by Equation (2.3-28):

$$T_1 - T_0 = \frac{5}{6}\eta\frac{V^2}{k}\left[1 + \frac{6}{5\,\mathrm{Br}}\right]\left[1 - \exp\left\{-\frac{48}{5}\,\mathrm{Ca}\right\}\right]$$

where:

$$\mathrm{Ca} = \frac{\alpha L}{\cos\beta V R^2}, \quad \mathrm{Br} = \frac{\eta V^2}{k(T_w - T_0)}$$

$R = \dfrac{2}{3}R_0$ (we take into account the radius variations from the center to the edge)

$V = \dfrac{Q}{2\pi R^2}$ (the flow is equally divided in the two channels)

T_w = wall temperature of the die .

- The flow in the triangular section is assumed to be a planar Poiseuille flow. The variation of temperature between point A and point C is given by Equation (2.3-66), i.e.:

$$T_2 - T_0 = \frac{99}{14}\eta\frac{V^2}{k}\left[1 + \frac{140}{99\,\mathrm{Br}}\right]\left[1 - \exp\left\{-\frac{560}{33}\,\mathrm{Ca}\right\}\right]$$

where:

$$\mathrm{Ca} = \frac{\alpha L \tan\beta}{V h^2}, \quad \mathrm{Br} = \frac{\eta V^2}{k(T_w - T_0)}, \quad V = \frac{Q}{Lh}$$

- The temperature increase in the rectangular slit is again given by Equation (2.3-66)

- *In the center zone:*

$$T_3 - T_2 = \frac{99}{140}\eta\frac{V^2}{k}\left[1 + \frac{140}{99\,\mathrm{Br}}\right]\left[1 - \exp\left\{-\frac{560}{33}\,\mathrm{Ca}\right\}\right]$$

where:

$$\mathrm{Ca} = \frac{\alpha l}{V h'^2}, \quad \mathrm{Br} = \frac{\eta V^2}{k(T_w - T_2)}, \quad V = \frac{Q}{Lh'}$$

- *Near the edges*

$$T_4 - T_1 = \frac{99}{140}\eta\frac{V^2}{k}\left[1 + \frac{140}{99\,\mathrm{Br}}\right]\left[1 - \exp\left\{-\frac{560}{33}\,\mathrm{Ca}\right\}\right]$$

with the Brinkman number given by $\mathrm{Br} = \dfrac{\eta V^2}{k(T_w - T_1)}$

Application

The polymer has the following properties:

viscosity, $\eta = 10^3 \, \text{Pa} \cdot \text{s}$
thermal conductivity, $k = 0.2 \, \text{W/m} \cdot {}^\circ\text{C}$
heat capacity, $\varrho c_p = 2 \cdot 10^6 \, \text{J/m}^3 \cdot {}^\circ\text{C}$

The dimensions of the die are:

total width, $2L = 1.40$ m
slit length, $l = 0.2$ m
angle of triangular section, $\beta = 2^\circ$
thickness of triangular section, $h = 1.5$ mm
initial radius of distribution channel, $R_0 = 15$ mm

The computations for the temperature distribution in the die for a polymer flow rate, $2Q$, equal to 700 kg/h are reported in the following two figures. It is shown in Figure 2.3-20, which corresponds to a thin die ($h' = h = 1.5$ mm) with a wall temperature maintained at 220° C, that the axial variation of the polymer temperature can be quite large. For a thicker die ($h' = 7$ mm) under the same conditions, the results reported in Figure 2.3-21 show much smaller axial variations of the polymer temperature, but a drastic temperature heterogeneity is observed at the exit. Figure 2.3-22 shows how polymer temperature homogeneity in a thicker die can be achieved by maintaining various parts of the die walls at different temperatures as indicated in the figure.

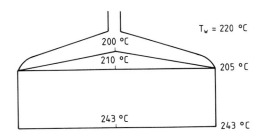

Figure 2.3-20   Temperature variations in a thin slit die

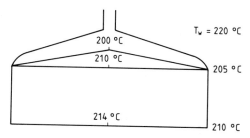

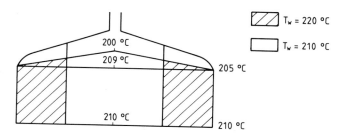

Figure 2.3-21   Temperature variations in a large slit die

Figure 2.3-22   Temperature variations in a large slit die – wall temperature controlled at two different temperatures

Chapter 3
Flow of Molten Polymers
in Various Geometries

3.1 Pressure build up in polymer flow

Only simple flow geometries have been studied in the first two chapters:
-   Poiseuille flow in tubes of constant diameter;
-   shear flow between parallel plates.

The pressure in these flow geometries remains constant or decreases linearly with the axial distance. In polymer processing, the flow geometries are frequently more complex. Examples of more complex flows are encountered in
-   calendering (Chap. 3.2);
-   extrusion dies (Chap. 3.3);
-   injection molds (Chap. 3.4);
-   compression and metering zones of an extruder (Chap. 4.4).

The pressure profiles obtained in these flow situations may be very different. In Chapter 3.1, we show several flow geometries in which pressure can be built up.

3.1-1 Qualitative analysis of three flow situations

a) The Rayleigh bearing

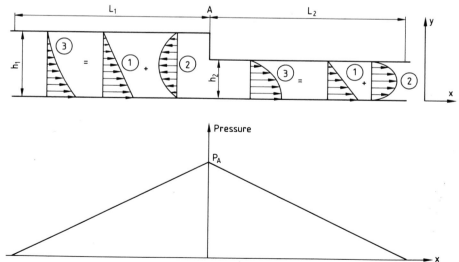

Figure 3.1-1   The Rayleigh bearing – mechanism for pressure build up

The flow geometry of the Rayleigh bearing is illustrated in Figure 3.1-1. It is a simple shear flow between parallel plates with a sudden restriction.

The linear velocity profile of type ① observed in simple shear flow does not respect here the mass conservation principle. One can well imagine that a pressure p_A is generated due to the sudden contraction, such that Poiseuille profiles of type ② are induced upstream as well as downstream. The net velocity profile of type ③ is concave downstream and convex upstream, thus respecting the principle of mass conservation.

b) Contact of two cylinders

The flow situation in calendering or flow between two cylinders is illustrated in Figure 3.1-2.

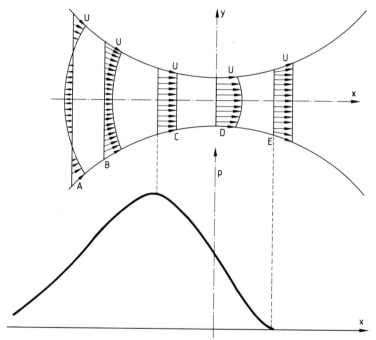

Figure 3.1-2 Flow between two cylinders

The fluid, dragged by the motion of the cylinders, flows at a rate equal to Q. It is reasonable to believe that there is a point C in the gap where the average velocity of the fluid is equal to the velocity of the cylinders, U. Upstream of point C, the average velocity of the fluid is less than that of the cylinders; the fluid is then accelerated by the presence of the cylinders and the pressure increases. However, downstream of point C, the average velocity of the fluid is greater than that of the cylinders; the fluid is then decelerated by the presence of the cylinders and the pressure decreases. It is further assumed that contact is lost at point E where the average fluid velocity is again equal to that of the cylinders. This point is symmetrical to point C, with respect to D, corresponding to the point of minimum gap.

It is clear from this simple analysis that the pressure increases from zero (or from a reference value) at the entrance to a maximum at point C, then decreases back to zero at the outlet.

c) The Reynolds bearing

The Reynolds or plane slider bearing represents a flow situation often encountered in mechanical lubrication, with thick oil films ($\approx 100\ \mu$). The problem is somewhat simplified by considering that the lower plate is moving with respect to the upper slider (Figure 3.1-3) with a velocity U. The lubricant flow rate is Q.

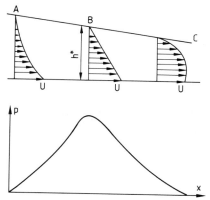

Figure 3.1-3   The Reynolds bearing

One may assume that the velocity profile is linear at point B as illustrated in Figure 3.1-3. Upstream of B, the velocity profile must then be concave to respect the flow continuity and the pressure increases with x. Downstream, the situation is opposite and the pressure decreases with x. A maximum in the pressure profile is then observed at point B.

We conclude from these three examples that the two necessary conditions to build up pressure in a flow geometry are:

– a decrease of the flow section affecting the mass balance;
– the existence of shear or drag flow due to the wall motion of the processing equipment.

3.1-2 Pressure generated in a Rayleigh bearing

At a reasonable distance from the sudden contraction, there is only one component of the velocity, u, which is a function of y (see Figure 3.1-1). The equation of motion in simple shear reduces to:

$$\frac{dp}{dx} = \eta \frac{d^2 u}{dy^2} \qquad (3.1\text{-}1)$$

Since u is a unique function of y and p is a unique function of x, the pressure gradient, dp/dx, is a constant given by $dp/dx = p_A/L_1$ in the upstream section or by $dp/dx = -p_A/L_2$ in the downstream section, assuming that the pressure is equal to zero at both the entrance and outlet.

Using the non-slip conditions at the bottom and upper plates, we can integrate Equation (3.1-1) to obtain:

upstream
$$u = U\left(1 - \frac{y}{h_1}\right) - \frac{1}{2\eta}\frac{p_A}{L_1}y(h_1 - y) \qquad (3.1\text{-}2)$$

downstream
$$u = U\left(1 - \frac{y}{h_2}\right) + \frac{1}{2\eta}\frac{p_A}{L_2}y(h_2 - y) \qquad (3.1\text{-}3)$$

The velocity profiles are different for the two zones of the bearing. The results are the sum of simple shear and Poiseuille flows. The flow rate per unit width (W) is for each zone:

upstream
$$\frac{Q}{W} = U\frac{h_1}{2} - \frac{1}{12\eta}\frac{p_A}{L_1}h_1^3 \qquad (3.1\text{-}4)$$

downstream
$$\frac{Q}{W} = U\frac{h_2}{2} + \frac{1}{12\eta}\frac{p_A}{L_2}h_2^3 \qquad (3.1\text{-}5)$$

Since the flow rate is conserved, the pressure p_A is readily obtained:

$$p_A = \frac{6\eta U(h_1 - h_2)}{h_1^3/L_1 + h_2^3/L_2} \qquad (3.1\text{-}6)$$

Numerical application

Let us consider the following conditons:

$$\eta = 5 \text{ kPa} \cdot \text{s}$$
$$U = 0.1 \text{ m/s}$$

$$h_1 = 1 \text{ mm}, \qquad L_1 = 0.1 \text{ m}$$
$$h_2 = 0.5 \text{ mm}, \qquad L_2 = 0.1 \text{ m}$$

From Equation (3-1-6), the pressure is:

$$p_A = 133 \text{ MPa}$$

This is a very high value.

Remark

This example does not correspond to real situations encountered in polymer processing. In Chapter 4.3, a more realistic case for extrusion is discussed. The Rayleigh bearing is, however, a flow geometry often encountered in lubrication. Taking the following conditions:

$$\eta = 0.1 \text{ Pa} \cdot \text{s}$$
$$U = 0.3 \text{ m/s}$$

$$h_1 = 0.5 \text{ mm}, \qquad L_1 = 0.05 \text{ m}$$
$$h_2 = 0.2 \text{ mm}, \qquad L_2 = 0.05 \text{ m}$$

we get:

$$p_A = 20 \text{ kPa}$$

If we assume that the pressure profile is linear (except in the vicinity of the contraction), the axial load created by the flow is given by the expression:

$$F = p_A \frac{(L_1 + L_2)W}{2} \tag{3.1-7}$$

For $W = 50$ mm, F is equal to 50 N.

3.1-3   Flow in a variable gap geometry

Let us consider the flow geometry shown in Figure 3.1-4, corresponding to a lubricated cylinder and plate system.

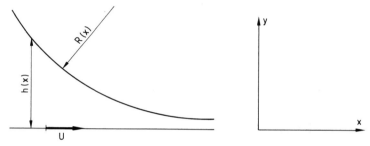

Figure 3.1-4   Lubricated cylinder and plate system

For Newtonian liquids, with negligible inertia, the Navier-Stokes equations reduce to:

$$\frac{\partial p}{\partial x} = \eta \left[\frac{\partial^2 u}{\partial x^2} + \frac{\partial^2 u}{\partial y^2} \right]$$
$$\frac{\partial p}{\partial y} = \eta \left[\frac{\partial^2 v}{\partial x^2} + \frac{\partial^2 v}{\partial y^2} \right] \tag{3.1-8}$$

and the equation of continuity becomes:

$$\frac{\partial u}{\partial x} + \frac{\partial v}{\partial y} = 0 \tag{3.1-9}$$

We can obtain approximate solutions of these equations for applications of interest.

a)   The lubrication approximations

Following STREETER (1961) and TANNER (1985), we propose the following:

– 1st *hypothesis:* if the gap does not vary rapidly with the axial distance, i.e. $dh/dx \ll 1$, then the term $\partial u/\partial x$ is negligible with respect to $\partial u/\partial y$;

– 2nd *hypothesis:* if the curvature of the body surfaces is not large, i.e. $h/R \ll 1$ and the radius of curvature expressed by $1/R = \partial^2 h/\partial x^2$, the term $\partial^2 u/\partial x^2$ is negligible with respect to $\partial^2 u/\partial y^2$ (see Appendix E).

– 3rd *hypothesis:* the Reynolds number is usually very small in polymer processing (Re $\ll 1$) so that the flow remains laminar.

These three hypotheses of hydrodynamic lubrication allows us also to neglect all terms containing the velocity component v in Equation (3.1-8). Hence Equation (3.1-8) can be written as:

$$\frac{dp}{dx} = \eta \left[\frac{\partial^2 u}{\partial y^2} \right] \tag{3.1-10}$$

while the continuity equation states that the axial flow rate per unit width, W, is constant, i.e.

$$\frac{Q}{W} = \text{constant} \tag{3.1-11}$$

The differential equation (3.1-10) is identical to that obtained for Poiseuille or shear flow between two parallel plates, but here the pressure gradient is no longer a constant since u is a function of x and y. The use of the hydrodynamic lubrication approximations is equivalent to assuming that, locally, the polymer flows between two parallel plates, or more generally in a constant cross section channel.

Remark

The hydrodynamic lubrication approximations are applicable when the angle made by the adjacent walls does not exceed 10° (BENIS, 1967). For larger angles, the errors made by using the approximations are no longer negligible. Nevertheless, the results are still useful for a qualitative evaluation of the phenomena.

b) The Reynolds equations

Equation (3.1-10) can be easily integrated with the following boundary conditions:

$$\text{B.C. 1)}: \quad \text{at } y = 0, \ u = U$$
$$\text{B.C. 2)}: \quad \text{at } y = h, \ u = 0$$

The velocity profile is then:

$$u(x,y) = U \left(1 - \frac{y}{h} \right) - \frac{1}{2\eta} \frac{dp}{dx} y(h-y) \tag{3.1-12}$$

This is the same type of velocity profile as encountered in the Rayleigh bearing, but here the pressure gradient varies with the axial distance, x. The flow rate is expressed by:

$$\frac{Q}{W} = \int_0^h u(x,y)\, dy = -\frac{h^3}{12\eta} \frac{dp}{dx} + U \frac{h}{2} \tag{3.1-13}$$

Since the flow rate is constant in the gap, we can write the three following equivalent expressions for the pressure gradient:

1st form:

$$\frac{dp}{dx} = \left[U\frac{h}{2} - \frac{Q}{W} \right] \frac{12\eta}{h^3} \qquad (3.1\text{-}14)$$

2nd form:

$$\frac{d}{dx}\left[-\frac{h^3}{12\eta}\frac{dp}{dx} + U\frac{h}{2} \right] = 0 \qquad (3.1\text{-}15)$$

3rd form:

We define h^* as the film thickness or gap for which the pressure is maximum, i.e. where $dp/dx = 0$ (point B in Figure 3.1-3). Then:

$$\frac{Q}{W} = \frac{Uh^*}{2} = -\frac{h^3}{12\eta}\frac{dp}{dx} + U\frac{h}{2}$$

which yields.

$$\boxed{\frac{dp}{dx} = 6\eta U \frac{h - h^*}{h^3}} \qquad (3.1\text{-}16)$$

Remark

This last form is correct only if the velocity of the plate, U, is constant within the contact zone.

Problem 3.1

3.1-A   Reynolds bearing

Consider the flow in a Reynolds bearing as illustrated in Figure 3.1-5. The lower plate is displaced at uniform velocity U and Equation (3.1-16) can be correctly used, by changing the variable x with z:

$$\frac{dp}{dz} = 6\eta U \frac{h - h^*}{h^3}$$

a)   Make the change of variable $z(h)$, and obtain the expression for the Reynolds equation as a unique function of h.

b)   Using the following boundary conditions:

$$\text{B.C. 1}) : \quad \text{at } z = 0, \ p = 0$$
$$\text{B.C. 2}) : \quad \text{at } z = L, \ p = 0$$

I)   obtain the expression for the pressure profile $p(h)$ as a function of the parameter h^*, as yet unknown;

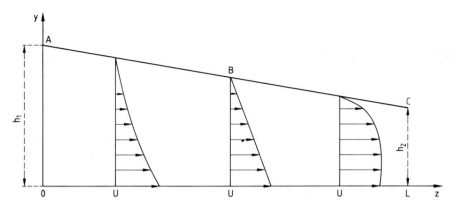

Figure 3.1-5   Reynolds bearing

II) find the expressions for h^*, for the flow rate Q, of the maximum pressure p_{max}, and the load F normal to the flow direction. W is the width of the Reynolds bearing.

c) Obtain the numerical values of h^*, Q, p_{max} and F for the following two cases:

I) Polymer:
$$\eta = 5 \text{ kPa} \cdot \text{s}$$
$$U = 0.1 \text{ m/s}$$
$$h_1 = 1 \text{ mm}, \quad h_2 = 0.5 \text{ mm}$$
$$L = 0.2 \text{ m}, \quad W = 50 \text{ mm}$$

II) Oil:
$$\eta = 0.1 \text{ Pa} \cdot \text{s}$$
$$U = 0.3 \text{ m/s}$$
$$h_1 = 0.5 \text{ mm}, \quad h_2 = 0.2 \text{ mm}$$
$$L = 0.1 \text{ m}, \quad W = 50 \text{ mm}$$

3.2   Calendering

3.2-1   General description

Calendering is the process used to produce sheets of thermoplastics and elastomers. As an illustration, we present below the fabrication process of polyvinyl chloride (PVC) sheets.

In contrast with extrusion or injection molding where all the solid and liquid transformations occur in the same machine, the manufacture of PVC sheets is done through an assembly of equipment as shown in Figure 3.2-1.

We will restrict this section to the phenomena encountered in calendering, and more specifically in the gap of the last two rolls. This is where the quality of the sheet (dimensions and surface finish) is controlled. The design and the proper operation of the calendering unit necessitates a good understanding of the phenomena involved in this processing. The flow pattern in the nip of the two rolls is quite complex as shown in Figure 3.2-2.

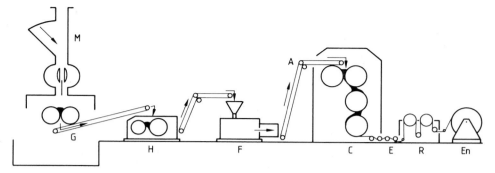

Figure 3.2-1   Manufacture of PVC sheets
  M: mixing of additives to PVC
  G: gelation
  H: homogeneisation
  F: filtration
  A: feed
  C: calendering
  E: extraction
  R: cooling
  E_N: windup

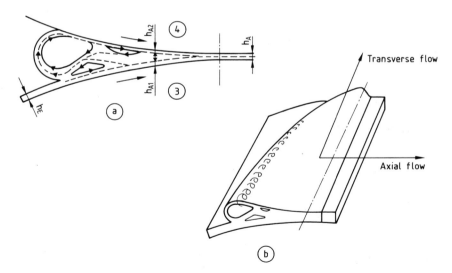

Figure 3.2-2   Flow in a calender showing the bank formation (from UNKRUER, 1972)

The flow proceeds as follows:

— the sheet in close contact with roll ③ (the lower roll in Figure 3.2-2) is fed as
   illustrated on the sketch a) of the figure;

- a part of the material, of thickness h_{A1}, goes directly through the nip along cylinder ③;
- the rest of the material, of thickness $h_E - h_{A1}$, is recirculated to form a bank;
- a portion of the bank material (h_{A2}) will flow in the nip along roll ④ (upper roll of Figure 3.2-2);
- the rest of the material ($h_E - h_{A1} - h_{A2}$) flows in the normal direction, thus increasing the width of the sheet.

We will simplify the analysis by assuming that the entry flow is symmetrical with respect to the plane of the gap as pictured in Figure 3.2-3. This is then similar to the situation described in Figure 3.1-2. We assume further that the polymer velocity at the exit (from point E and downstream) is equal to that of the lower roll.

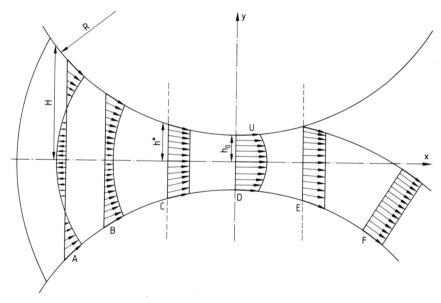

Figure 3.2-3   Approximate flow in calendering

3.2-2 Symmetrical Newtonian analysis

We now present the model originally developed for different boundary conditions by ARDICHVILLI (1938), GASKELL (1950), MCKELVEY (1962), and TASKERMAN-KROZER et al. (1975).

The hydrodynamic lubrication approximations are valid:

- The surface of the rolls is assumed to be parabolic in the flow region, i.e.

$$h = h_0 \left[1 + \frac{x^2}{2Rh_0} \right]$$ (3.2-1)

where $h(x)$ is half of the polymer thickness and h_0 is half of the gap. The gap being very small compared to the radius of the rolls, the derivative becomes:

$$\left| \frac{dh}{dx} \right| = \left| \frac{x}{R} \right| = \sqrt{\frac{2(h - h_0)}{R}} \ll 1 \qquad (3.2\text{-}2)$$

– The radius of curvature is simply the radius of the rolls. hence:

$$\frac{h}{R} \ll 1 \qquad (3.2\text{-}3)$$

The Navier-Stokes equations reduce to Equation (3.1-10), i.e.:

$$\frac{dp}{dx} = \eta \left[\frac{\partial^2 u}{\partial y^2} \right]$$

and from the boundary conditions shown in Figure 3.2-3, we obtain a Reynolds equation quite similar to those obtained in Section 3.1-3:

$$\frac{dp}{dx} = 3\eta U \left[\frac{h - h^*}{h^3} \right] \qquad (3.2\text{-}4)$$

where $2h^*$ represents the gap at point C where the pressure is maximum; it is also equal to the thickness of the sheet at the exit.

a) Spread height

Spread height is defined as the ratio of the sheet thickness in contact with the roll at point F and the gap between the two rolls (point D):

$$r = \frac{h^*}{h_0} \qquad (3.2\text{-}5)$$

Defining a new variable by:

$$a = \pm \sqrt{\frac{h}{h_0} - 1} = \frac{x}{\sqrt{2Rh_0}} \qquad (3.2\text{-}6)$$

we may write the Reynolds equation as:

$$\frac{dp}{da} = 3\eta U \frac{\sqrt{2Rh_0}}{h_0{}^2} \left[\frac{a^2 - a^{*2}}{(a^2 + 1)^3} \right] \qquad (3.2\text{-}7)$$

with

$$a^* = \sqrt{\frac{h^*}{h_0} - 1} = \sqrt{r - 1} \qquad (3.2\text{-}8)$$

Equation (3.2-7) can be integrated from $-a_H = -\sqrt{H/h_0 - 1}$ to a, assuming a zero pressure at the entrance (H is the bank height):

$$p(a) = 3\eta U \frac{\sqrt{2Rh_0}}{h_0{}^2} \int\limits_{-a_H}^{a} \frac{a^2 - a^{*2}}{(a^2 + 1)^3} da \qquad (3.2\text{-}9)$$

The expression for a^* is obtained by taking the pressure equal to zero at the exit, i.e.:

$$p(a^*) = 0 = \int\limits_{-a_H}^{a^*} \frac{a^2 - a^{*2}}{(a^2 + 1)^3} da \qquad (3.2\text{-}10)$$

Results of the spread height as a function of the ratio of bank height to gap clearance are reported in Figure 3.2-4. As in most operations the bank is large compared to the gap ($H/h_0 > 10$), it is clear from the figure that the spread height is constant, with a ratio slightly above 1.2.

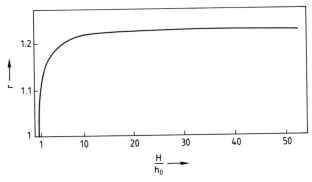

Figure 3.2-4   Spread height as a function of the bank to nip ratio

b) Roll separating force and torque exerted on the roll

Knowing the value of the recovery ratio, hence of a^*, we can calculate the pressure profile by integrating equation (3.2-9). Figure 3.2-5 reports pressure profiles in the nip region calculated for the following conditions:

Rolls radius : $R = 0.3$ m
Gap distance : $2h_0 = 0.4$ mm
Rolls velocity : $U = 0.15$ m/s
Polymer viscosity : $\eta = 1$ kPa $\cdot$ s

We observe that the pressure profiles are of the same shape as drawn qualitatively in Figure 3.1-2. For large enough polymer banks (as encountered in industry) the pressure peak is almost independent of the bank height.

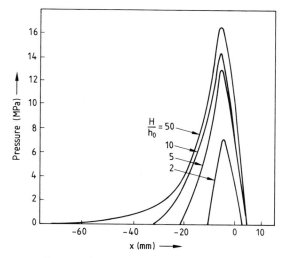

Figure 3.2-5 Pressure profiles in calendering for several values of bank to nip ratio

The separating force acting on the rolls can now be calculated by integrating the pressure profile:

$$F = W \int_{\text{contact}} p(x)\,dx = 6\eta RUW \frac{1}{h_0} \int_{-a_H}^{a^*} \int_{a^*}^{a'} \frac{a^2 - a^{*2}}{(a^2 + 1)^3}\,da\,da' \qquad (3.2\text{-}11)$$

where W is the sheet width. Using the equilibrium value for a^* deduced from Figure 3.2-4 and assuming that $a_H = \infty$ (this is reasonable since the pressure does not increase as soon as the bank height is sufficiently large), we obtain:

$$F = 1.23 \frac{\eta RUW}{h_0} \qquad (3.2\text{-}12)$$

and the torque exerted on the roll is:

$$C = 1.62\eta RUW \sqrt{\frac{2R}{h_0}} \qquad (3.2\text{-}13)$$

These two expressions are of special interest to the design engineer as they express:

- the elastic deformation of the rolls, hence the curvature of the plastic sheets produced by calendering;
- the power required to rotate the calendering unit.

Using the data on which Figure 3.2-5 is based, we get the following numerical values for a calendering width, W, equal to 1 m:

$$\text{Separating force, } F = 276.8 \text{ kN}$$
$$\text{Resisting torque, } C = 4.0 \text{ kN} \cdot \text{m}$$

3.2-3 Asymmetrical Newtonian analysis

The assumption in Section 3.2-2, that the flow in the bank as illustrated in Figure 3.2-2 can be made symmetrical as done in Figure 3.2-3 is somewhat unrealistic. Removing this assumption, but considering a two-dimension flow (i.e. neglecting transverse flow), the Navier-Stokes equations for this problem have been solved in terms of the stream and vorticity functions by AGASSANT and ESPY (1985), and in terms of velocity and pressure by MITSOULIS et al. (1985). The graphical results were shown in Chapter 1.4. We observe that the calculated streamlines in the bank are very close to those obtained experimentally by stopping the rolls, sampling in the bank, cooling and cutting it in thin slices, then polishing the solid samples. The photograph of Figure 3.2-6a shows the experimental streamlines compared to the computed ones of Figure 3.2-6b.

a)

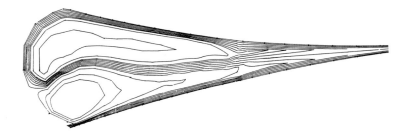

b)

Figure 3.2-6   Streamlines in a calendering bank
a)   Photograph of the flow field in the bank
b)   Calculated streamlines

Figure 3.2-7 shows that the pressure profile calculated on the axis of the bank is quite similar to that obtained under the assumptions of hydrodynamic lubrication. The following operating conditions and property were used:

Rolls radius :	$R = 0.1125$ m
Gap distance :	$2h_0 = 0.245$ mm
Rolls velocity :	$\Omega = 9.5$ rev/min.
Polymer viscosity :	$\eta = 1$ kPa $\cdot$ s

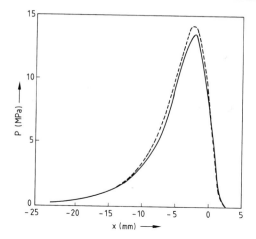

Figure 3.2-7 Pressure profile on the bank axis
—— hydrodynamic lubrication assumptions,   ----- finite-elements method

3.2-4 Shear-thinning fluids

The results of the previous sections are based on approximate values of the viscosity. In general the polymer viscosity decreases greatly with shear rate. This is known as shear thinning, accounted for in Chapter 1.5 by using the power-law expression. The total stress tensor is given, for a generalized power-law fluid, by the following expression:

$$\boldsymbol{\sigma} = -p\boldsymbol{\delta} + 2K|\dot{\gamma}|^{n-1}\dot{\boldsymbol{\varepsilon}} \qquad (3.2\text{-}14)$$

Models for the calendering of shear-thinning polymers have been developed by MCKELVEY (1962), PEARSON (1966), CHONG (1968), BRAZINSKI et al. (1970), and AGASSANT and AVENAS (1977), all using the power-law expression for the viscosity. ALSTON and ASTILL (1973) have made use of the hyperbolic tangent viscosity Equation (1.5-5) and KIPARISSIDES and VLACHOPOULOS (1976) have compared the results obtained by a finite-elements method for the two rheological models.

a) Equation of motion

The rate-of-deformation tensor for planar shear flow is given in the most general case by the following:

$$[\dot{\varepsilon}] = \begin{bmatrix} \dfrac{\partial u}{\partial x} & \dfrac{1}{2}\left[\dfrac{\partial u}{\partial y} + \dfrac{\partial v}{\partial x}\right] & 0 \\[2ex] \dfrac{1}{2}\left[\dfrac{\partial u}{\partial y} + \dfrac{\partial v}{\partial x}\right] & \dfrac{\partial v}{\partial y} & 0 \\[2ex] 0 & 0 & 0 \end{bmatrix} \qquad (3.2\text{-}15)$$

Using the hydrodynamic lubrication assumptions, we neglect the terms in v and its derivatives with respect to the terms containing u and derivatives of u, i.e.

$|\partial v/\partial y| = |\partial u/\partial x| \ll |\partial u/\partial y|$. This is equivalent to assuming the existence locally of simple shear flow and the rate-of-deformation tensor reduces to:

$$[\dot{\varepsilon}] = \begin{bmatrix} 0 & \frac{1}{2}\left[\frac{\partial u}{\partial y}\right] & 0 \\ \frac{1}{2}\left[\frac{\partial u}{\partial y}\right] & 0 & 0 \\ 0 & 0 & 0 \end{bmatrix} \qquad (3.2\text{-}16)$$

and the generalized rate of deformation is:

$$\dot{\gamma} = \sqrt{2\sum_{ij} \dot{\varepsilon}_{ij}^2} = \left|\frac{\partial u}{\partial y}\right| \qquad (3.2\text{-}17)$$

The expression for the stress tensor then reduces to:

$$[\sigma] = \begin{bmatrix} -p & K\left|\frac{\partial u}{\partial y}\right|^{n-1}\frac{\partial u}{\partial y} & 0 \\ K\left|\frac{\partial u}{\partial y}\right|^{n-1}\frac{\partial u}{\partial y} & -p & 0 \\ 0 & 0 & -p \end{bmatrix} \qquad (3.2\text{-}18)$$

and the x-component of the equation of motion for negligible mass and inertia effects is:

$$\frac{dp}{dx} = K\frac{\partial}{\partial y}\left[\left|\frac{\partial u}{\partial y}\right|^{n-1}\frac{\partial u}{\partial y}\right] \qquad (3.2\text{-}19)$$

b) Generalized Reynolds equation

Equation (3.2-19) must be integrated in two parts, since the sign of the velocity gradient changes with the axial position (see Figure 3.2-2):

-   Downstream of point C, $\dfrac{\partial u}{\partial y} < 0$ for $y > 0$

-   Upstream of point C, $\dfrac{\partial u}{\partial y} > 0$ for $y > 0$

Then *downstream* of point C, Equation (3.2-19) may be written as:

$$-\frac{dp}{dx} = K\frac{\partial}{\partial y}\left[-\frac{\partial u}{\partial y}\right]^n$$

and integrated with the non slip conditions on the rolls to obtain the expression for the velocity profile:

$$u = \frac{n}{n+1}\left[\frac{1}{K}\left(-\frac{dp}{dx}\right)\right]^{1/n}[y^{1+1/n} - h^{1+1/n}] + U \qquad (3.2\text{-}20)$$

and the expression for the flow rate is:

$$Q = \frac{2n}{n+1}W\left[\frac{1}{K}\left(-\frac{dp}{dx}\right)\right]^{1/n}h^{2+1/n} + 2UhW \qquad (3.2\text{-}21)$$

In terms of h^* (see Figure 3.2-3), $Q = 2WUh^*$ and:

$$\frac{dp}{dx} = -K\left[\frac{2n+1}{n}U\right]^n \frac{(h^* - h)^n}{h^{2n+1}}$$

(3.2-22)

Upstream of point C, the corresponding expression is:

$$\frac{dp}{dx} = K\left[\frac{2n+1}{n}U\right]^n \frac{(h - h^*)^n}{h^{2n+1}}$$

(3.2-23)

These two results can be written in the following single equation, which we call the generalized Reynolds equation:

$$\boxed{\frac{dp}{dx} = K\left[\frac{2n+1}{n}U\right]^n \frac{|h - h^*|^{n-1}(h - h^*)}{h^{2n+1}}}$$

(3.2-24)

3.2-5 Integrated generalized Reynolds equation

We make the same change of variables as done previously and integrate Equation (3.2-23) using the same boundary conditions, i.e. the pressure is assumed to be equal to zero at the inlet and outlet of the contact.

We observe that the spread height is independent of the bank height as soon as the bank is large enough. Also the spread height is a weak function of shear thinning as illustrated by Figure 3.2-8.

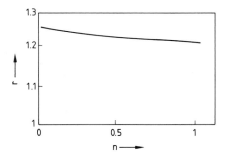

Figure 3.2-8 Variation of the spread height with the shear-thinning index n for large banks

On the other hand, Figure 3.2-9 shows that the pressure profile varies greatly with the shear-thinning index. It follows that the calendering force and torque are strongly dependent on the value of n, as reported in Figure 3.2-10. The results of

Figures 3.2-9 and 3.2-10 have been calculated for the following conditions:

Rolls radius : $R = 0.3$ m
Rolls velocity : $U = 0.15$ m/s
Gap : $2h_0 = 0.4$ mm
Bank : $H/h_0 = 50$
Consistency index : $K = 10^4$ Pa $\cdot$ s^n

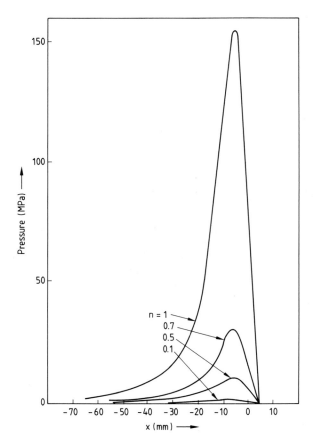

Figure 3.2-9   Pressure profile in the calendering nip as a function of the shear-thinning index n

As an example, for a polymer with a K value equal to 10^4 Pa $\cdot$ s^n, we obtain a calendering force equal to $3 \cdot 10^6$ N if the polymer is Newtonian ($n = 1$) and only $7 \cdot 10^4$ N for a typical shear-thinning polymer ($n = 0.3$).

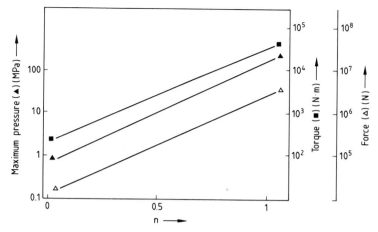

Figure 3.2-10   Variations of the maximum pressure, of the separating force, and of the torque with the shear-thinning parameter n (based on the same data as in Figure 3.2-9)

3.2-6 Thermal effects in calendering

The flow of a high viscosity polymer between the rollers of a calendering unit can generate a considerable amount of heat. In most cases, when processing PVC for example, the operating temperature (180–210° C) is very close to the temperature at which the walls are controlled. It is then appropriate to examine the evolution of the mean temperature, by increments as shown in Figure 3.2-11, from the bank entrance to the exit from the contact.

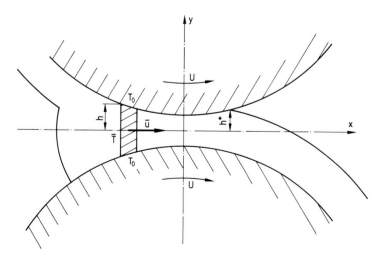

Figure 3.2-11   Solution by increments in calendering

A thermal energy balance on a differential volume element $2hW\Delta x$ yields the following differential equation:

$$\varrho c_p h \bar{u} \left(\frac{d\bar{T}}{dx} \right) = -h_c(\bar{T} - T_0) + \eta \int_0^h \left(\frac{\partial u}{\partial y} \right)^2 dy \tag{3.2-25}$$

where $\bar{u}$ is the average velocity. The heat transfer coefficient, h_c, can be evaluated from the numerical results obtained by SAILLARD (1982) for the shear-heating of polymer flow between two parallel plates (Chapter 2.3), $\mathrm{Nu} = h_c h/k = 10$.

a) Uncoupled solution

Let us consider the velocity profile to be independent of the temperature. Then solutions of Equation (3.2-25) can be obtained for different velocity profiles. For example, the velocity profile for a Newtonian fluid is:

$$u(x,y) = U \left[\frac{3}{2} \frac{h - h^*}{h^3} (y^2 - h^2) + 1 \right] \tag{3.2-26}$$

In this case, the integration can be performed analytically and the result in terms of a, defined by Equation (3.2-6) is:

$$\bar{T}(a) = T_0 + \mu'' + \frac{\lambda a + \mu}{a^2 + 1} + \frac{\lambda' a + \mu'}{(a^2 + 1)^2}$$
$$- \left[\mu'' + \frac{\lambda a_H + \mu}{a_H^2 + 1} + \frac{\lambda' a_H + \mu'}{(a_H^2 + 1)^2} \right] \exp\{B(\tan^{-1} a_H - \tan^{-1} a)\} \tag{3.2-27}$$

where a_H is the value of a evaluated at the bank entrance and:

$$\lambda' = \frac{r^2 A}{4 + B^2/4}, \quad \lambda = \frac{-2rA + 3\lambda'}{2 + B^2/2}, \quad \mu' = \frac{B}{4}\lambda', \quad \mu = \frac{B}{2}\lambda, \quad \mu'' = \frac{A + \lambda}{B}$$

$$A = \frac{3\eta U \sqrt{2Rh_0}}{\varrho c_p rh_0^2}, \quad B = \frac{\mathrm{Nu} \sqrt{2Rh_0}}{\varrho c_p rh_0^2}, \quad \text{and} \quad r = \frac{h^*}{h_0},$$

is the spread height.

Figure 3.2-12 shows the average temperature profile of the polymer as a function of its position in the calendering unit. The high values of $\bar{T}$ in the case of PVC would normally result in thermal degradation. This is not observed experimentally; hence shear-heating is less important than calculated by Equation (3.2-27).

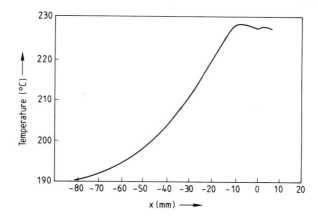

Figure 3.2-12   Mean temperature profile in the gap of a calender of radius $R = 0.275$ m, and of gap $2h_0 = 0.5$ mm. The rolls velocity is equal to 0.15 m/s and the wall temperature is at $190°$ C

b) Coupled solution

In practical situations, the viscosity of the polymer decreases sharply as the temperature increases (see Appendix C and Figure 3.2-13). The flow and pressure fields are hence affected and the equations of motion and of thermal energy have to be solved simultaneously, for example, by an increment method as done by AGASSANT and AVENAS (1977) and by AGASSANT (1980). The equations are:

$$\frac{dp}{dx} = 3\eta U \frac{h - h^*}{h^3}$$

$$\frac{d\bar{T}}{dx} = -\frac{h_c}{\varrho c_p \bar{u} h}(\bar{T} - T_0) + \frac{\eta}{\varrho c_p \bar{u} h} \int_0^h \left(\frac{\partial u}{\partial y}\right)^2 dy$$

and

$$\eta = K(\bar{T})|\dot{\bar{\gamma}}|^{n(\bar{T})-1}$$

The viscosity data of a PVC as a function of shear rate for three different temperatures are reported in Figure 3.2-13.

The computed pressure profiles reported in Figure 3.2-14 show that the pressure in the gap of the calender is slightly lower in the non-isothermal case compared with the isothermal shear-thinning case.

The temperature profiles computed for the temperature-dependent viscosity data are shown in Figure 3.2-15. The temperature increments are considerably less than reported before and this is more in line with the observed thermal stability of PVC.

Figure 3.2-16 reports the variation of the force and torque with increasing calender velocity. The separation between the two rollers is constant and equal to 0.4 mm, the wall temperature is maintained at $190°$ C and the polymer temperature at the entrance of the bank is equal to $195°$ C.

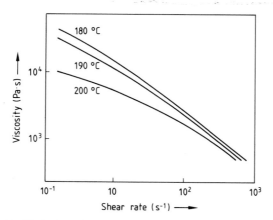

Figure 3.2-13   Variation of the viscosity of a PVC as a function of shear rate and temperature

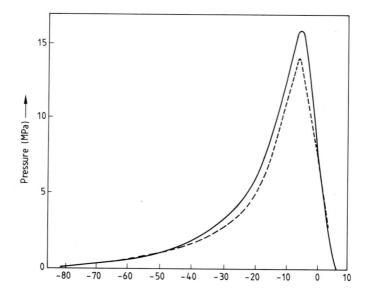

Figure 3.2-14   Pressure profiles in the gap of a calender for the conditions reported in Figure
3.2-12 and viscosity data of Figure 3.2-13
—— isothermal shear-thinning case ----- non-isothermal shear-thinning case

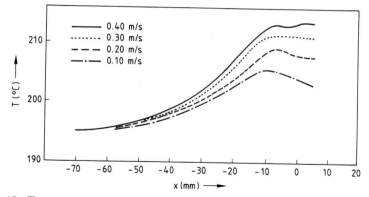

Figure 3.2-15   Temperature profiles in the gap of a calender for several values of the roll velocity

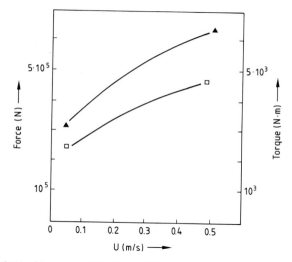

Figure 3.2-16   Variation of the force (□) and torque (▲) with increasing velocity of the calender

Remarks

In fact, as shown by FINSTON (1951), BEKIN et al. (1975), TORNER (1974) and KIPARRISIDES and VLACHOPOULOS (1978), the mean temperature may mask large temperature differences between the center zone of the flow where the shear rate is very small and wall regions where shear rates can be very large. Figure 3.2-17 shows details of temperature variations, calculated by KIPARRISIDES and VLACHOPOULOS (1978).

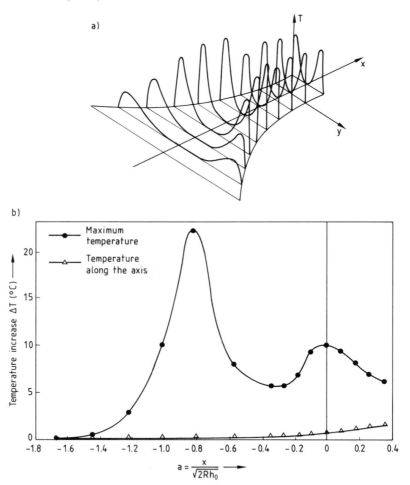

Figure 3.2-17 a) Temperature profiles. b) Evolution of the maximum temperature, ———•——— and of the center zone temperature, ———△——— (from KIPARRISIDES and VLACHOPOULOS, 1978)

On the other hand, the temperature difference between the polymer and the roll can be very large in other processes as, for example, in the calendering of polymethyl methacrylate (wall temperature at 100° C and polymer temperature at 250° C). The approximate solution based on the mean temperature is no longer valid in such a case: a solution of the coupled equations of motion and of energy must be considered on a point by point basis. An example of isotherms and velocity profiles computed by AGASSANT (1985) are reported in Figure 3.2-18. The rapid cooling of the polymer in the vicinity of the calender walls (Figure 3.2-18a) results in large modifications of the velocity profiles, as shown in Figure 3.2-18b.

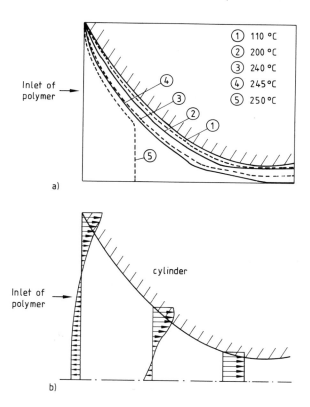

Figure 3.2-18   Calendering of polymethyl methacrylate
a)   Isotherms
b)   Velocity profiles

3.3   Extrusion dies

In this section, we first give a brief description of the most commonly used dies in the plastics industry. Then we present the general equations that govern the polymer flow and temperature in extrusion dies. Finally, specific examples related to the tubing, wire-coating, sheeting and film blowing dies are given in Sections 3.3-3 to 3.3-6 respectively.

3.3-1 Description of extrusion dies

The polymer, initially plasticated in a single screw (see Chapter 4) or twin screw extruder, flows under pressure through a die which sets the shape of the product. The main extrusion dies are.

a) Sheeting die

The sheeting die (also called coat-hanger die) is used to fabricate large sheets of polymer. The molten polymer pumped by the extruder screw flows first through a manifold "a", of decreasing cross section from A to B, to enter into a narrow slit "b". A straining bar "c" and a relaxation channel "d" allow for a more uniform polymer distribution in the lips "e". Obviously, the construction of the die must be such that the extruded polymer sheet is of constant thickness and uniform temperature (see Chapter 2.3).

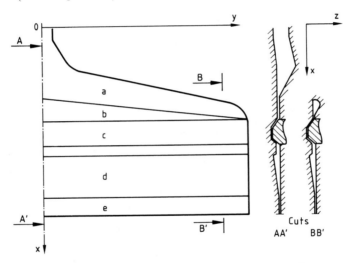

Figure 3.3-1   Sketch of a sheeting die (only one half of the die is shown)

b) Tubing die

For the production of plastic tubes, the polymer is fed by the extruder around a mandrel that is attached to the main body of the die by spider legs. The design must be such that tubes of uniform wall thickness can be extruded with minimum residual effects on the mechanical properties due to the presence of spider legs (weldline effects).

c) Wire-coating die

The die geometries used for the coating of telephone wires a few millimeters in diameter are different from those used for submarine telecommunication cables, the diameter of which could attain 0.1 m. In general, the wire or cable is fed by

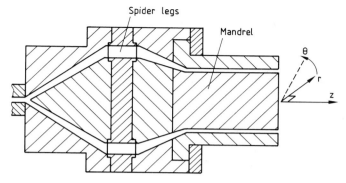

Figure 3.3-2   Sketch of a typical tubing die

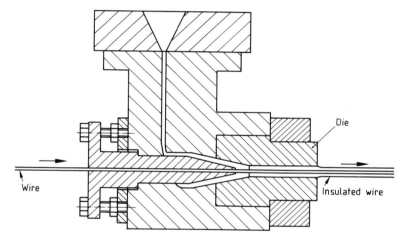

Figure 3.3-3   Sketch of a wire-coating die

a guider perpendicularly to the extruder. The polymer is distributed around the guider by a tubular die and flows along the wire in the cylindrical channel. The major problems encountered are the eccentricity of the wire with respect to the insulation (i.e. problems related to the non-uniform flow of the polymer in the die) and the rupture of the wire due to the very large forces exerted by the polymer on the wire.

d) Film-blowing die

In fabrication of the plastic film, a thin walled tube is first extruded and then blown by air under pressure. Instead of using regular tubing dies (of Figure 3.3-2), dies as the one illustrated in Figure 3.3-4 are preferred, in order to eliminate weldline effects. The polymer is fed from a central cylindrical channel through channels B into helical distribution channels H; then the polymer flows in the tubular channel D.

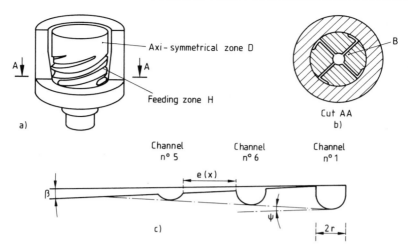

Figure 3.3-4   Sketch of a film-blowing die
a)   three-dimensional view
b)   cut view of the distribution zone
c)   cut view of the helical channel

e) Profile dies

It is not possible to present here a general sketch of profile dies, as the die
geometries vary from one application to another. The difficulty is no longer related
to flow homogeneity in the die, but rather to the securing of a shape of pre-
established and constant dimension. This requires good knowledge and control of
the extrudate swell (see Chapter 6) and of the cooling rate.

3.3-2 General flow equations

The flow of a polymer in extrusion dies of complex geometries as described above
cannot, in general, be rigorously analyzed to account for complex rheology and for
thermal effects. And this is true even with access to high performance computers.
However, we observe that in most applications, the polymer flows in a thin layer,
i.e. that the flow is either unidirectional or bi-directional. This allows us to make
use of the following approximations (see VERGNES, 1985, and VERGNES and
AGASSANT, 1986).

a) Hydrodynamic lubrication approximations

In various cases, the flow is unidirectional and the cross section varies slowly
in the flow direction. This is the case for most axi-symmetrical flows such as
encountered in tubing dies (Figure 3.3-2) and in the tubular zone of the wire-
coating dies (Figure 3.3-3). It is then possible to use the hydrodynamic lubrication
approximations as presented in Chapter 3.1. The Reynolds equation is for a
Newtonian polymer:

$$\frac{dp}{dx} = -\frac{12\eta Q}{h(x)^3} \qquad (3.3\text{-}1)$$

Analytical solutions are obtained in various cases, as shown in Problems 3.3-A and 3.3-B for the flow in conical and dihedral channels.

The flow of shear-thinning polymers is more difficult to analyze, particularly in more complex die geometries, for which numerical methods have to be used. A few cases are presented in Sections 3.3-3 and 3.3-4.

b) Confined flow approximations

In other cases, the flow is bi-directional and the tranverse flow section varies slowly. It is then possible to generalize the hydrodynamic lubrication approximations as follows:

– we assume that the velocity component w is negligible with respect to the components u and v;

– the derivatives of components u and v with respect to x and y are negligible compared to the derivatives with respect to z. Figure 3.3-5 illustrates a typical geometry where the flow is confined.

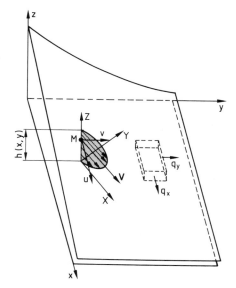

Figure 3.3-5 Typical geometry of confined flow

For Newtonian polymers, the Navier-Stokes equations become:

$$\frac{\partial p}{\partial x} = \eta \left[\frac{\partial^2 u}{\partial z^2} \right]$$

$$\frac{\partial p}{\partial y} = \eta \left[\frac{\partial^2 v}{\partial z^2} \right] \tag{3.3-2}$$

These two equations are independent and using the non-slip conditions at the walls, we may write the two following independent flow rate expressions (per unit width):

$$q_x(x, y) = \frac{-h^3(x, y)}{12\eta} \frac{\partial p}{\partial x}$$

$$q_y(x, y) = \frac{-h^3(x, y)}{12\eta} \frac{\partial p}{\partial y}$$

(3.3-3)

Making flow rate balances on a finite differences grid, we obtain a map of the pressures and velocities of the overall flow (TADMOR et al., 1974).

In contrast, the equations for shear-thinning polymers are no longer independent:

$$\frac{\partial p}{\partial x} = K \frac{\partial}{\partial z} \left[\left[\left| \frac{\partial u}{\partial z} \right|^2 + \left| \frac{\partial v}{\partial z} \right|^2 \right]^{(n-1)/2} \frac{\partial u}{\partial z} \right]$$

$$\frac{\partial p}{\partial y} = K \frac{\partial}{\partial z} \left[\left[\left| \frac{\partial u}{\partial z} \right|^2 + \left| \frac{\partial v}{\partial z} \right|^2 \right]^{(n-1)/2} \frac{\partial v}{\partial z} \right]$$

(3.3-4)

$$\frac{\partial p}{\partial z} = 0$$

It is, therefore, no longer possible to obtain separate expressions for the two flow rates q_x and q_y. The difficulty can be overcome by using, at each point, a coordinate system for which an axis coincides with the mean velocity vector of the fluid at that point (see Figure 3.3-5). This approach has been used by VERGNES (1979) and by FENNER and NADIRI (1979). In this coordinate system, the equations (3.3-4) become:

$$\frac{\partial p}{\partial X} = K \frac{\partial}{\partial Z} \left[\left| \frac{\partial U}{\partial Z} \right|^{n-1} \frac{\partial U}{\partial Z} \right], \qquad \frac{\partial p}{\partial Y} = \frac{\partial p}{\partial Z} = 0$$

(3.3-5)

and the flow rate vector is given by:

$$q = -\frac{2n}{2n + 1} \left(\frac{1}{K} \right)^{1/n} \left[\frac{h(x, y)}{2} \right]^{(2n+1)/n} |\nabla p|^{(1-n)/n} \nabla p$$

(3.3-6)

The components in the regular x, y, z coordinates are then:

$$q_x(x, y) = -\frac{2n}{2n + 1} \left(\frac{1}{K} \right)^{1/n} \left[\frac{h(x, y)}{2} \right]^{(2n+1)/n}$$

$$\times \left[\left(\frac{\partial p}{\partial x} \right)^2 + \left(\frac{\partial p}{\partial y} \right)^2 \right]^{(1-n)/2n} \frac{\partial p}{\partial x}$$

$$q_y(x, y) = -\frac{2n}{2n + 1} \left(\frac{1}{K} \right)^{1/n} \left[\frac{h(x, y)}{2} \right]^{(2n+1)/n}$$

$$\times \left[\left(\frac{\partial p}{\partial x} \right)^2 + \left(\frac{\partial p}{\partial y} \right)^2 \right]^{(1-n)/2n} \frac{\partial p}{\partial y}$$

(3.3-7)

A finite differences method can be used to solve the problem as in the Newtonian case. Such a solution is presented in Section 3.3-5 for a sheeting die.

c) Other geometrical approximations

Other approximations may be used:

- Unwinding of a three-dimensional geometry to obtain a planar geometry. This is done, for example, to calculate the flow in the distribution zone of the wire-coating die as shown in Figure 3.3-6, or of the film-blowing die.

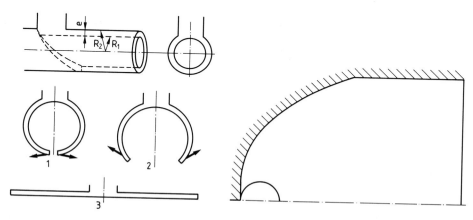

Figure 3.3-6   Unwinding the distribution zone of the wire-coating die

- Uncoupling of the complex bi-directional flow into two independent unidirectional flows. Problems 3.3-C and 3.3-D illustrate two simplified cases based on this concept.

d) Average temperature approximation

Thermal effects are of importance in die flows, but the polymer temperature usually remains close to that of the die walls. This allows us to use the average polymer temperature over the flow cross section.

- For *uni-directional flows* of shear-thinning fluids, a heat balance in terms of the average temperature is (see Chapter 2.3):

$$\varrho c_p U h \frac{d\bar{T}}{dx} = -2h_c(\bar{T} - T_w) + K \int\limits_0^h \left(\frac{\partial u}{\partial y}\right)^{n+1} dy \qquad (3.3\text{-}8)$$

where U is the average velocity in the flow direction, h_c is the heat transfer coefficient that can be calculated from correlations of the Nusselt number, Nu (see Chapter 2.3), and T_w is the temperature of the die wall.

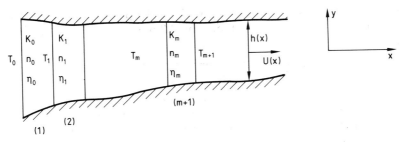

Figure 3.3-7   Solution by the increment method

For temperature-dependent viscosities, the equations of motion and energy are coupled and no analytical solutions can be obtained. We can, however, apply the increment method which is based on the following:

- The initial temperature T_0 is known.

- The Newtonian viscosity η_0 or the power-law parameters K_0 and n_0 are used for the first increment.

- The Reynolds equation is solved to determine the pressure and velocity profile in the increment.

- The thermal energy equation is then solved to calculate the average temperature, T_1, which is used for the next increment.

- New values of η_1 or K_1 and n_1 are used for the second increment and the process is repeated until the end of the geometry is reached.

— For *bi-directional flows*, the energy equation in terms of the average temperature may be written as:

$$\varrho c_p \left[U \frac{\partial \bar{T}}{\partial x} + V \frac{\partial \bar{T}}{\partial y} \right] h$$

$$= -2 h_c (\bar{T} - T_w) + K \int_0^h \left[\left(\frac{\partial u}{\partial y} \right)^2 + \left(\frac{\partial v}{\partial z} \right)^2 \right]^{(n+1)/2} dz \quad (3.3-9)$$

where U and V are the components of the average velocity in the x and y directions of flow. We can now proceed with an iterative method according to the algorithm of Figure 3.3-8.

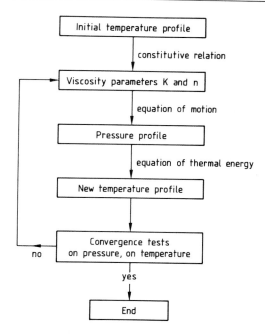

Figure 3.3-8 Algorithm for the solution of a bi-directional flow with temperature-dependent viscosity parameters

3.3-3 Tube die

If we exclude possible effects due to the spider legs that hold the mandrel to the main body, the flow is axi-symmetrical (Figure 3.3-2), the hydrodynamic lubrication approximations are always applicable and downstream the spider legs, the flow can be considered uni-directional. Such a tubular flow has been analyzed by FREDRIKSON and BIRD (1958), BIRD, STEWART and LIGHTFOOT (1960), LOSSON (1974) and SAILLARD et al. (1984), among others.
The Navier-Stokes equations for a *Newtonian polymer* reduce to:

$$\frac{dp}{dz} = \frac{\eta}{r} \frac{\partial}{\partial r} \left(r \frac{\partial w}{\partial r} \right) \tag{3.3-10}$$

Using the non-slip conditions on the mandrel and die surfaces, Equation (3.3-10) can be integrated readily to obtain the following expressions for the velocity and flow rate (Figure 3.3-9):

$$w(r, z) = \frac{1}{4\eta} \left(\frac{dp}{dz} \right) \left[(r^2 - r_i^2) - (r_e^2 - r_i^2) \frac{\ln(r/r_i)}{\ln(r_e/r_i)} \right] \tag{3.3-11}$$

$$Q = -\frac{\pi}{8\eta} \left(\frac{dp}{dz} \right) (r_e^2 - r_i^2) \left[(r_i^2 + r_e^2) - \frac{r_e^2 - r_i^2}{\ln(r_e/r_i)} \right] \tag{3.3-12}$$

This particular expression for the Reynolds equation (see Chapter 3.1) can be integrated with respect to z to obtain the pressure profile $p(z)$:

For a shear-thinning polymer which obeys the power-law expression, the equation of motion reduces to:

$$\frac{dp}{dz} = \frac{K}{r} \frac{\partial}{\partial r} \left[\left(r \frac{\partial w}{\partial r} \right) \left| \frac{\partial w}{\partial r} \right|^{n-1} \right] \tag{3.3-13}$$

The absolute value within Equation (3.3-13) forces us to integrate differently for negative and positive values for $\partial w / \partial r$. Let us define r^* as the radius at which the velocity goes through extremes (see Figure 3.3-9).

- The Reynolds equation is now given by:

$$\frac{dp}{dz} = \frac{-2K(Q/2\pi)^n}{\left[\int\limits_{r_i}^{r^*} r \int\limits_{r_i}^{r} (r^{*2}/r - r)^{1/n} \, dr \, dr + \int\limits_{r^*}^{r_e} r \int\limits_{r}^{r_e} (r - r^{*2}/r)^{1/n} \, dr \, dr \right]^n} \tag{3.3-14}$$

- The energy equation in terms of the average temperature is:

$$\varrho c_p Q \frac{d\bar{T}}{dz} = 2\pi k \, \mathrm{Nu} \, \frac{r_e}{r_e - r_i} (T_w - \bar{T}) - Q \frac{dp}{dz} \tag{3.3-15}$$

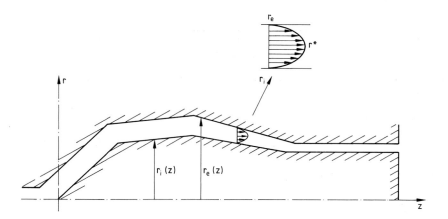

Figure 3.3-9   Sketch of a tubing die

- With the help of the increment method discussed above, we obtain the axial pressure and temperature profiles illustrated in Figure 3.3-10.

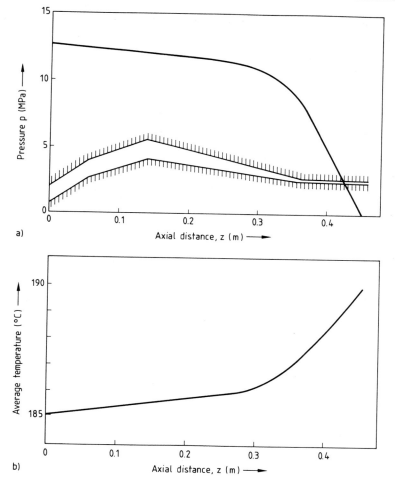

Figure 3.3-10   Axial pressure and temperature variations in a tubing die of geometry shown in
Figure 3.3-9
a)   Pressure profile
b)   Temperature profile

Both the increase in temperature due to viscous dissipation and the pressure
drop are important mostly in the last section of the die, where the shear rate is
comparatively much larger.

3.3-4   Wire-coating die

Flows in wire-coating dies have been analyzed by FENNER and WILLIAMS (1967),
FENNER (1970), CASWELL and TANNER (1978), FENNER and NADIRI (1979) and
VERGNES (1979, 1981). Two distinct zones are observed in Figure 3.3-11:
– ˙an axi-symmetrical zone
–   a distribution zone around the guider

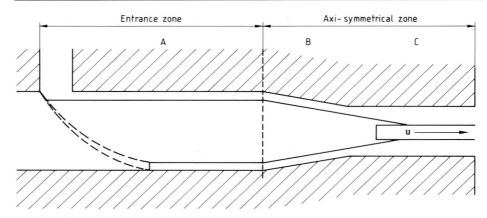

Figure 3.3-11   Sketch of a wire-coating die

a) Axi-symmetrical zone

In this zone, we distinguish two types of flow:
–   Tabular flow is observed in part B of the die and the analysis is identical to that discussed in Section 3.3-3 for tubing dies.
–   In part C of the die, the polymer is sheared between the fixed die wall and the wire which is displaced at high speed. Equation (3.3-10) may be more difficult to integrate as the velocity profile is not known a-priori and can take anyone of the different forms shown in Figure 3.3-12.

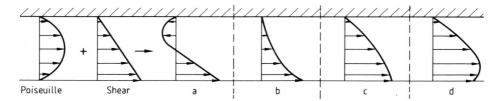

Figure 3.3-12   Various velocity profiles observed in the shear zone of a wire-coating die

One can calculate the axial pressure profile and the shear stress exerted on the die and guider walls and on the wire. Typical results are shown in Figure 3.3-13. The shear force acting on the wire is obtained by integrating the shear stress over the surface. In extreme cases, as in coating telephone cables, the force can be large enough to break the wire.

b) Distribution zone around the guider

Around the guider, the flow field at first appears to be completely three-dimensional, hence extremely difficult to solve. In fact, however, as illustrated by Figure 3.3-6, unwinding the die channel allows us to reduce the problem to a bi-directional or confined flow represented in Figure 3.3-5.

A solution by a finite differences method as discussed briefly in Section 3.3-2 is presented in Figure 3.3-14. A map of the pressure and velocity field is shown in

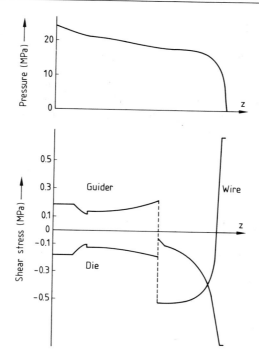

Figure 3.3-13 Axial pressure profile and shear stresses in a wire-coating die

terms of isobars and streamlines. If all the streamlines have the same separation at the die exit, the flow rate is uniform and the polymer is well distributed by the die. If the streamlines are not equidistant, then the flow is not uniform and the wire could be eccentric with respect to the plastic insulation.

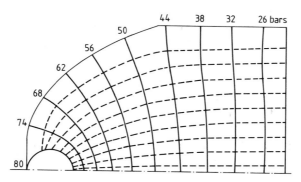

Figure 3.3-14 Isobars and streamlines in the distribution zone of a wire-coating die
——— isobars, ----- streamlines

3.3-5 Sheeting die

The flow in sheeting dies has been analyzed by many authors: ITO (1970), KLEIN and KLEIN (1973), MATSUBARA (1979), and VERGNES et al. (1980 et 1984). The geometry is of the same type as encountered in the distribution zone of the wire-coating die, but the thickness of the cross section, h, is not constant.

Figure 3.3-15 reports results obtained for a polymer obeying the power-law expression. As shown, the power-law index plays an important role on the flow distribution at the exit of the die.

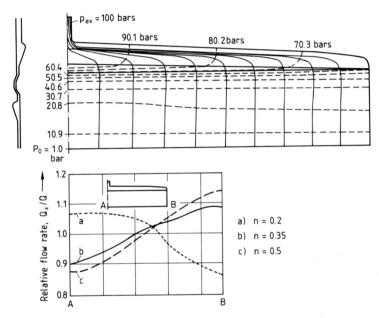

Figure 3.3-15   Pressure distribution and relative flow rate in a sheeting die

The temperature profile in the die is also of interest, as shown in Figure 3.3-16. When the polymer is fed to the die at 200° C and the die walls also are maintained at 200° C, the temperature variations inside the die are small but the flow distribution at the die exit is far from being uniform, as illustrated by curve b. The flow distribution becomes considerably less satisfactory when the wall temperature is controlled at 180° C (curve a). A more uniform distribution is obtained for die walls maintained at 220° C (curve c of the figure). Obviously the flow distribution is highly sensitive to temperature control, as discussed in Chapter 2.3.

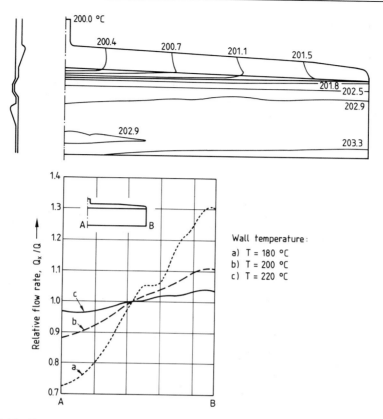

Figure 3.3-16 Temperature variations inside a sheeting die. Influence on the flow distribution.

3.3-6 Film-blowing die

The flow in film blowing dies has been analyzed by PROCTER (1972), HELMY and
WORTH (1980), SAILLARD and AGASSANT (1984), VERGNES et al. (1984).
The analysis of the flow geometry illustrated in Figure 3.3-4 appears to be, a-priori,
quite complex. The problem can be simplified considerably through the following
approximations:

–   The helical distribution channel is unwound to obtain the geometry illustrated
    in Figure 3.3-17:
–   The helical geometry is approximated by a channel of circular cross section
    and the leak flow between two adjacent channels is approximated by the flow
    between two parallel plates (see Figure 3.3-18).

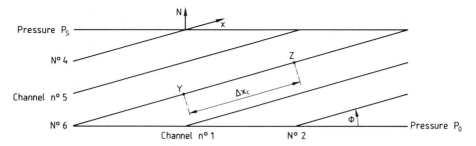

Figure 3.3-17   Six channel feed zone in a film blowing die – The unwinding is shown for half a circumference

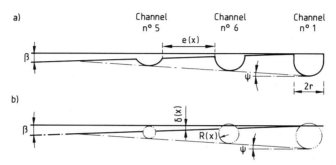

Figure 3.3-18   Longitudinal view of a film die mandrel
a)   exact geometry
b)   approximate geometry

–   Further we assume that the flow in the die is the sum of two independent flows:
   •   Flow along the axis of the unwound channel:

$$q(x) = \frac{n\pi}{3n+1}\left(\frac{1}{2K}\right)^{1/n} R(x)^{1/n+3} \left|\frac{\partial p}{\partial x}\right|^{1/n} \qquad (3.3\text{-}16)$$

   •   Leakage flow between two adjacent channels

$$q_L(x) = \frac{n}{2(2n+1)}\left(\frac{1}{2K}\right)^{1/n} \delta(x)^{1/n+2} \left|\frac{\partial p}{\partial N}\right|^{1/n} \qquad (3.3\text{-}17)$$

–   Assuming that the isobars of Figure 3.3-17 are horizontal lines, we can obtain the pressure gradient $\partial p/\partial N$ as a function of $\partial p/\partial x$ for the unwound channel.
–   Flow balances at any position can be applied as illustrated in Figure 3.3-19. For a position above a restriction, the flow rate IN has to be equal to the flow rate OUT. For a channel element, the sum of the flow rate IN and the leakage flow from the preceding channel must balance the flow rate OUT and the leakage flow to the subsequent channel, i.e.

$$q(x + \Delta x) = q(x) - q_L(x)\,\Delta x \cos\Phi + q_L(x - \Delta X_C)\,\Delta x \cos\Phi \qquad (3.3\text{-}18)$$

Using the expressions (3.3-16 and 17) in the flow balance (3.3-18), we get a non-linear set of algebraic equations in which the pressure is the only unknown.

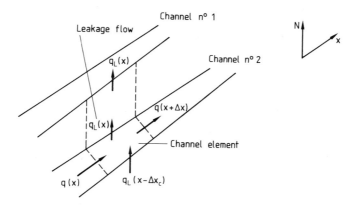

Figure 3.3-19   Flow balances in the channel and above the restriction of a film blowing die

The results are reported in Figure 3.2-20. Part a) shows the pressure profile along the axis of the channel, but the more interesting result is the tangential distribution of the flow rate at the perimeter of the die exit, shown in part b) of the figure. Variations close to 20% about the average value are observed.

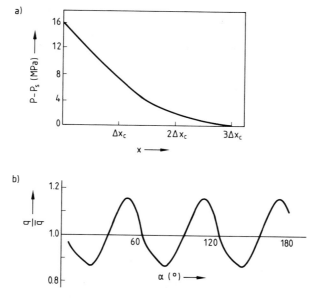

Figure 3.3-20   Pressure profile (a) and tangential distribution of the exit flow rate (b) in a film blowing die

Problems 3.3

3.3-A   Flow in a conical channel

Consider the conical channel illustrated in Figure 3.3-21. The axial variation of $R(z)$ is small, i.e.

$$\frac{R_0 - R_1}{L} \ll 1$$

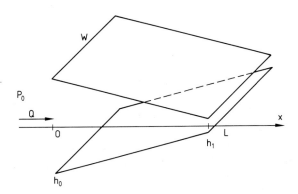

Figure 3.3-21   Flow in a conical channel

a) Summarize the hypotheses of hydrodynamic lubrication and obtain the relation between $dp(z)/dz$ and $R(z)$ in terms of the flow rate.
b) Integrate the previous relation in terms of R and obtain the expression for Q as a function of the pressure at the entrance, p_0 (assume that the pressure is equal to zero at the exit). What would be the radius of an equivalent circular channel of constant cross section? Simplify for the case where $R_1 \ll R_0$.

3.3-B   Flow in a dihedron

Consider the flow in the dihedron illustrated in Figure 3.3-22, for which h varies slightly with x, i.e. $(h_0 - h_1)/L \ll 1$.

Figure 3.3-22   Flow in a dihedron

a) Answer the questions a) and b) of Problem 3.3-A, replacing the variable R by h.
b) What is the expression for the pressure profile? Illustrate the profiles for the following conditions:

$$W = 1\,\text{m}, \quad L = 0.1\,\text{m}, \quad h_0 = 1\,\text{mm}, \quad h_1 = 0.2\,\text{mm}$$
$$\eta = 1\,\text{kPa}, \quad p_0 = 10^7\,\text{Pa}$$

3.3-C Flow in a sheeting die – First part

Consider the sheeting die of a simplified geometry shown in Figure 3.3-23. The feed channel is of circular cross section of radius R and length L. The lips are of length l and of variable thickness $h(x)$ with:

$h(0) = h_0$ on the feed side

$h(L) = h_L$ on the opposite side at the end of the channel.

We assume that the expression for $h(x)$ is such that the flow rate per unit width, $q = Q/L$, is uniform, i.e. independent of x.

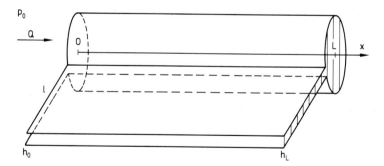

Figure 3.3-23 First example of a sheeting die

   Because of the pressure drop in the channel, $p(x)$ is a decreasing function of x. A uniform flow rate, q, has to be obtained by increasing $h(x)$ with the distance x. The pressure is equal to p_0 at the entrance and zero at the exit of the die.

a) Obtain the relation p_0 vs. Q in terms of h_0, l and L.
b) Obtain the expression for the local pressure $p(x)$, assuming that the Poiseuille relation is valid locally. What is the geometrical condition such that flow remains possible all along the die width?
c) Determine the expression for $h(x)$ in terms of the geometrical parameters of the die. Give the expression for the ratio h_L/h_0.

d) Numerical applications
Use the following values:

$$L = 0.5 \text{ m}, \quad l = 0.1 \text{ m} \quad \text{and} \quad R = 10 \text{ mm}$$

What is the maximum possible value for h_0?

Obtain h_L for $h_0 = 1$ mm
for $h_0 = 2$ mm
for $h_0 = h_{0\text{MAX}}$.

3.3-D   Flow in a sheeting die – Second part

In the second part we examine the flow in a more realistic sheeting die, as illustrated in Figure 3.3-24.

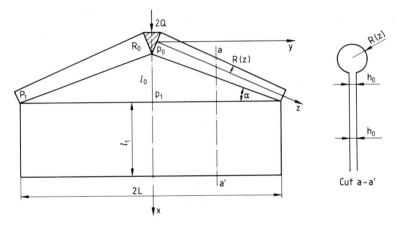

Figure 3.3-24   Simplified geometry of a sheeting die

The feed channel is of circular cross section with a radius R, decreasing with z. The polymer flows from this channel into a triangular zone of constant thickness, h_0, and then to a rectangular channel of length, l_1, also of thickness h_0.

The function $R(z)$ has to be such that the flow rate per unit width at the exit is constant (i.e. independent of y). The pressure at the entrance is p_0 and the pressure at the exit may be taken as zero. The flow is isothermal and the polymer is assumed to be a Newtonian fluid. Also the flow field in the feed channel is assumed to be undisturbed by leakage to the die lips, where the flow is assumed to be in the x-direction only.

a) Establish a relationship between the pressure, p_0, the flow rate, Q, the polymer viscosity, η, and the geometrical parameters, α, h_0, L and l_1 such that the flow rate per unit die width is constant.

b) Obtain the expression for the pressure profile, $p(z)$, in the feed channel.

c) Find the expression for $R(z)$ of the feed channel. What is the value of R_0?

d) Numerical applications
    Use the following values:

$$L = 0.5 \text{ m}, \quad l_0 = 0.1 \text{ m} \quad h_0 = 1 \text{ mm}$$
$$\text{and} \quad h_0 = 2 \text{ mm}$$

3.4   Flow of polymers in molds

3.4-1 Injection molding

Injection molding is the process used to manufacture objects of complex shape in large quantity. This is a cyclic process and its main steps are described in Figure 3.4-1.

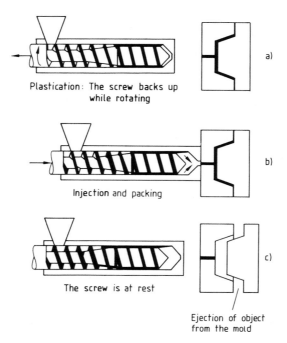

Figure 3.4-1 Injection molding cycle

- In step a, the polymer is melted in a plasticating extruder similar to those presented in Chapter 4.

- In step b, the molten polymer is injected under pressure into a cold mold. Once the mold is filled, the pressure is maintained until the polymer is solidified. During packing, additional molten polymer is injected to compensate for changes in polymer density during cooling.

- In step c, the object is cooled in the mold and then ejected from the mold.

It is not possible to reduce the large variety of molds into a small number of simple geometries. However, we can distinguish two main types of molding operation:

- *Single cavity molding*. This is used to manufacture objects of complex shape and of large dimensions. Frequently, the polymer is injected radially from a sprue, as illustrated in Figure 3.4-2.

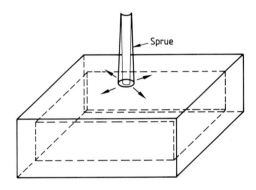

Figure 3.4-2   Single cavity injection molding

- *Multi-cavity molding.* This is used to manufacture objects of simpler shape and smaller volume and the injection proceeds as shown in Figure 3.4-3. The polymer is fed initially through one or more runners before filling the various cavities. Through the use of restriction gates or runners of different sizes, the filling time of all the cavities can be balanced.

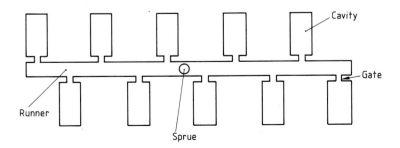

Figure 3.4-3   Multi-cavity injection molding

Injection molding problems are of two types:

- During the filling stage, the hot polymer must fill the cold mold completely before it solidifies (incomplete filling is called a short shot). Moreover, the shear rate and the viscous dissipation should not be large enough to avoid polymer degradation.
- As the polymer solidifies in the mold, its density increases and if no additional polymer were injected, the object would shrink and its shape would possibly change due to non uniform cooling (an effect known as warpage). The packing and cooling stages should be such that shrinkage and warpage of the manufactured objects are minimized, with final dimensions as close as possible to the design values.

We will restrict the analysis in this chapter to the filling stage. The modelling is complex because:

- molds are frequently of complex shape;
- the flow is unsteady with filling time of the order of one second;
- thermal effects are important. The polymer is injected at very high rates, resulting in large viscous dissipation effects. Moreover, large transverse temperature gradients are observed between the cold mold walls (close to room temperature) and the injected polymer (with temperature of the order of 200–300° C). Hence, it is necessary to solve the coupled equations of motion and of thermal energy on a point by point basis;
- finally, polymers are usually complex rheologically, i.e. their viscosity is shear-thinning and they exhibit elastic properties which may play an important role in transient flows. Nevertheless, in this chapter, polymers will be assumed to be inelastic with a Newtonian or power-law viscosity.

Various simple flow geometries have been analyzed by many authors in the literature:

- KAMAL and KENIG (1972), WINTER (1975), BERGER and GOGOS (1973), WU et al. (1974), and VINCENT (1984) have examined the case of radial flow from a central injection point, as illustrated in Figure 3.4-4.

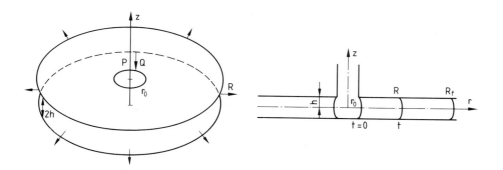

Figure 3.4-4   Radial flow between two parallel disks

- WHITE (1975), BROYER et al. (1975), VAN WIJNGAARDEN et al. (1982), and PHILIPON et al. (1985), among others, have analyzed the flow between parallel and non-parallel plates.
- WILLIAMS and LORD (1975) have studied the flow through circular channels.

In the case of molds of complex shape, two kinds of methods are encountered:
- The mold geometry may be laid flat and then described through a series of simpler geometries: radial flow, flow between parallel plates and flow in circular channels. Several commercial computer programs now available are based on this concept. Examples are Moldflow from General Electric, FILL 2 from CEMEF and CISI Ingenierie, (AGASSANT et al., 1988), Kalmul from I.K.V. (Aachen), Carmold from Cornell University (New-York), ABC flow from ATOCHEM, BILLION and CEMEF (ALLES et al., 1986), and many others.
- The mold geometry may be considered as a tridimensional thin layer and a mesh at the mid surface is introduced. Mechanical and thermal balance equations are written on this mesh and solved using finite elements, finite differences or finite volume methods. Several commercial computer programs are now developped by General Electric, Cornell University, Louvain La Neuve University (Belgium) and by C.L.I.P. (Club des Logiciels de l'Industrie Plastique - France).

Successive models for the injection in a disk-shaped mold are presented in Sections 3.4-2 to 3.4-4. The analysis of a multi-cavity mold is reserved for the problem section.

3.4-2 Isothermal Newtonian flow in a disk-shaped mold

Let us assume that at time $t = 0$ the polymer starts to flow in the mold illustrated in Figure 3.4-4 at a constant flow rate, Q. We will examine how the injection pressure varies with time.

a) Postulates

At some reasonable distance from the injection point, we will assume that the flow is axi-symmetrical and purely radial (see Figure 3.4-4). The velocity components in cylindrical coordinates are:

$$u = u(r, z), \quad v = w = 0 \tag{3.4-1}$$

and the rate-of-deformation tensor is given by:

$$[\dot{\varepsilon}] = \begin{bmatrix} \dfrac{\partial u}{\partial r} & 0 & \dfrac{1}{2}\dfrac{\partial u}{\partial z} \\ 0 & \dfrac{u}{r} & 0 \\ \dfrac{1}{2}\dfrac{\partial u}{\partial z} & 0 & 0 \end{bmatrix} \tag{3.4-2}$$

The continuity equation is given for incompressible fluids by $tr[\dot{\varepsilon}] = \partial u/\partial r + u/r = (\partial(ur)/\partial r)/r = 0$. Hence $u = A(z)/r$ and the rate-of-deformation tensor can be written as:

$$[\dot{\varepsilon}] = \begin{bmatrix} -\frac{1}{r^2}A(z) & 0 & \frac{1}{2r}\frac{dA}{dz} \\ 0 & \frac{1}{r^2}A(z) & 0 \\ \frac{1}{2r}\frac{dA}{dz} & 0 & 0 \end{bmatrix} \qquad (3.4\text{-}3)$$

The stress tensor for a Newtonian fluid is given by:

$$[\sigma] = \begin{bmatrix} -p - \frac{2\eta}{r^2}A(z) & 0 & \frac{\eta}{r}\frac{dA}{dz} \\ 0 & -p + \frac{2\eta}{r^2}A(z) & 0 \\ \frac{\eta}{r}\frac{dA}{dz} & 0 & -p \end{bmatrix} \qquad (3.4\text{-}4)$$

b) Equation of motion

The components of the equation of motion, neglecting inertial and gravitational terms, are:

r-component:

$$-\frac{\partial p}{\partial r} + \frac{\eta}{r}\frac{d^2A(z)}{dz^2} = 0$$

θ-component:

$$\frac{\partial p}{\partial \theta} = 0 \qquad (3.4\text{-}5)$$

z-component:

$$\frac{\partial p}{\partial z} = 0$$

As the pressure is a unique function of r, the first eqation can be written as:

$$\frac{1}{\eta}r\frac{dp}{dr} = \frac{d^2A(z)}{dz^2} = C = \text{const.} \qquad (3.4\text{-}6)$$

Since C is a constant, both sides of the equation can be integrated separately to obtain:

$$A(z) = \frac{C(z^2 - h^2)}{2}$$

and

$$u(r, z) = \frac{C(z^2 - h^2)}{2r} \qquad (3.4\text{-}7)$$

c) Velocity profile

The constant C can be expressed in term of the flow rate. We will assume that the flow rate is constant for the whole filling cycle (in modern injection machines, the flow rate can be varied at will during the filling stage, in particular to obtain a constant front velocity in molds of variable cross section). Hence

$$Q = 4\pi r \int_0^h u(r, z)\, dz = -\frac{4\pi C h^3}{3} \qquad (3.4\text{-}8)$$

As a consequence, the velocity profile is:

$$u(z) = -3Q\frac{(z^2 - h^2)}{8\pi rh^3} \tag{3.4-9}$$

d) Pressure profile

Integrating the first member of Equation (3.4-6) yields:

$$p(r) = -C\eta \ln \frac{R}{r} \tag{3.4-10}$$

This result is obtained by making use of the boundary condition, $p(R) = 0$, where R is the radial position of the melt front, at time t. Eliminating C with the help of Equation (3.4-8), we get:

$$p(r) = \frac{3\eta Q}{4\pi h^3} \ln \frac{R}{r} \tag{3.4-11}$$

For a constant flow rate, the radial position R of the melt front with time is expressed by the following:

$$t = 2\pi h(R^2 - r_0{}^2)/Q \tag{3.4-12}$$

Thus, the filling time is obtained as a particular case:

$$t_F = 2\pi h(R_F^2 - r_0{}^2)/Q \tag{3.4-13}$$

where R_F and r_0 are respectively the radius of the disk-shaped mold and of the sprue.

The injection pressure is obtained from Equation (3.4-11) evaluated at $r = r_0$:

$$p_I = \frac{3\eta Q}{4\pi h^3} \ln \frac{R}{r_0} \tag{3.4-14}$$

Example 3.4-1

Let us consider the following numerical values:

$$\eta = 6.4 \text{ kPa} \cdot \text{s}$$
$$R_F = 0.15 \text{ m}, \quad r_0 = 5 \text{ mm}$$
$$h = 1 \text{ mm}$$
$$Q = 5 \ 10^4 \text{ mm}^3/\text{s}$$

The results for the injection pressure as a function of the melt front position are shown in Figure 3.4-5 (the results for shear-thinning cases are discussed in the next section). A very large injection pressure is required to inject the Newtonian polymer. This is unrealistic and, as polymers are shear thinning, the viscosity in the mold will be considerably less that 6.4 kPa·s. As shown in the next section, the required injection pressure will be considerably lower. The corresponding results for the pressure profile in the mold at various times are shown in Figure 3.4-6.

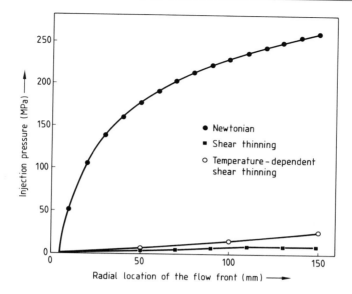

Figure 3.4-5   Injection pressure as a function of melt front position in a disk-shaped mold

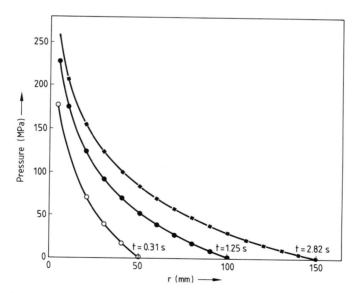

Figure 3.4-6   Radial pressure profile in a disk-shaped mold – Newtonian case

3.4-3 Filling of a shear-thinning fluid in a disk-shaped mold

Using the same postulates as in the previous case, the stress tensor for power-law polymers is:

$$[\sigma] = \begin{bmatrix} -p - \frac{2K}{r^2}A(z)\dot{\gamma}^{n-1} & 0 & \frac{K}{r}\frac{dA}{dz}\dot{\gamma}^{n-1} \\ 0 & -p + \frac{2K}{r^2}A(z)\dot{\gamma}^{n-1} & 0 \\ \frac{K}{r}\frac{dA}{dz}\dot{\gamma}^{n-1} & 0 & -p \end{bmatrix} \qquad (3.4\text{-}15)$$

where the rate of deformation is expressed by:

$$\dot{\gamma} = \frac{1}{r}\sqrt{4\frac{A^2}{r^2} + \left(\frac{dA}{dz}\right)^2}$$

Referring to the Newtonian case, the term A/r is negligible with respect to the term dA/dz and A/r^2 can be neglected compared with d^2A/dz^2 (this is equivalent to assuming that the elongational part of the flow is weak with respect to the shear part). The r-component of the equation of motion reduces to:

$$\frac{r^n}{K}\frac{dp}{dr} = -\frac{d}{dz}\left(-\frac{dA}{dz}\right)^n = C \qquad (3.4\text{-}16)$$

Both members of the equation can be integrated separately to obtain the expressions for the velocity and pressure profiles.

a) Velocity profile

$$u(r,z) = \frac{A(z)}{r} = -\frac{n}{n+1}\frac{(-C)^{1/n}}{r}(z^{1+1/n} - h^{1+1/n}) \qquad (3.4\text{-}17)$$

The flow rate is obtained as in the Newtonian case:

$$Q = \frac{4\pi n}{2n+1}(-C)^{1/n}h^{2+1/n} \qquad (3.4\text{-}18)$$

Eliminating C, the velocity profile is then:

$$u(r,z) = -\frac{2n+1}{n+1}\frac{Q}{4\pi r h^{2+1/n}}(z^{1+1/n} - h^{1+1/n}) \qquad (3.4\text{-}19)$$

b) Pressure profile

The integration of the first member of Equation (3.4-16) leads to:

$$p(r) = -\frac{CK}{1-n}(r^{1-n} - R^{1-n}) \qquad (3.4\text{-}20)$$

Eliminating C with the help of result (3.4-18), we get:

$$p(r) = \frac{K}{h(1-n)} \left[\frac{(2n+1)Q}{4\pi nh^2} \right]^n (R^{1-n} - r^{1-n}) \qquad (3.4\text{-}21)$$

and, as a particular case, the injection pressure is:

$$p_I = \frac{K}{h(1-n)} \left[\frac{(2n+1)Q}{4\pi nh^2} \right]^n (R^{1-n} - r_0^{1-n}) \qquad (3.4\text{-}22)$$

The radial position of the melt front is still given by Equation (3.4-12).

Example 3.4-2

We use the same values as in Example 3.4-1, except for the viscosity which is replaced by the following power-law parameters:

$$K = 6.4 \text{ kPa} \cdot \text{s}^n \quad \text{and} \quad n = 0.39$$

The results for the injection pressure as a function of the melt front position are compared with those for the Newtonian case in Figure 3.4-5. As expected, the required pressure is considerably lower for the shear-thinning polymer. This is a well-known result, already established in Chapter 1.5. We also observe in the figure that the pressure profile is considerably more linear when filling a disk-shaped mold with a shear-thinning polymer than when the polymer is Newtonian.

We present in Figure 3.4-7 the pressure profile in the mold as a function of time. Here again, a comparison with results of Figure 3.4-6 shows that the pressure decrease of the shear-thinning polymer with radial position is considerably more linear.

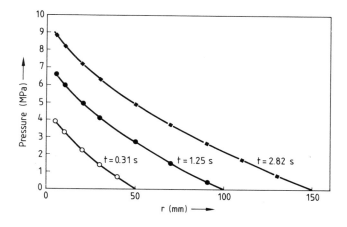

Figure 3.4-7 Radial pressure profile in a disk-shaped mold – Shear-thinning polymer

3.4-4 Filling of a shear-thinning temperature dependent fluid in a disk-shaped mold

The following analysis is taken from VINCENT (1984), and PHILIPON et al. (1985). We assume that both the heat capacity and the heat conductivity are independent of the temperature. The thermal energy equation is:

$$\varrho c_p \left[\frac{\partial T}{\partial t} + u \frac{\partial T}{\partial r} \right] = k \left[\frac{\partial^2 T}{\partial z^2} + \frac{1}{r} \frac{\partial}{\partial r} \left(\frac{\partial T}{\partial r} \right) \right] + K \dot{\gamma}^{n+1} \qquad (3.4\text{-}23)$$

The consistency index, K, is taken to be a function of T, hence to vary with the spatial position (r, z). The power-law index, n, is assumed to be constant.

We will further assume, as we did in Chapter 2.3, that heat conduction in the flow direction $\left(\frac{k}{r} \frac{\partial}{\partial r} \left(r \frac{\partial T}{\partial r} \right) \right.$ term) is negligible with respect to heat convection $(\varrho c_p u \, \partial T / \partial r$ term). Finally, we neglect the elongational part of the flow, as was done in the previous section. The rate of deformation then reduces to:

$$\dot{\bar{\gamma}} = \dot{\gamma} = \frac{1}{r} \left| \frac{\partial A}{\partial z} \right| \qquad (3.4\text{-}24)$$

Equation (3.4-23) is simplified to:

$$\varrho c_p \left[\frac{\partial T}{\partial t} + u \frac{\partial T}{\partial r} \right] = k \frac{\partial^2 T}{\partial z^2} + K(r, z) \left[\frac{1}{r} \left| \frac{\partial A}{\partial z} \right| \right]^{n+1} \qquad (3.4\text{-}25)$$

A major difficulty in solving Equation (3.4-25) arises from the unsteady nature of the flow. As illustrated in Figure 3.4-8, two zones need to be examined.

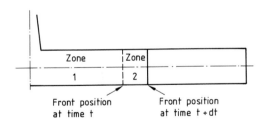

Figure 3.4-8 Unsteady filling of a disk-shaped mold

- In zone 1 which is filled completely with polymer at time t, the equation can be solved by an implicit finite differences method.
- In zone 2, which is filled between the interval t and $t + dt$, we will take into account the fountain flow effect, as discussed by TADMOR (1974). We assume that in zone 2 the polymer flows at a uniform temperature equal to the average melt front temperature at time t.

Another difficulty stems from the non separability of the pressure and velocity components in the equation of motion. Through the thermal effects, K depends on r and z and the r-component of the equation of motion is now:

$$r^n \frac{dp}{dr} = -\frac{\partial}{\partial z}\left[K(r,z)\left| - \frac{dA}{dz}\right|^n\right] \qquad (3.4\text{-}26)$$

Each member of the equation is now a function of r and integration is not possible without a knowledge of the temperature profile. Thus, the equations of motion and of energy are coupled through the dependence of the viscosity (or parameter K) on the temperature. An Arrhenius type of dependence (see Appendix C) can be introduced or we can make direct use of the viscosity data obtained at various temperatures. The following algorithm can be used to write a computer program for the analysis of the filling stage of a mold.

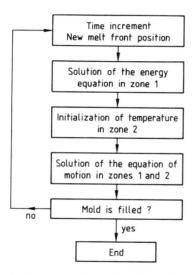

Figure 3.4-9   Algorithm for the calculation of the filling of a mold

Example 3.4-3

We have tested this numerical scheme in terms of filling the disk-shaped mold of the previous sections. The following properties corresponding to a polyamide have been used.

$$K_0 = 6.4 \text{ kPa} \cdot \text{s}^n, \quad n = 0.39 \text{ at } T_0 = 300\,^\circ\text{C}$$

activation energy : $E/R = 6220\,^\circ\text{K}$ (see Appendix C)
thermal diffusivity : $\alpha = 6.5\ 10^{-8}\ \text{m}^2/\text{s}$
heat capacity : $\varrho c_p = 3.5\ \text{MJ}/^\circ\text{C} \cdot \text{m}^3$

The results are reported and compared with the isothermal filling case in Figures 3.4-5 and 3.4-10. It is interesting to note that the injection pressure is larger when thermal effects are taken into account, whereas smaller pressures were generated in non-isothermal calendering (Chapter 3.2). This is to be expected since the walls of the mold are maintained at a temperature much lower than that of the injected polymer.

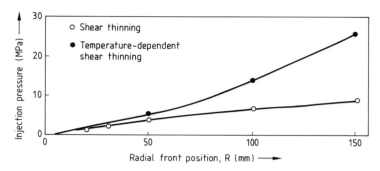

Figure 3.4-10   Thermal effects on the injection pressure in a disk-shaped mold

The cooling effect of the walls is quite visible on Figure 3.4-11 which shows a few velocity profiles near the melt front for two flow rates: 50 mL/s and 150 mL/s. Due to the large increase of the viscosity, the velocity in the vicinity of the walls is very small. Two typical results for the temperature profiles are shown in Figure 3.4-12. Except for a layer near the walls, which decreases in thickness with the flow rate, the flow may be considered as isothermal.

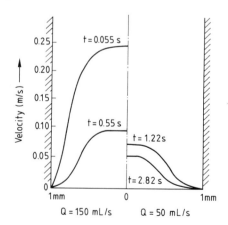

Figure 3.4-11   Velocity profiles near the melt front in a disk-shaped mold

These results are not necessarily a general illustration of what may happen in most filling. Figure 3.4-13 compares the temperature profiles at two radial positions in the mold for two different polymers: the previous polyamide and a PVC. Due to its very large viscosity, the contribution of viscous dissipation to the temperature profile for the PVC is quite large in the high shear rate zone, i.e. at a given distance from the wall. An equivalent result is shown in Figure 3.3-14,

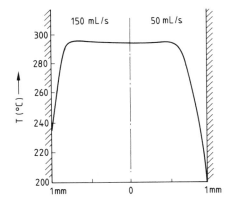

Figure 3.4-12   Temperature profiles near the melt front in a disk-shaped mold for two flow rates

where the temperature is plotted as a function of radial position at the end of the filling stage. For both polymers, the temperature of the core remains relatively constant until the mold is completely filled whereas near the wall surface, the temperature drops rapidly. At 0.3 mm from the surface, the temperature of the polyamide still drops, but less rapidly; in contrast, the temperature of the PVC increases very quickly above the core temperature before decreasing at the end of the mold.

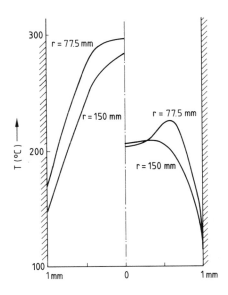

Figure 3.4-13   Temperature profiles at two radial positions in a disk-shaped mold
On the left side: polyamide; on the right side: PVC

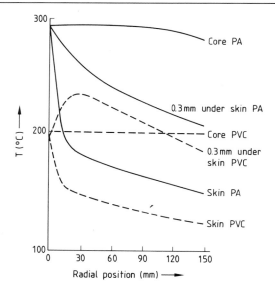

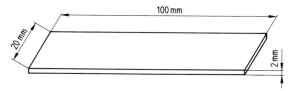

Figure 3.4-14 Temperature profiles at the end of filling in a disk-shaped mold
——— polyamide, - - - - - PVC

Problem 3.4

3.4-A Injection molding in a twin rectangular cavity mold

Let us consider the rectangular cavity with dimensions given in Figure 3.4-15.
Injection is done in a twin cavity mold, as illustrated in Figure 3.4-16.

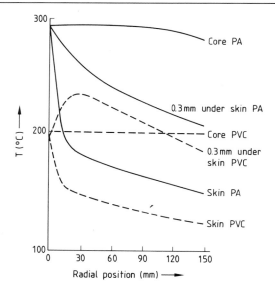

Figure 3.4-15 Dimensions of the rectangular cavity

The molding of such articles takes place in a cooled mold, where thermal
effects are necessarily important. However, we will assume that the viscosity of
the polymer is constant and equal to 1 kPa·s, and so look for first order solutions
for the filling time and the temperature profile.

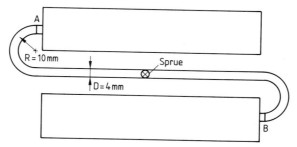

Figure 3.4-16   Sketch of the twin cavity mold

The runners consist of circular channels, 4 mm in diameter. Entrance effects from the runner to the rectangular cavities, at point A and B, can be neglected.

a) Obtain the expressions relating the flow rate and the pressure drop:
   I)   for the runners;
   II) for the flat cavities (the thickness of the cavities is very small with respect to their width, so that the side wall effects can be neglected).
b) If the injection pressure in maintained constant at 100 MPa, calculate
   I)   the filling time of the runners (up to points A and B);
   II) the filling time of the flat cavities (in this case, the pressure $p_A(t) = p_B(t)$ at points A and B has to be determined as a function of time).
   III) Is the overall filling time reasonable?
c) We wish to determine the temperature profile in the mold. For a first approximation, we assume that the temperature is not a function of time (quasi steady-state assumption) and we make use of the equation of energy in terms of the average temperature developed in Chapter 2.3. The wall temperature of the mold is maintained at 20°C and the polymer is injected at 200°C. The thermal properties of the polymer are:

thermal conductivity: $k = 0.2$ W/m $\cdot$ °C

heat capacity: $\varrho c_p = 2$ MJ/m$^3 \cdot$ °C

   I)   Obtain the values of the Cameron number as functions of time (different Cameron numbers for the runners and the flat cavities have to be calculated).
   II) Plot the average temperature in the runners and in the cavities as a function of the distance in the flow direction, at the characteristic filling times obtained in b).

Chapter 4
Polymer Extrusion

4.1 General description of a single screw extruder

4.1-1 The different zones of an extruder

In a plasticating extruder, we distinguish three different zones which correspond to three different physical states of the polymer. These zones, as illustrated in Figure 4.1-1, are

I) the feed zone, where the polymer is completely in the solid state, in granular or powder form, more or less compacted;

II) the plasticating zone, where melting occurs and where both solid and liquid states are present;

III) the metering zone, where the totally molten polymer is pumped towards the extruder die.

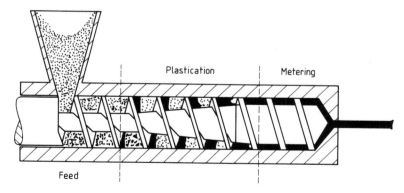

Figure 4.1-1 Single screw plasticating extruder

The diameter of the screw shaft always increases from the feed section to the die, for the entire length in the case of an universal screw or for only one part, as illustrated in Figure 4.1-2. In this case, the three zones described above correspond approximately to the three different screw geometries shown in the figure.

This type of screw is normally designed so that the compression zone coincides with the plasticating zone. This however, is not achieved for all extrusion conditions.

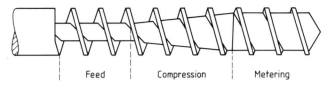

Feed   Compression   Metering

Figure 4.1-2   Extruder screw geometry

4.1-2   Screw characteristics

The main characteristics of a screw in relation to the extruder barrel are shown in Figure 4.1-3:

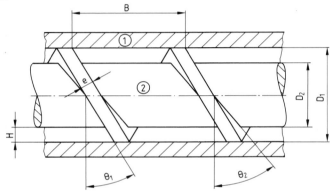

Figure 4.1-3   Details of an extruder screw;   ① extruder barrel, ② screw shaft

The main dimensions are:

    internal diameter of the extruder barrel: D_1
    diameter of the screw shaft: D_2
    channel depth: $H(D_1 = D_2 + 2H)$
    screw pitch: B

The flight angle varies with the radial distance:

    at the top of the flights, θ_1 is given by:

$$\tan \theta_1 = \frac{B}{\pi D_1} \qquad (4.1\text{-}1)$$

on the shaft, θ_2 is given by:

$$\tan \theta_2 = \frac{B}{\pi D_2} \qquad (4.1\text{-}1\text{a})$$

for any radial position r, $\theta(r)$ is given by:

$$\tan \theta(r) = \frac{B}{2\pi r} \qquad (4.1\text{-}1\text{b})$$

The pitch of the screw for most extruders designed to process plastics is equal to the barrel diameter; hence:

$$B = D_1 \quad \text{and} \quad \theta_1 = 17° \, 40' \, .$$

The angle θ_2 however may be different, for example:

$$D_2 = \frac{2}{3} D_1 \quad \text{and} \quad \theta_2 = 26° \, .$$

The flight thickness, e, may also vary with the radial position:

$$e = e_1 \quad \text{at the top of the flight,}$$

$$e = e_2 \quad \text{at the bottom of the flight.}$$

4.1-3 Screw geometry

Let us consider a helix of pitch B and diameter D as illustrated in Figure 4.1-4. The flight thickness is e and θ is the angle of the flight with respect to the screw shaft axis. The axial distance between the two adjacent helices is B' which is given by

$$B' = B - \frac{e}{\cos \theta} \tag{4.1-2}$$

The section included between the two helices $abcd$ may be unwound into a parallelogram $a'b'c'd'$. The helix length for one turn is:

$$Z = \frac{\pi D}{\cos \theta} = \frac{B}{\sin \theta} \tag{4.1-3}$$

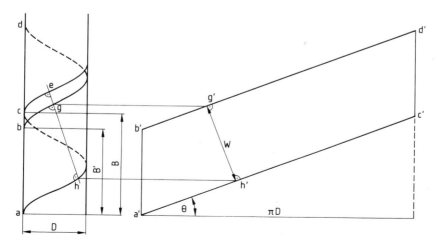

Figure 4.1-4   Helix geometry

The intercept of any plane normal to both helices and the cylindrical surface of the screw is an ellipsoidal arc gh which can be unwound into a segment $g'h'$. This segment is perpendicular to $a'c'$ and $b'd'$ and its length is

$$W = gh = g'h' = B' \cos\theta = B\cos\theta - e \qquad (4.1\text{-}4)$$

This is referred to as the screw channel width, which may vary with the distance from the screw axis. Such variations are shown in Figure 4.1-5. For any position, r:

$$W(r) = B\cos\theta(r) - e(r) \qquad (4.1\text{-}5)$$

So that:

at the top of the flights : $W_1 = B\cos\theta_1 - e_1$
at the bottom of the flights : $W_2 = B\cos\theta_2 - e_2$

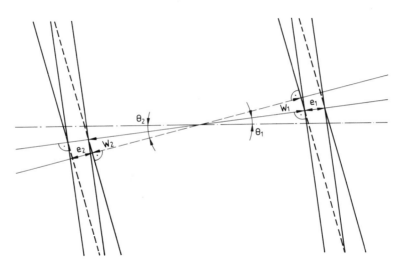

Figure 4.1-5 Variations of the channel width with radial distance
—— top of the flights ----- bottom of the flights

It is not possible to unwind the screw channel into a rectangular form without creating distortions, since the helix angle varies with the radial position r and no plane is orthogonal to both the top and the bottom of the flights. However, the helical screw channel may be approximated by a rectilinear channel of rectangular cross section, of width W and height H, as shown in Figure 4.1-6. This approximation considerably simplifies the analysis of the extrusion process and facilitates understanding the important transport phenomena that take place during extrusion. The approximation is valid for values of H small compared to that of D_2.

Rectangular Cartesian coordinates can be used to describe the flow field with z, being the longitudinal direction (along the direction of the channel), x and y being transverse directions (normal to the channel direction).

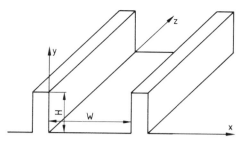

Figure 4.1-6   Unwinding of the screw channel

4.1-4 Hypothesis on the rotational speed

It is difficult to describe the flow field in an extruder using fixed coordinates with respect to the barrel. It is more practical to consider that the screw is stationary and that the barrel is rotating with an inverse rotational speed. This is equivalent to setting the origin of the axes on the screw.

The only difference between these situations is the stress distribution generated by the centrifugal forces acting on the polymer. These forces play an insignificant role on the flow of solid, but may alter the flow field of a liquid. The flow between two coaxial cylinders (Couette flow) illustrated in Figure 4.1-7 is a good example of the effects of centrifugal forces.

If the inner cylinder is rotating, the radial pressure gradient due to the centrifugal force is maximal at the inner cylinder. This tends to destabilize the main flow field and generate secondary flows. These secondary flows or vortices appear at a critical angular velocity, Ω_c, given by the following expression proposed by TAYLOR (1923) for Newtonian fluids:

$$\Omega_c = 40 \frac{\nu}{\sqrt{R \Delta R^3}} \qquad (4.1\text{-}6)$$

where R is the radius of the outer cylinder, ΔR is the gap and ν is the fluid kinematic viscosity.

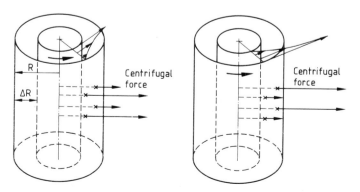

Figure 4.1-7   Centrifugal forces in Couette flow

When the outer cylinder is rotating, the radial pressure gradient (due to the centrifugal force) is now maximum at the outer cylinder. This stabilizes the main flow and vortices appear at a critical rotational speed considerably larger (≈ 100 times) than given by formula (4.1-6). The reader should refer to SCHLICHTING (1955) for more information.

Let us now consider a large extruder with a screw of radius R equal to 0.20 m and screw channel height $H(= \Delta R)$ equal to 50 mm. The application of relation (4.1-6) to a polymer of viscosity η equal to 10^3 Pa $\cdot$ s gives a critical speed, Ω_c equal to 8×10^4 radiants/s. This is far beyond the range of the rotational speed of commercial extruders. Centrifugal forces are therefore negligible in extrusion of plastic materials and the hypothesis of the barrel rotating is quite acceptable.

4.1-5 Relative velocity of the barrel

The assumption of a stationary screw with that of a rectilinear unwound screw channel leads to the situation described in Figure 4.1-8.

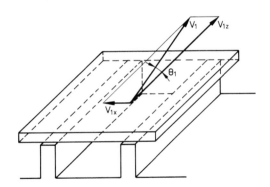

Figure 4.1-8   Relative motion of the barrel with respect to the screw

The barrel must be unwound into a plane which moves at an angle θ_1 with respect to the rectangular channel; V_1 is the linear velocity of the barrel, i.e. $V_1 = \Omega D_2/2$, V_{1z} is the velocity component along the channel axis and V_{1x} is the transverse component normal to the channel axis. These components are given by:

$$V_{1z} = V_1 \cos \theta_1$$
$$V_{1x} = V_1 \sin \theta_1 \qquad (4.1\text{-}7)$$

We say that the relative transverse motion of the barrel is from the front to the back of a flight. Finally, in the modelling of the extruder, the flow field will be decomposed into:

I)   a longitudinal flow, with a velocity component in the z-direction;

II) a transverse flow with x and y-components of the velocity field (i.e. components normal to the channel axis).

4.1-6 Reference extruder

In the next section of Chapter 4, we will frequently refer to an extruder, of which the main dimensions are presented in Figure 4.1-9.

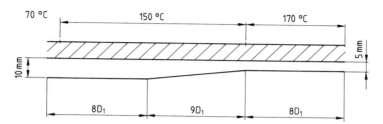

Figure 4.1-9 Dimensions of the reference extruder

Inside diameter of barrel	D_1	$= 120$ mm
Screw diameter metering zone	D_{2m}	$= 110$ mm
feed zone	D_{2f}	$= 100$ mm
Screw pitch	B	$= D_1$
Screw length	L	$= 25\,D_1$
– metering zone	L_m	$= 8\,D_1$
– compression zone	L_c	$= 9\,D_1$
– feed zone	L_f	$= 8\,D_1$
Flight thickness	e	$= 5$ mm
Screw rotational speed	Ω	$= 60$ rpm

4.2   Feed zone of an extruder

4.2-1   Solid conveying in the feed zone

The solid polymer, which is in a powder or granular form, initially has a certain cohesion and it can develop internal friction. Then, the polymer powder or granule is rapidly compacted, or broken, as pressure and temperature are applied. It behaves like a non-deformable solid, of helicoïdal shape, sliding in the gap between the barrel and the screw.

Before getting into the details of the mathematics, we can consider two limiting cases:

I)   the polymer sticks perfectly to the screw and slides on the barrel. The screw seals itself progressively with time and after a few minutes of operation, the output is nil;

II) the polymer slides perfectly on the screw and sticks to the barrel. The output is then very large, but the torque required to rotate the screw can attain a very high value and, in some cases, the system can block.

These two situations illustrate the important role of the coefficient of friction between the polymer and the metallic surfaces of the screw and barrel.

4.2-2 Metal – polymer friction

The friction coefficient between two solid bodies is given by Coulomb's law:

$$\sigma = fp \qquad (4.2\text{-}1)$$

where σ is the shear stress exerted in the opposite direction with respect to the relative velocity of the two solid bodies, p is the contact pressure, and f is the coefficient of friction which is assumed to be independent of the relative velocity of the two bodies. The physical meaning of this law is the following:

- the true contact area, a, is only a fraction of the apparent contact area, a_0;

- the shear stress, σ_0, developed by friction across the true contact area is independent of the pressure and velocity;

- the apparent shear stress, $\sigma = a\sigma_0/a_0$, depends on p since the true contact area can be considered proportional to p as illustrated in Figure 4.2-1.

$$p_2 > p_1 \quad \Rightarrow \quad a_2 > a_1 \quad \Rightarrow \quad \sigma_2 = \frac{a_2}{a_0}\sigma_0 > \sigma_1 = \frac{a_1}{a_0}\sigma_0$$

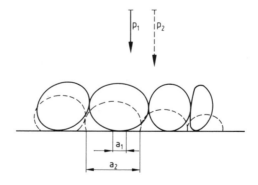

Figure 4.2-1   Physical interpretation of Coulomb's law

Coulomb's law does not hold for pressures, p, larger than the yield stress of the polymer. In that case, the real contact area becomes equal to the apparent contact area; hence it does not vary anymore with the applied pressure and the friction stress is no longer proportional to p. We then take the friction stress as a constant equal to the yield stress of the polymer.

The values of the friction coefficient for various polymers in contact with a polished steel are given in Table 4.1

Table 4.1. Friction coefficient between a polymer and polished steel (TADMOR and KLEIN, 1970, p. 70)

Polymer	f
Polyamide	0.25
Polytetrafluoroethylene	0.04
Polyethylene	0.15
Polystyrene	0.5
Polyvinylchloride	0.5

CHUNG (1970) has proposed another model to characterize the friction between a polymer and a metal:

$$\sigma = cu^{\alpha} \qquad (4.2\text{-}2)$$

where c and α are two parameters which depend on the polymer-metal system.

4.2-3 Principle of Archimedes' screw

The real situation of a helicoïdal-shape solid submitted to the friction of the barrel, screw, and flight surfaces is too complex to hope to obtain an exact solution of the problem. We will, however, present six simplified models which illustrate different physical aspects. These models, which describe more or less the reality, were first proposed by DARNELL and MOL (1956) and later refined by TADMOR and KLEIN (1970).

Model 1

The solid is considered to be a parallelepiped of surface S, submitted to an uniform pressure p. Friction is exerted on two parallel plates as illustrated in Figure 4.2-2.

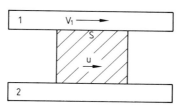

Figure 4.2-2 Friction between two plates

The velocity of the solid is u, smaller than the velocity of the upper plate, $V_1(0 < u < V_1)$; plate 1 exerts a force F_1 in the direction $V_1(F_1 = f_1 Sp)$ whereas plate 2 exerts a force F_2 in the opposite direction ($F_2 = f_2 Sp$). Then,

- if $f_1 > f_2$, u increases to attain V_1, and the solid moves with the plate 1;
- if $f_1 < f_2$, u decreases to zero and the solid remains stationary as plate 2;
- if $f_1 = f_2$, u has an arbitrary value between 0 and V_1.

In the last case, if the friction coefficient varies slightly (due, for instance, to temperature variations) the solid will successively adhere to either plate; the solid displacement will be irregular and the situation unstable.

Model 2

The solid is in a rectangular rectilinear channel (unwound screw channel). Friction is exerted on the bottom of the channel and by the upper plate (barrel) as illustrated in Figure 4.2-3. The velocity of the upper plate, V_1, is at an angle θ with the channel axis. The friction exerted by the flights walls is neglected.

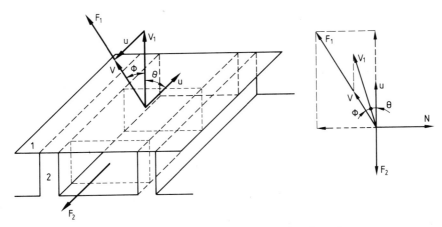

Figure 4.2-3 Friction exerted on the polymer in a rectangular channel

The velocity of the solid is $u(0 \leq u \leq V_1 \cos \theta)$ and the relative velocity V of the barrel with respect to the solid is at an angle ϕ with the vector $V_1(0 < \phi < \pi/2 - \theta)$. The barrel exerts a force in the direction of V of magnitude equal to $F_1 = f_1 S p$. The bottom of the channel exerts a force in the opposite direction to u, of magnitude $F_2 = f_2 S p$.

A balance of forces leads to a non-zero normal opposite force, N, on the back flight.

– Longitudinal equilibrium

$$F_1 \cos(\theta + \phi) = F_2$$

hence

$$f_1 \cos(\theta + \phi) = f_2 \tag{4.2-3}$$

– Transverse equilibrium

$$F_1 \sin(\theta + \phi) = N \tag{4.2-4}$$

The angle ϕ and u can be obtained from Equation (4.2-3).

– If $f_2/f_1 < \cos \theta$, $\phi = \cos^{-1}(f_2/f_1) - \theta$ and

$$u = V_1 \frac{\sin\theta \sin\phi}{\cos\phi - f_2/f_1 \cos\theta} \qquad (4.2\text{-}5)$$

– If $f_2/f_1 \geq \cos\theta$, then $\cos(\theta + \phi) > \cos\theta$, which is in contradiction with the condition $0 < \phi < \pi/2 - \theta$; hence, $u = 0$ and ϕ is unspecified.

When non zero, the velocity u is stable. Indeed, let us consider that the velocity is changed from u to u', then

– if $u' > u$, $F_1 \cos(\theta + \phi') < F_1 \cos(\theta + \phi) = F_2$, the solid is decelerated, and u' decreases;

– if $u' < u$, $F_1 \cos(\theta + \phi') > F_1 \cos(\theta + \phi) = F_2$, the solid is accelerated, and u' increases.

Conclusion of Model 2

In conclusion, although the model is far from being realistic, it gives useful information. The prediction from Model 2 of a stable solid velocity u (unstable velocity predicted by Model 1) is of importance. This illustrates the Archimedes' screw principle which can convey solid materials such as powder, granular, sand, gravel, etc.

The condition for which ϕ is now zero shows that the solid velocity increases as the ratio f_2/f_1 decreases, and as special cases:

– if $f_2 = f_1$, $u = 0$;
– if $f_2 = 0$, $u = V_1 \cos\theta$.

The solid output can be increased by lowering the friction coefficient on the screw and increasing the coefficient on the barrel. In many cases, grooves parallel to the screw axis are made in the barrel to increase the friction on the barrel. Also, cooling of the barrel in the feed zone avoids melting of the polymer, thus ensuring a good friction coefficient on the barrel surface.

Nevertheless, we note the model predicts that solid transport is impossible if $f_1 = f_2$. Even if the surfaces of the barrel and of the screw are originally of different nature, they can become quite similar with use due to wear, material deposit, sealing of grooves etc. Then, according to Model 2, the solid output will fall to zero. In fact, this is not the case and solid flow is possible even if the friction coefficients are the same. This illustrates that the assumption of a rectilinear channel is a crude approximation for the feed zone. The following models will show that it is necessary to include torque calculations to get more realistic results.

4.2-4   Influence of channel depth on the solid velocity

Model 3

In Model 3, the solid of an annular shape is submitted to the friction of two co-axial cylinders as shown in Figure 4.2-4.

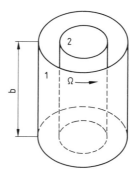

Figure 4.2-4   Friction between two co-axial cylinders

The surfaces of both cylinders of height b are $S_1 = \pi D_1 b$ and $S_2 = \pi D_2 b$. The velocity of the outer cylinder is $V_1 = \Omega D_1/2$ where Ω is its angular velocity. We consider that the linear velocity of the solid is $u(0 < u < V_1)$. The solid is then submitted to two torques:

- in the direction of Ω:

$$f_1 S_1 p D_1/2 = \pi f_1 b p D_1{}^2/2 \qquad (4.2\text{-}6)$$

- in the opposite direction

$$f_2 S_2 p D_2/2 = \pi f_2 b p D_2{}^2/2 \qquad (4.2\text{-}7)$$

- If $f_1 D_1{}^2 > f_2 D_2{}^2$, u increases to the value of V_1;
- If $f_1 D_1{}^2 < f_2 D_2{}^2$, u decreases to zero;
- If $f_1 D_1{}^2 = f_2 D_2{}^2$, u is unspecified, between 0 and V_1.

This shows that the influence of the barrel surface is more important in a cylindrical geometry, due to the contact surface and to the lever arm through which the friction forces are transmitted to the solid. In particular, if $f_1 = f_2$, the inner cylinder slides over the solid which adheres to the outer cylinder.

A qualitative criterion for the flow of the solid can be obtained by combining Models 2 and 3:

$$\frac{f_2}{f_1}\left(\frac{D_2}{D_1}\right)^2 < \cos\theta \qquad (4.2\text{-}8)$$

For a pitch equal to the diameter, $\theta = 17° 40'$ (see Section 4.1), $\cos\theta \approx 0.95$ and even if $f_1 = f_2$, the solid will flow provided that $D_2/D_1 < 0.975$. This condition is easily met in the feed zone. This model also shows that the feeding efficiency increases as the ratio D_2/D_1 decreases.

Model 4

The true helicoïdal geometry is considered in Model 4. The friction from the flights is neglected and the pressure is assumed to be uniform and equal to p. The situation is described in Figure 4.2-5.

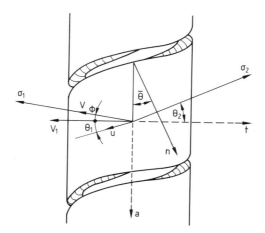

Figure 4.2-5 Helical flow of solid polymer

As shown in Section 4.1, the angle of the flights varies from θ_1 at the level of the barrel to θ_2 on the screw body. We consider that the solid flows in the screw channel at velocity u; the relative velocity V of the barrel with respect to the solid is at an angle ϕ with the vector V_1. At any point, the barrel exerts a shear stress σ_1 in the direction V whereas the friction of the screw yields a shear stress σ_2 which is in the opposite direction to u, at an angle θ_2 with t:

$$\sigma_1 = f_1 p, \quad \sigma_2 = f_2 p$$

A balance of the forces leads to a normal force n (per unit surface) on the back flight. The resulting force and torque (along and with respect to the screw axis) due to the various frictions must be equal to zero at equilibrium. If S_1 and S_2 are the contact surfaces of the barrel and screw, we obtain at equilibrium,

  — forces in screw axis direction:

$$\sigma_1 \sin \phi \, S_1 + \sigma_2 \sin \theta_2 \, S_2 = \int_f n \cos \theta \, ds \qquad (4.2\text{-}9)$$

  — torques:

$$\sigma_1 \cos \phi \, S_1 \frac{D_1}{2} = \sigma_2 \cos \theta_2 \, S_2 \frac{D_2}{2} + \int_f n \sin \theta \frac{D}{2} \, ds \qquad (4.2\text{-}10)$$

where $\int_f$ refers to the integral over the surface of a back flight; ds is the surface element. We will assume that:

$$\int_f n \cos\theta \, ds = N \cos\bar{\theta}, \quad \int_f n \sin\theta \, D ds = N \sin\bar{\theta}\bar{D}$$

where $N = \int_f n \, ds$, $\bar{\theta}$ and $\bar{D}$ are respectively the mean angle and mean diameter of the flights ($\bar{D} = (D_1 + D_2)/2$, $\tan\bar{\theta} = B/\pi\bar{D}$).

If we consider a simple turn of the screw:

$$S_1 = \frac{B}{\sin\theta_1} W_1, \quad S_2 = \frac{B}{\sin\theta_2} W_2$$

and Equations (4.2-9, 4.2-10) become:

$$f_1 \frac{\sin\phi}{\sin\theta_1} W_1 + f_2 W_2 = \frac{N}{pB} \cos\bar{\theta} \tag{4.2-11}$$

$$f_1 \frac{\cos\phi}{\sin\theta_1} W_1 \frac{D_1}{\bar{D}} = f_2 \frac{\cos\theta_2}{\sin\theta_2} W_2 \frac{D_2}{\bar{D}} + \frac{N}{pB} \sin\bar{\theta} \tag{4.2-12}$$

Eliminating N we obtain:

$$\boxed{\begin{aligned} &\cos\phi = K \sin\phi + M \\[2mm] &\text{with } K = \frac{\bar{D}}{D_1} \tan\bar{\theta} = \tan\theta_1 \\[2mm] &\text{and } M = \frac{f_2 W_2 \bar{D}}{f_1 W_1 D_1} \sin\theta_1 \left[\tan\bar{\theta} + \frac{D_2}{\bar{D}} \cot\theta_2 \right] \end{aligned}} \tag{4.2-13}$$

The graphical solution of Equation (4.2-13) in Section 4.2-9 shows that a solution exists only if $M < 1$ and that ϕ increases as M decreases. The criterion for the flow of solid in the channel is then:

$$\frac{f_2 W_2 \bar{D}}{f_1 W_1 D_1} \sin\theta_1 \left[\tan\bar{\theta} + \frac{D_2}{\bar{D}} \cot\theta_2 \right] < 1 \tag{4.2-14}$$

and for maximum values of ϕ and u, the ratios f_2/f_1 and D_2/D_1 must be minimum. We obtain the same results as with Models 2 and 3. In particular, if f_2 is equal to zero, u is maximum:

$$M = 0 \rightarrow \phi = \frac{\pi}{2} - \theta_1 \rightarrow u = V_1 \cos\theta_1$$

Equation (4.2-14) shows that even if f_1 and f_2 are equal, the flow of solid is possible due to the relative depth of the channel.

Remark

For small angles θ, the qualitative criterion (4.2-8) obtained from Model 3 is verified.

$$\tan \bar{\theta} + \frac{D_2}{\bar{D}} \cot \theta_2 \approx \frac{D_2}{\bar{D}} \cot \theta_2$$

$$\frac{W_2}{W_1} = \frac{B \cos \theta_2 - e_2}{B \cos \theta_1 - e_1} \approx \frac{\cos \theta_2}{\cos \theta_1}$$

Hence:

$$M \approx \frac{f_2}{f_1} \frac{\cos^2 \theta_2}{\cos \theta_1} \frac{D_2}{D_1} \frac{\sin \theta_1}{\sin \theta_2} = \frac{f_2}{f_1} \left(\frac{D_2}{D_1} \right) \cos \theta_2 \frac{\tan \theta_1}{\tan \theta_2}$$

$$= \frac{f_2}{f_1} \left(\frac{D_2}{D_1} \right)^2 \cos \theta_2$$

Therefore $M < 1$ is equivalent (for small angles) to

$$\frac{f_2}{f_1} \left(\frac{D_2}{D_1} \right)^2 < 1$$

4.2-5   Flow rate – Optimum screw angle

The flow rate in the feed zone is the product of the solid velocity, projected on the screw axis, times the surface of the solid bed taken perpendicularly to the screw axis (see Figure 4.2-6):

$$Q = u \sin \theta_1 \left[\frac{\pi}{4} \left(D_1{}^2 - D_2{}^2 \right) - \frac{\bar{e}H}{\sin \bar{\theta}} \right] \tag{4.2-15}$$

Since

$$\tan \phi = \frac{u \sin \theta_1}{V_1 - u \cos \theta_1} \tag{4.2-16}$$

Equation (4.2-15) can be written as,

$$Q = \Omega \frac{D_1}{2} \frac{\tan \phi \tan \theta_1}{\tan \phi + \tan \theta_1} \left[\frac{\pi}{4} (D_1{}^2 - D_2{}^2) - \frac{\bar{e}H}{\sin \bar{\theta}} \right] \tag{4.2-17}$$

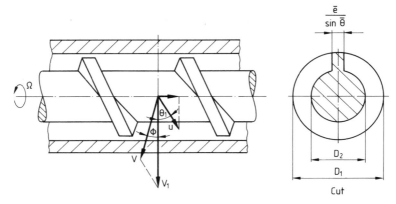

Figure 4.2-6   Flow of solid polymer in the feed zone

The solid flow rate is proportional to the quantity q given by

$$q = \frac{\tan \phi \, \tan \theta_1}{\tan \phi + \tan \theta_1} \qquad (4.2\text{-}18)$$

which varies from 0 to $\sin \theta_1 \cos \theta_1$ as ϕ varies from 0 to $\pi/2 - \theta_1$. For f_2 equal to zero, ϕ is equal to $\pi/2 - \theta_1$ and q is maximum when θ_1 is equal to 45°. In real situations, f_2 is different from zero and the optimal value of θ_1 can be quite different. To obtain this optimum, we must consider the friction exerted by the walls of the flights. This is done in Model 5.

Model 5

We now include the friction exerted by the walls of the flights, assuming the same friction coefficient f_2 as for the main screw body. The shear stresses on the flights are:

– first, the friction exerted by the mean pressure p of the solid on the back and forward flights: $\sigma_2 = f_2 p$;

– second, the opposing friction on the back flight only: $f_2 n$ (n is the opposing normal force defined above).

If S_f is the wall surface of a flight, a force balance yields:

$$\sigma_1 \sin \phi \, S_1 + \sigma_2 \sin \theta_2 \, S_2 + 2\sigma_2 \cos \bar{\theta} \, S_f = \int_f n(\cos \theta - f_2 \sin \theta) \, ds \ (4.2\text{-}19)$$

$$\sigma_1 \cos \phi \, S_1 D_1 = \sigma_2 \cos \theta_2 \, S_2 D_2 + 2\sigma_2 \sin \bar{\theta} \, S_f + \int_f n(\sin \theta + f_2 \cos \theta) \, ds \\ (4.2\text{-}20)$$

These two equations can be combined to obtain:

$$\cos \phi = K \sin \phi + M$$

$$\text{with } K = \frac{D \sin \bar{\theta} + f_2 \cos \bar{\theta}}{D_1 \cos \bar{\theta} - f_2 \sin \bar{\bar{\theta}}}$$

$$\text{and } M = \frac{W_2}{W_1} \frac{f_2}{f_1} \sin \theta_1 \left(K + \frac{D_2}{D_1} \cot \theta_2 \right)$$

$$+ 2 \frac{H}{W_1} \frac{f_2}{f_1} \sin \theta_1 \left(K + \frac{\bar{D}}{D_1} \cot \bar{\theta} \right)$$

$$(4.2\text{-}21)$$

This result is too complex to allow a direct physical interpretation. It has been used by DARNELL and MOL (1956) to obtain the value of the screw angle for maximum flow rate. They have considered that:

$$f_1 = f_2 = f$$

$$\frac{W_2}{W_1} = \frac{\cos \theta_2}{\cos \theta_1} \quad \text{(negligible flight thickness)}$$

$$H \ll W_2$$

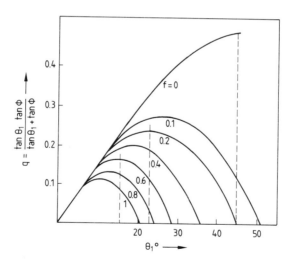

Figure 4.2-7 Flow rate as a function of the screw angle – The same values of friction coefficient for the screw and the barrel have been used

Figure 4.2-7 shows that the optimum angle θ_1 is a decreasing function of f. For most polymers, f falls between 0.25 and 0.5; these values correspond to an optimum approximately equal to 20°. This result explains why the pitch is usually made equal to the screw diameter ($\theta_1 = 17° 40'$). Figure 4.2-7 finally shows that this last choice corresponds to $q \approx 0.2$ for any value of the friction coefficient.

Hence the following approximate expression for the flow rate is obtained:

$$Q = 0.2 \left[\frac{\pi}{4} (D_1{}^2 - D_2{}^2) - \frac{\bar{e}H}{\sin \bar{\theta}} \right] V_1 \qquad (4.2\text{-}22)$$

Remark

If f_2 is equal to zero, the flow rate is given by

$$Q = 0.29 \left[\frac{\pi}{4} (D_1{}^2 - D_2{}^2) - \frac{\bar{e}H}{\sin \bar{\theta}} \right] V_1 \qquad (4.2\text{-}23)$$

The flow rate is then considerably less than the maximum value ($q_{max} = 0.5$ for $\theta_1 = 45°$).

4.2-6 Effect of back pressure

In all the models presented so far, the pressure is assumed to be constant. In most practical situations, a pressure profile is developed in the feed zone. The effect is presented in Model 6.

Model 6

We reconsider Model 2, but with two values for the pressure p_1, at the entrance and p_2 at the exit of the feed zone. This is illustrated in Figure 4.2-8.

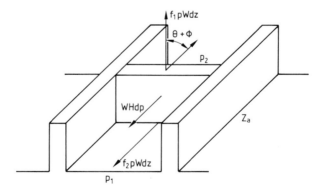

Figure 4.2-8   Effect of the pressure gradient in the feed zone

We assume that the pressure is uniform across a flow section of the channel and varies from p_1 to p_2 over the distance Z_a. For a volume element of thickness dz, we have:

$$p f_1 W \, dz \, \cos(\theta + \phi) = p f_2 W \, dz + W H \, dp \qquad (4.2\text{-}24)$$

or

$$\cos \phi = \tan \theta \sin \phi + \frac{1}{\cos \theta} \frac{f_2}{f_1} + \frac{H}{f_1 \cos \theta} \frac{d \ln p}{dz} \qquad (4.2\text{-}25)$$

If the velocity of the solid is constant, ϕ and $d\ln p/dz$ are independent of z and p varies exponentially with z. Then, from the boundary conditions, we have:

$$\frac{d\ln p}{dz} = \frac{1}{Z_a} \ln\left(\frac{p_2}{p_1}\right) \qquad (4.2\text{-}26)$$

Combining Equations (4.2-25) and (4.2-26) we get:

$$\cos\phi = \tan\theta \sin\phi + \frac{1}{\cos\theta} \frac{f_2}{f_1} + \frac{H}{f_1 \cos\theta} \frac{1}{Z_a} \ln\frac{p_2}{p_1} \qquad (4.2\text{-}27)$$

Here again, this expression is of the same type as obtained from Models 4 and 5.

The pressure effect increases the value of the coefficient M, hence decreases ϕ, u and the flow rate. It is not possible to give a simple expression for the flow rate as a function of pressure. Numerical or graphical calculations have to be performed.

The results of all the six models can be combined to obtain the general expression for ϕ (TADMOR and KLEIN, 1970)

$$\cos\phi = K\sin\phi + 2\frac{H}{W_1}\frac{f_2}{f_1}\sin\theta_1\left(K + \frac{\bar{D}}{D_1}\cot\bar{\theta}\right)$$

$$+ \frac{W_2}{W_1}\frac{f_2}{f_1}\sin\theta_1\left(K + \frac{D_2}{D_1}\cot\theta_2\right)$$

$$+ \frac{\bar{W}}{W_1}\frac{H}{Za}\frac{1}{f_1}\sin\bar{\theta}\left(K + \frac{\bar{D}}{D_1}\cot\bar{\theta}\right)\ln\frac{p_2}{p_1} \qquad (4.2\text{-}28)$$

$$\text{where} \quad K = \frac{\bar{D}}{D_1}\frac{\sin\bar{\theta} + f_2\cos\bar{\theta}}{\cos\bar{\theta} - f_2\sin\bar{\theta}}$$

(Z_a is measured at the barrel). The effect of back pressure is to decrease the flow rate significantly, as soon as the term $(H/Z_a)\ln p_2/p_1$ is no longer negligible with respect to f_2.

4.2-7 Recent developments

BROYER and TADMOR (1972), LOVEGROVE and WILLIAMS (1974) have proposed refinements for the effect of pressure based on the following:
- the solid is highly compressible (due to entrapped air);
- the pressure in the solid bed varies with position relative to barrel, screw, flights and axis of the channel (this is an elastic solid and not a liquid, which would transmit the pressure in all directions).

TADMOR and BROYER (1972) and PEIFFER (1981) have examined thermal effects, namely heat generation due to friction between the polymer granules and the barrel, and heat diffusion into the solid bed.

Finally, CHUNG (1975) has assumed that the solid polymer is coated by a thin layer of polymeric liquid. The Coulomb friction coefficient is then replaced by a viscosity term.

4.2-8 Conclusion

The different models that we have examined explain how solid particles can be transported with an Archimedes' screw. They also explain some of the technological choices used in extruder design:

- Grooving and cooling the barrel increase the value of the friction coefficient f_1, hence the value of ϕ and the flow rate of the solid polymer in the feed zone (FRANZKOCH and MENGES, 1978; RAUTENBACH and PEIFFER, 1982; RAUWENDAAL and FERNANDEZ, 1984).
- In most cases, the same steel is used for the screw and barrel construction. Hence the values of f_1 and f_2 are about the same and the choice of a screw angle of 17° 40' (this corresponds to the pitch equal to the screw diameter) appears to be optimum for most polymers. Is this choice also optimum when the friction on the barrel is much larger than that on the screw?

4.2-9 Remarks on the solution of Equation 4.2-21

One has to solve the following equation:

$$\cos \phi = K \sin \phi + M \qquad (4.2\text{-}21)$$

with $0 < \phi < \pi/2 - \theta$

The following graphical solution can be easily obtained:

Point A of coordinates $x = \cos \phi$, $y = \sin \phi$ is located on a circular arc $(0 < \phi < \pi/2 - \theta)$. The solution of Equation (4.2-21) is the intersection of this arc with the equation $y = Kx + M$. K and M are always positive (because $f_2 \cot \overline{\theta} \simeq 3$). The intersection exists only if $M < 1$ which means that:

$$\sin \phi = \frac{\sqrt{1 + K^2 - M^2} - KM}{1 + K^2}$$

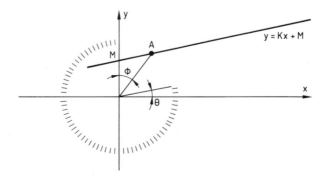

Figure 4.2-9   Graphical solution of Equation (4.2-21)

4.3   The plastication of polymer

4.1-3   Phenomenological considerations

One may believe at first glance that plastication (or fusion) of polymer granules in an extruder occurs at random, molten polymer appearing progressively at the interfaces of the granules and at the boundaries between metallic surfaces and the solid polymer particles. This is suggested by Figure 4.3-1.

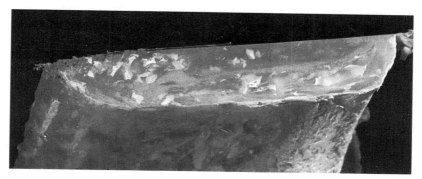

Figure 4.3-1   Random plastication

Systematic experiments conducted on extrusion show on the contrary that plastication occurs in a well ordered manner. These experiments have consisted in stopping the extruder after steady-state conditions have been attained, cooling rapidly and then sampling the polymer contained in the screw channel. The polymer which is still in granular form can easily be distinguished from the molten polymer. The first observations of that type have been made by MADDOCK (1959) by removing the screw from the extruder; other experiments have been conducted on extruders with barrels that can be opened.

The observations have shown the following:

– the polymer first melts at the barrel surface, creating a liquid film at the interface, as illustrated in Figure 4.3-2;

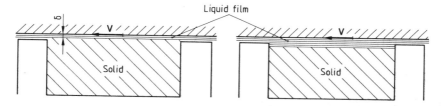

Figure 4.3-2   Liquid film formation at the barrel surface

– the film increases progressively and then a liquid pocket is created along the back flight (see Figure 4.3-3). The pocket increases progressively toward the front flight. The solid particles, on the other hand, form a continuous bed along the front flight and the bed width decreases along the axis of the screw channel (illustrated in Figure 4.3-4 for an unwound channel).

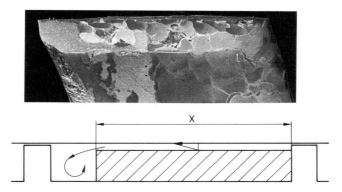

Figure 4.3-3   Formation of liquid pocket along the back flight

above: photograph,   *below:* sketch

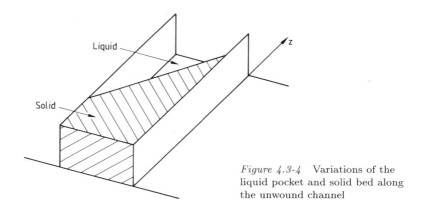

Figure 4.3-4   Variations of the liquid pocket and solid bed along the unwound channel

The experiments show two successive mechanisms for the plastication:

– a delay zone during which a liquid film at the interface between the barrel and the polymer is formed;

– a zone of plastication along the back flight, where the molten polymer is progressively accumulated.

First, a qualitative interpretation of the mechanisms is presented; models are developed in the next sections.

a) Delay zone

KACIR and TADMOR (1972) have proposed the following description of the delay
zone. Even in the zone where the barrel is cooled, plastication of the polymer
may occur near the interface of the barrel and the solid polymer due to heat
generated by friction. On the other hand, as soon as the polymer particles reach
the zone of the barrel which is heated at a temperature above the melting point
(or the transition temperature for amorphous polymers), plastication occurs at
the surface of the barrel. From that point, a liquid film is progressively formed
between the barrel surface and the solid bed. The following phenomena occur:

–   Molten polymer diffuses through the solid particles; and depending on whether
    or not the molten polymer wets the metallic surface, a liquid film will be formed
    more or less rapidly. If the wetting properties are not good, drops of molten
    polymer will diffuse by capillary forces in the pores of the solid bed as shown
    in Figure 4.3-5.

Figure 4.3-5   Diffusion of molten polymer in pores of the polymer solid bed

–   Molten polymer flows back through the clearance between the screw flights and
    the barrel. This is possible whenever the film thickness of the molten polymer
    is smaller than the clearance and whenever a positive pressure gradient is
    developed along the screw axis. The mechanisms described in Section 4.2
    for the solid transportation remain valid, provided that the Coulomb friction
    coefficient, f_1, is replaced by a viscous friction term $\sigma = \eta V/\delta$.

In summary, the delay to fusion is longer when:

–   the clearance between the screw flights and the barrel is large;
–   the wetting properties of the metallic surfaces are poor.

b) Initiation of plastication

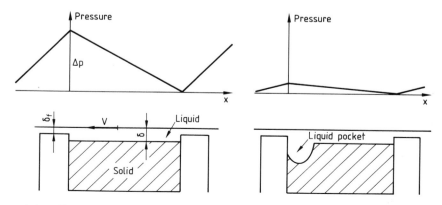

Figure 4.3-6   Pressure build-up along the back flight

As the layer of the molten polymer becomes larger than the clearance between the back flight and the barrel, a pressure profile is developed in the liquid film, according to the mechanism of the Rayleigh bearing discussed in Section 3.1. As illustrated in Figure 4.3-6, if the pressure is large enough to deform the solid bed, a liquid pocket is formed along the back flight. The pressure is then reduced. The modelling of the initiation process is discussed in Section 4.3-2 and it is the object of Problem 4.3-A.

c) Back flight plastication

According to Figure 4.3-3, the mechanism may be described by the following:

- the liquid pocket pushes the solid bed towards the front flight and the barrel. The section of the solid bed practically remains rectangular with a decreasing width with time;
- the solid particles are molten through heat conduction and shear friction at the interface of the barrel and solid bed. A liquid film of thickness less than 1 mm is formed;
- the liquid generated in the film feeds the liquid pocket, due to the relative velocity of the barrel, so that the film thickness may remain constant.

Remark

The fundamental aspect of this mechanism is that the film thickness remains very small throughout the plastication zone. This is better understood using the following analogy suggested by TADMOR and KLEIN (1970).

The case of a solid placed against the moving hot wall of the barrel is analogous to a rod of wax pushed against a hot rotating cylinder as shown in Figure 4.3-7.

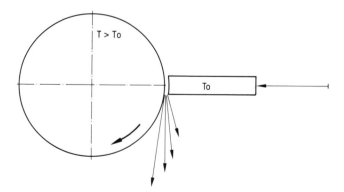

Figure 4.3-7 Analogy for polymer plastication

In this experiment, it is observed that the wax rod melts at a constant rate and that the film thickness, δ, between the rod and the rotating cylinder remains constant. Indeed, a balance on the film is the following:

- a flow rate, q_E, proportional to δ, is projected away from the cylinder surface;
- a flow rate, q_I, is fed to the film at the interface by fusion of the wax. This flow rate is inversely proportional to δ, since heat conduction through the

film and heat generated by viscous dissipation on the film are both inversely proportional to δ;

— the equality between q_E and q_I necessarily leads to a single possible value for δ and moreover this value is stable since

- if δ is accidentally decreased, q_E decreases and q_I increases; then the film thickness, δ, will increase to attain the initial value;
- on the contrary, if δ is momentarily increased, q_E increases and q_I decreases, then the film thickness will decrease to the initial value.

Such experiments have been conducted and modeled for solid polymers by MOUNT and CHUNG (1978) and by McCLELLAND and CHUNG (1983).

The modelling of the back flight plastication is discussed in Section 4.3-3 and it is the object of Problem 4.3-B.

4.3-2 Initiation of plastication

As soon as the film thickness at the barrel surface becomes larger than the clearance between the flights and the barrel, a pressure is built-up on the back flight. The variation of the pressure in a plane (P) perpendicular to the screw axis can be calculated from Problem 3.1A; in particular, the pressure build-up on the back flight is given by:

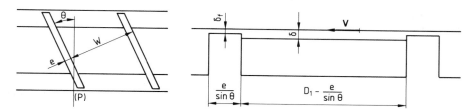

Figure 4.3-8   Pressure build-up on the back flight

$$\Delta p = \frac{6\eta V}{\sin\theta}\, \frac{\delta - \delta_f}{\delta_f^3/e + \delta^3/W} \tag{4.3-1}$$

Figure 4.3-9 shows that the pressure increases considerably as soon as δ is slightly larger than δ_f. It is easy to imagine that the pressure may exceed the value of p_c, the pressure necessary to deform the solid bed. A liquid pocket is then formed (WEY, 1984).

The figure also shows that the maximum pressure built-up decreases as the clearance, δ_f, increases. It could become inferior to the value of p_c, necessary to deform the solid bed. Under those conditions, the polymer plastication on the back flight is no longer possible.

Various other modes of plastication have indeed been observed (CHUNG, 1976; EDMONDSON and FENNER, 1975). A progressive plastication by conduction from

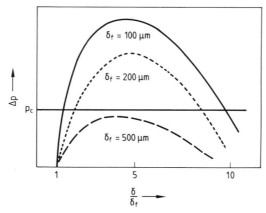

Figure 4.3-9 Pressure profile along the back flight

the barrel and the screw has been observed by LINDT (1976). This mode is illustrated in Figure 4.3-10(a). A plastication on the front flight is possible when the clearance is large enough. The leak flow becomes non negligible and the liquid is accumulated on the front flight as shown in Figure 4.3-10(b) (MENGES and KLENK, 1967).

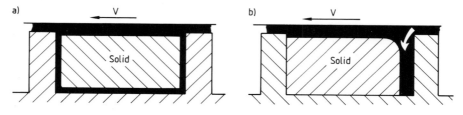

Figure 4.3-10 Other modes of plastication

4.3-3 Modelling of the back flight plastication

We discuss in this section the model proposed by TADMOR and KLEIN (1970) to analyze the back flight plastication. As soon as the liquid film is formed, we can calculate in the plane (P) of Figure 4.3-8 the thickness, δ, of the film at the interface between the barrel and the polymer bed, and then the fractions of liquid and solid polymer.

We define a new variable, V_s, which is the velocity of the solid bed towards the barrel, assumed to be uniform. Figure 4.3-11 shows the main variables of the problem.

a) Plastication rate

The plastication rate can be obtained from the following balances:
- a mass balance at the solid-liquid interface:

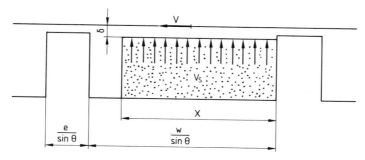

Figure 4.3-11   Modelling of back flight plastication

$$V_s \varrho_s X' = \frac{V \varrho_l \delta}{2} \tag{4.3-2}$$

where ϱ_s and ϱ_l are respectively the specific mass of the solid and of the liquid polymer.

— a heat balance at the same interface:

$$\underbrace{\varrho_s V_s [c_{ps}(T_m - T_s) + \lambda]}_{\substack{\text{Heat required to} \\ \text{bring the solid} \\ \text{from } T_s \text{ to } T_m}} = \underbrace{\frac{k_l(T_1 - T_m)}{\delta}}_{\substack{\text{heat of} \\ \text{fusion}}} \underbrace{\phantom{\frac{k_l(T_1 - T_m)}{\delta}}}_{\substack{\text{heat} \\ \text{conduction}}} + \underbrace{\eta \frac{V^2}{2\delta}}_{\substack{\text{viscous} \\ \text{dissipation}}} \tag{4.3-3}$$

where c_{ps} is the heat capacity per unit mass of the solid, λ, its latent heat of fusion per unit mass, k_l is the thermal conductivity of the liquid; T_1, T_m and T_s are respectively the wall temperature of the barrel, the melting temperature and the temperature of the solid polymer.

The expression for δ is obtained by eliminating V_s from both equations:

$$\delta = \frac{2k_l(T_1 - T_m) + \eta V^2}{V_{1x}\varrho_l[c_{ps}(T_m - T_s) + \lambda]} X \tag{4.3-4}$$

where X is the width of solid bed in the screw channel and V_{1x} is the relative velocity component of the barrel in the direction normal to the channel axis (see Figure 4.1-8). Details of the intermediate steps are given in the solution of Problem 4.3.B.

The expression for the plastication rate per unit length of unwound channel can now be derived from the following:

$$\omega = \frac{\varrho_l \delta V_{1x}}{2} = \phi \sqrt{X} \tag{4.3-5}$$

$$\phi = \sqrt{\frac{V_{1x}\varrho_l[k_l(T_1 - T_m) + \eta V^2/2]}{2[c_{ps}(T_m - T_s) + \lambda]}} \tag{4.3-6}$$

The parameter ϕ is a characteristic of the overall plastication efficiency for given thermo-mechanical conditions. The numerator represents the source terms, heat conduction and viscous dissipation; the denominator contains the heat required for the plastication: sensible heat and latent heat of fusion.

b) Optimum plastication rate

The plastication rate, proportional to ϕ, can be increased by first increasing the heat source terms, i.e.:

- using higher wall temperature of the barrel up to the maximum possible without thermally degrading the polymer;
- increasing the rotational speed of the screw.

We note that the ratio of the heat terms in the numerator of Equation (4.3-6) is the Brinkman number, defined in Section 2.3:

$$\text{Br} = \frac{\eta V^2}{k_l(T_1 - T_m)} \tag{4.3-7}$$

Two limiting cases can be examined from the expression for the temperature profile in the liquid film (Problem 4.3-B). These are shown in Figures 4.3-12:

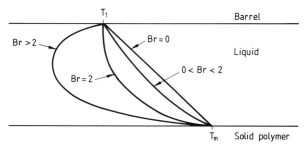

Figure 4.3-12   Temperature profile in the liquid film

- *The Brinkman number is less than 2.* Viscous dissipation is not very important and the temperature anywhere in the film is less than that of the barrel;
- *The Brinkman number is greater than 2.* Viscous dissipation becomes important and the liquid temperature will exceed the wall temperature.

If a polymer cannot be submitted to a temperature higher than T_1, even for a short period of time, the velocity of the screw has to be restricted:

$$\text{Br} < 2 \;\;\rightarrow\;\; V < \sqrt{\frac{2k_l(T_1 - T_m)}{\eta}} \tag{4.3-8}$$

Example 4.3-1

Let us consider the following properties and conditions:

$$k_l = 0.2 \text{ W/m} \cdot °\text{C}$$
$$\eta = 10^3 \text{ Pa} \cdot \text{s}$$
$$T_1 - T_m = 100° \text{ C}$$

Then Equation (4.3-8) gives:

$$V < 0.20 \text{ m/s}$$

This is the magnitude of the linear speed of extruder screws above which shear heating in the liquid film becomes important. This corresponds to speeds used in so-called isothermal extrusion. In this case, the Brinkman number is of the order of 1 $(0.1 < \text{Br} < 10)$ and the heat transferred by conduction and the heat generated by viscous dissipation are of the same magnitude.

On the contrary, if the polymer can be submitted to a temperature above T_1 for a short period of time, then the screw can be operated at a velocity greater than the limiting value corresponding to Br = 2. This is the case of the so-called adiabatic extrusion for which V may attain values above 1 m/s, with a corresponding Brinkman number higher than 25. The heat transferred by conduction is then negligible with respect to the heat generated by viscous forces.

The plastication rate, or ϕ, can be increased by lowering the heat required for the plastication:

- to bring the polymer from T_s to T_m (melting point);
- to melt the polymer:

The first term is often more important than the second. For example, in the case of a polyamide (Nylon 6):

$$c_{ps}(T_m - T_s) \approx 4 \times 10^5 \text{ J/kg}$$
$$\lambda = 8 \times 10^4 \text{ J/kg} \quad \text{(for 20 to 30\% of crystallinity)}$$

For amorphous polymers, the heat of fusion is equal to zero.

It is obvious that preheating of the polymer granules or powder is effective in lowering the heat requirement for plastication, hence in increasing ϕ and the plastication rate of the extruder.

c)   Effect of flight clearance

The expression for the plastication rate obtained above does not include the effect of the clearance, δ_f, between the screw flights and the barrel. But the value of δ, the liquid film thickness, is always less than 1 mm, which is not much larger than the clearance, δ_f, of the order of 0.1 mm or more.

We now examine the effect of the flight clearance on the plastication. Defining the film thickness by δ', the previous two balances become:

- the heat balance is the same, i.e.:

$$\varrho_s V_s [c_{ps}(T_m - T_s) + \lambda] = \frac{k_l(T_1 - T_m) + \eta V^2/2}{\delta'} \qquad (4.3\text{-}3a)$$

– the mass balance is now given approximately by:

$$\omega' = \varrho_s V_s X' = \frac{V \varrho_l (\delta' - \delta_f)}{2} \qquad (4.3\text{-}9)$$

since a liquid film of thickness δ_f flows directly above the back flight and does not contribute to the liquid pocket.

Comparing these two balances to the previous ones, we obtain:

$$\delta'(\delta' - \delta_f) = \delta^2 \qquad (4.3\text{-}10)$$

where δ would be the film thickness if the flight clearance was equal to zero (Equation 4.3-10 is valid only if $\delta' > \delta_f$). Hence:

$$\frac{\delta'}{\delta} = \sqrt{1 + \frac{\delta_f{}^2}{4\delta^2}} + \frac{\delta_f}{2\delta} \qquad (4.3\text{-}11)$$

The result is shown in Figure 4.3-13; we notice the following limits:

– If $\delta_f \ll \delta$, $\delta' \approx \delta$
– If $\delta_f \gg \delta$, $\delta' \approx \delta_f$

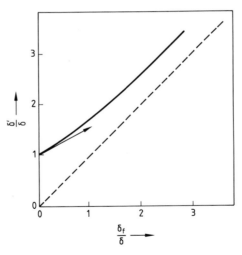

Figure 4.3-13 Variation of δ'/δ vs. δ_f/δ

The consequences are the following:

– There is always a thickness δ' to the liquid film greater than the clearance δ_f. Let us assume, for example, that δ_f is equal to the value of δ obtained previously, then $\delta' = 1.2\delta_f$.

- At the end of the plastication zone, the value of δ tends to zero as the thickness δ' approaches the value of δ_f.
- Also, the negative influence of the flight clearance on the plastication rate is greater as δ_f is larger compared to δ. The plastication rate is given by:

$$\frac{\omega'}{\omega} = \frac{\delta' - \delta_f}{\delta} = \sqrt{1 + \frac{\delta_f^2}{4\delta^2}} - \frac{\delta_f}{2\delta} \qquad (4.3\text{-}12)$$

and the result is shown in Figure 4.3-14.

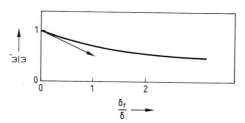

Figure 4.3-14   Influence of flight clearance on plastication rate

For the previous case, with $\delta_f = \delta$, $\omega'/\omega = 0.62$; the observed flow rate is less than 2/3 of the theoretical maximum value. This illustrates once more that the performances of an extruder rapidly decrease with wear of the screw.

4.3-4 Length of the plastication zone

The plastication rate per unit length is by definition equal to the variation of the solid polymer flow rate, i.e.:

$$\omega = -\frac{d}{dz}[\varrho_s u H X] = \phi\sqrt{X} \qquad (4.3\text{-}13)$$

where ϕ is defined in Equation (4.3-6). We now examine the influence of the screw geometry on the plastication rate.

a) Constant depth channel

In the simplest case where

- $\varrho_s(z) = $ constant (the solid's compressibility is neglected)
- $u(z) = $ constant (the solid's velocity is assumed to be constant)
- $H(z) = $ constant (there is no compression, constant depth channel)

Equation (4.3-13) can be readily integrated to obtain:

$$\frac{X}{W} = \left(1 - \frac{\phi z}{2\varrho_s u H \sqrt{W}}\right)^2 \qquad (4.3\text{-}14)$$

and the length required for complete plastication $(X/W = 1)$ is:

$$Z_t = \frac{2\varrho_s u H \sqrt{W}}{\phi} = \frac{2G}{\phi\sqrt{W}} \qquad (4.3\text{-}15)$$

where G is the mass flow rate.

In Problem 4.3-B, the value of Z_t for the reference extruder of Section 4.1 is found to be equal to 6.31 m. This value corresponds to 16 helix turns.

For given thermo-mechanical conditions, the length required for the plastication zone increases linearly with the flow rate. Hence, if the flow rate can possibly be increased by a reduction of the pressure drop across the die, the plastication zone is then not long enough to assure the complete melting of the polymer. On the other hand, it is obvious that for a given flow rate the length Z_t decreases as the parameter ϕ increases.

b) Effect of compression

Obviously, it is important to restrict the total length of an extruder. The length of the plastication zone can be decreased by using compression zones, as illustrated in Figures 4.1-1 and 4.1-2. In such compression zones, the depth varies from H_a to H_p on the length Z_c, i.e.:

$$H(z) = H_a - Az \qquad (4.3\text{-}16)$$

where $A = (H_a - H_p)/Z_c$. The effect of the compression on the plastication rate is the following. The decrease of the depth of the solid bed results in an increase of its width, hence of X and of the plastication rate. The rate per unit length being larger, the total length required is then shorter. Figure 4.3-15 reports the curves $X(z)$, for various values of the parameters A, which are solutions of the following equation:

$$-\varrho_s u \frac{d}{dz}\left[(H_a - Az)\,X\right] = \phi\sqrt{X} \qquad (4.3\text{-}17)$$

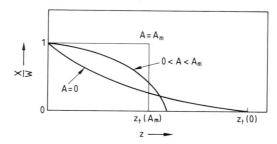

Figure 4.3-15   Solid bed profiles for three different compression factors

When A is equal to zero (no compression), the solid bed profile is parabolic and the required plastication length is given by $Z_t(0)$. As A increases, the curve becomes less concave and then convex and more important is the decrease of the plastication length, given by:

$$Z_t(A) = Z_t(0) \left[1 - \frac{A Z_t(0)}{4 H_a} \right] \tag{4.3-18}$$

In Problem 4.3-B, $Z_t(A)$ is found to be equal to 4.90 m, compared to 6.31 m in absence of compression. Hence plastication is completed in 12.4 helix turns. This is larger than the 9 turns for the compression of the reference extruder (Section 4.1-6).

The curve X/W exhibits a limiting behaviour for $A_m = 2H_a/Z_t(0)$. In this case, the contraction of the solid bed is done on an infinitely short distance and $Z_t(A_m) = Z_t(0)/2$. If A is greater than A_m, there is no solution for $X(z)$ with values between 0 and W. The mechanism of plastication described here does not apply anymore. The polymer, plasticated in an increment dz of the screw channel, cannot be accumulated in the back of the flight. It is believed then that a large pressure gradient is built-up, which modifies completely the simple shear flow assumed so far in the liquid film. However, no satisfactory modelling of such a complex flow has been yet proposed.

c) Types of screw

Two types of screw are widely used for most applications:

– Long compression type with 8 to 15 helix turns for the compression (or the total length in the case of the universal screw). The value of the compression factor is small. Typical dimensions are:

$$H_a = 10 \text{ mm}$$
$$H_p = 5 \text{ mm}$$
$$Z_c = 3.6 \text{ m}$$
$$A \approx \frac{1}{700}$$

With such a screw, A is always less than A_m and the back flight plastication mechanism always applies, provided that the clearance between the barrel and the flights is not too large (see Section 4.3-3).

– Short compression type, with 1 or 2 helix turns of compression. The compression factor is large, typically: $A \approx 1/10$. The values of A is larger than A_m, at least for large flow rates.

If back flight plastication has been initiated in the first zone of the screw (of constant depth channel), X may increase in the compression zone and plastication may proceed in the usual way. If not, plastication will proceed under a different mechanism. It could be very rapid with very large shear-heating effects since the controlling mechanism of a liquid film of constant thickness δ is not effective anymore.

These considerations are useful guides for the selection of an extruder screw. Long compression screws are adapted to the processing of polymers sensitive to shear-heating, either because of their limited thermal stability (e.g. PVC, ABS), or because of their very high viscosity. Short compression screws are used for processing polymers of low viscosity in the molten state (e.g. polyamides, polyolefines, polymethylmethacrylate, ...).

Remark

One of the screw's characteristics is its compression ratio, which is the ratio of the flow section in the feed zone to the flow section in the metering zone. This ratio is not much different from H_a/H_p. As discussed above, it is clear that the plastication mechanism for two screws having the same compression ratio but different plastication lengths could be totally different.

4.3-5   Recent theories

DONOVAN (1971) has proposed the following improvements for the TADMOR and KLEIN (1970) model:

- the temperature variation in the solid bed is taken into account:
- the velocity of the solid bed is not taken as a constant, but it is assumed to increase as plastication proceeds. Indeed, STREET (1961) has shown that for some polymers the velocity u of the solid bed increased from the beginning to the end of the plastication. EDMONDSON and FENNER (1975) and SHAPIRO et al. (1976) have proposed models to account for the solid bed acceleration. Since then, extensive experimental work has confirmed the new hypothesis (COX and FENNER, 1980; FUKASE et al., 1982).
- MOUNT and et al. (1982) have made use of a shear-thinning and temperature dependent viscosity and LINDT (1981) considered a layer of molten polymer along the screw shaft.

4.3-6   Conclusions

The plastication process described in this section is of a two-fold interest:
- On the one hand, the combined convection of the solid and liquid increases the plastication rate. If plastication proceeded by heat conduction alone, the residence time would become excessively long.
- On the other hand, since the powder or granular polymer forms a single helical block, air and dissolved gases can easily be removed through the porous bed towards the feed hopper.

The models developed to describe the plastication process show the influence of key parameters on the efficiency of the extrusion process:
- *The clearance between the tip of the flights and the barrel.* With wear of the screw, the rate of plastication not only decreases, but the mechanism itself is modified. A given extruder may be correctly operating until wear becomes important enough that the mechanism of back flight plastication is modified. Then, suddenly, partly unmelted polymer is extruded.

- *The compression factor.* For screws with a short compression zone, plastication may proceed through a different mechanism compared to that for screws with a long compression zone.
- *The Brinkman number.* At high Brinkman numbers (i.e. when heat generated by viscous forces is important with respect to heat conduction), the temperature in the liquid layer between the solid bed and the barrel wall may become very large. This could explain why corrosion in extruders has been observed in the compression zone and not at the end of the screw where the wall temperature and the mean temperature of the polymer are usually maximum.
- *Preheating of the polymer granules increases the plastication rate.*

The understanding of the plastication mechanism has resulted in important technological progresses. In the fifties, it was current practice to enhance shear effects by adding mixing zones (see Figure 4.3-16) in the hope of increasing the throughput of extruders. Since the mechanism of back flight plastication is well understood, instead of mixing and tearing off solid granules, it is clear that the convection process described above should be favored. Mixing heads, however, remain of interest for the outlet zone where all the polymer is in the liquid state.

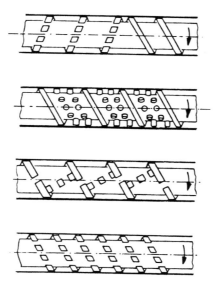

Figure 4.3-16   Mixing devices on extruder screws

Special screw designs have been developed to enhance back flight plastication.

The Maillefer screw

The screw developed by MAILLEFER (1959) includes on its entire length a regular flight of constant pitch equal to its diameter, plus a supplementary flight in the compression zone. The pitch of this flight is such that there is one turn less than the main flight in the compression zone.

From a transverse cut, the supplementary flight is seen to uniformly scan the width of the main channel, from the front to the back of the main flight along the channel (see Figure 4.3-17). Hence, the position of the supplementary flight is very close to that of the boundary between the solid bed and the liquid pocket for a regular screw. Indeed, we observe that the solid particles remain in the space located in front of the supplementary flight, whereas the liquid is dragged by the barrel towards the pocket located behind (see Figure 4.3-18).

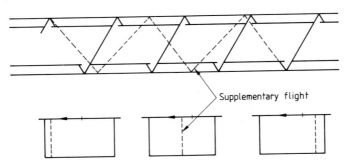

Figure 4.3-17   Simplified sketch of the Maillefer screw

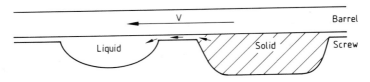

Figure 4.3-18   Channel of the Mailllefer screw

The supplementary flight acts as a solid boundary between the solid bed and the liquid pocket. Its main function is to prevent solid particles to be pulled in the liquid pocket. Throughout the compression zone, the solid bed is kept in front of the supplementary flight with little possibility of breaking up. Moreover the shearing in the thin liquid layer of the gap between the tip of the supplementary flight and the barrel wall ensures the plastication of any small particles that could be dragged with the melt.

The Maillefer screw is, however, not flexible; a given screw geometry is suitable for a polymer under specific operating conditions. Many modifications of the Maillefer screw have been proposed by INGEN HOUSZ and MEIJER (1981) and by ELBIRLI et al. (1983). Finally, we notice that the Maillefer system is now more frequently used as mixing device in the metering zone of the screw.

The Cohen screw

The body of the Cohen screw is such that the depth of the channel decreases from the front to the back of the flight as illustrated in Figure 4.3-19. This screw has two main advantages: I) the solid bed is forced against the barrel through a "corner effect", and II) the rupture of the bed is avoided. The plastication process is enhanced by such a geometry (it is worth mentioning that the channel of classical screws in the compression zone is inclined in the opposite direction; this is not favoring the upkeep of the solid bed). Moreover the Cohen screw is much more flexible than the Maillefer screw.

Figure 4.3-19   The Cohen screw

Problems 4.3

4.3-A   Initiation of back flight plastication

The plastication process in an extruder is initiated by the formation of a thin liquid layer of thickness δ at the interface of the solid bed and the barrel wall. First, the liquid flows through the clearance, δ_f, and then hits the side of the back flight. As in the situation encountered in the Rayleigh bearing (see Section 3.1), a pressure is built-up at the flight. This extra pressure Δp will deform the solid bed, thus creating a liquid pocket at the back of the flight.

The geometry of the screw is defined in Figure 4.3-20. This geometry can be unwound and represented by Figure 4.3-21 in a plane perpendicular to the screw axis. The pressure is necessarily the same at points A and C which are geometrically the same (see Figure 4.3-20).

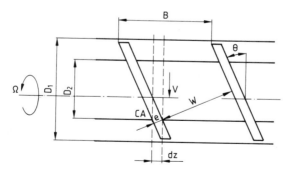

Figure 4.3-20   Sketch of a screw pitch

Calculate the value for the extra pressure Δp as a function of δ/δ_f for the following conditions:

$$D_1 = 120 \text{ mm}, \quad D_2 = 100 \text{ mm}$$
$$e = \quad 5 \text{ mm}, \quad B = D_1$$
$$\Omega = 60 \text{ rev/min}$$
$$\delta_f = 0.1 \text{ mm}$$
$$\eta = 10^3 \text{ Pa} \cdot \text{s}$$

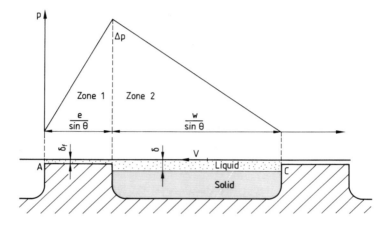

Figure 4.3-21   Mechanism of the Rayleigh bearing in extrusion

4.3-B   Back flight plastication

Let us consider the process of back flight plastication as illustrated in Figure 4.3-22. The axis x' is taken in the direction of the relative velocity V and W' and X' are respectively the width of the channel and of the solid bed in the x' coordinates.

The solid bed is pushed against the barrel wall and its upper surface is displaced at the velocity V_s, uniform with respect to x' and constant with time. The polymer melts at the upper surface, is dragged by the barrel and then is accumulated at the back of the flight, forming a vortex flow.

In this problem, we will examine the phenomena encountered in the liquid film between the solid bed and the barrel. The analysis is based on the following hypotheses:

– the metallic wall of the barrel is maintained at a constant and known temperature T_1;
– the solid bulk temperature is T_s, assumed to be known and constant;
– the melting temperature is T_m, assumed to be well defined as in the case of crystalline polymers;
– the film thickness is δ, not a function of x'; the flow in the film is uniform with respect to x' and under steady-state conditions; the velocity profile is then linear, given by:

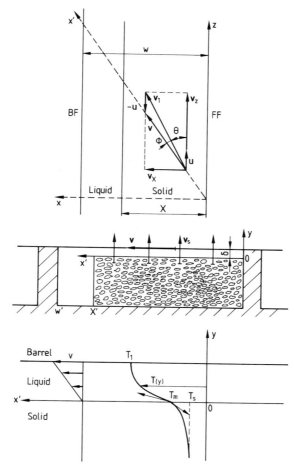

Figure 4.3-22   Back flight plastication

$$u(y) = V\frac{y}{\delta}$$

with the origin of the y-axis taken at the solid-liquid interface;
— finally, the temperature profiles in the liquid film and in the solid bed are assumed to be uniform with respect to x' and under steady-state conditions (hypothesis of equilibrium regime in the liquid film).

The following properties and conditions are assumed to be known:

- the densities of the solid (ϱ_s) and of the liquid (ϱ_l),
- the heat conductivities of the solid (k_s) and of the liquid (k_l),
- the heat capacities per unit mass of the solid (c_{ps}) and of the liquid (c_{pl}),
- the latent heat of fusion of the solid, λ,
- the liquid's viscosity, η, evaluated at the mean temperature of the film and at the corresponding shear rate,
- the width X of the solid bed at a given time,
- the dimensions of the extruder and the flow rate.

a) Obtain the relative values of u (velocity of the solid) and of V_z. What is then the value of the angle ϕ (see Figure 4.3-21)?

b) Show that the problem has two unknowns which can be determined from two equations obtained from balances of two physical quantities of the liquid film.

c) Write down the equation of heat transfer for the liquid film. Obtain the expression for the heat flux from the liquid film to the liquid-solid interface. Under what conditions is shear-heating important in the liquid film?

d) Write down the equation of heat transfer for the solid bed. Show that the temperature in the solid is close to T_s at a short distance from the interface. Obtain the expression for the heat flux from the liquid-solid interface to the solid body.

e) Write down the heat balance for the interface.

f) Determine the two unknowns of this problem and calculate the rate of plastication (kg/s). How does this rate depend on X?

For the numerical applications, the dimensions of the reference extruder of Section 4.1-6 are to be considered. The extruded polymer is a low density polyethylene with the following properties:

$$T_m = 110° \text{ C}$$
$$\lambda = 12 \text{ kJ/kg} \quad (50\% \text{ of crystallinity})$$
$$\varrho_l = 812 \text{ kg/m}^3, \quad \varrho_S = 900 \text{ kg/m}^3$$
$$k_l = k_s = 0.2 \text{ W/m} \cdot °\text{C}$$
$$c_{pl} = c_{ps} = 2.5 \text{ kJ/kg} \cdot °\text{C}$$
$$\eta = 10^3 \text{ Pa} \cdot \text{s}$$

And the operating conditions are:

$$Q = 300 \text{ kg/h}$$
$$\Omega = 60 \text{ rev/min}$$
$$T_1 = 150° \text{ C}, \quad T_s = 20° \text{ C}.$$

4.4   Flow of molten polymer in extruders

The flow of molten polymer successively occurs, as illustrated in Figure 4.4-1, in the compression and in the metering or pumping zone. In the compression zone, the flow section between the barrel and the screw decreases in the direction of the screw channel and the solid and the liquid flows are considered. In the metering zone, the flow section is constant and the entire polymer is normally in the molten state.

However, in some cases, the situation may be more complex: the plastication may be only completed in the metering zone or on the contrary, before the end of the compression zone.

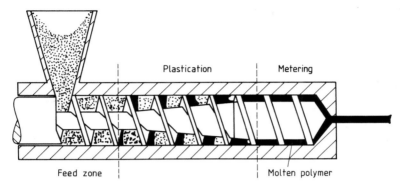

Figure 4.4-1   Flow of molten polymer in an extruder

4.4-1 Metering zone

The metering or pumping zone of an extruder is used to bring the molten polymer to the pressure required to obtain a given flow rate in the die. It has been analyzed by FENNER (1970) and by TADMOR and KLEIN (1970).

a) Geometry

If we unwind the channel of the screw and make the assumption that the barrel is rotating around the screw, we obtain a shear flow for a rectangular cavity as illustrated in Figure 4.4.2

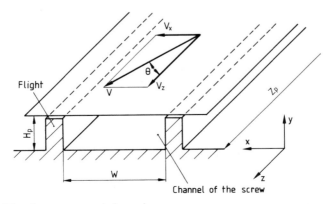

Figure 4.4-2   Flow in an unwound channel

The direction of the shear flow is at an angle θ with respect to the channel axis. Considering the usual dimensions of extruders (see the reference extruder of Section 4.1-6), we will assume that:

$$Z_p \gg W \gg H_p$$

and in our first analysis, we will further assume that the clearance between the tip of the flights and the barrel wall, δ_f, is nil.

b) Equation of motion

Except for the zone in the vicinity of the die entrance, the flow is uniform with respect to z. Hence the vector $\boldsymbol{u}$ is independent of z:

$$\boldsymbol{u} = \begin{vmatrix} u\,(x,y) \\ v\,(x,y) \\ w(x,y) \end{vmatrix}$$

Then, the Navier-Stokes equations for a flow of a Newtonian liquid under negligible inertial and gravitational forces reduce to:

$$\frac{\partial p}{\partial x} = \eta \left[\frac{\partial^2 u}{\partial x^2} + \frac{\partial^2 u}{\partial y^2} \right] = f(x,y)$$

$$\frac{\partial p}{\partial y} = \eta \left[\frac{\partial^2 v}{\partial x^2} + \frac{\partial^2 v}{\partial y^2} \right] = g(x,y) \qquad (4.4\text{-}1)$$

$$\frac{\partial p}{\partial z} = \eta \left[\frac{\partial^2 w}{\partial x^2} + \frac{\partial^2 w}{\partial y^2} \right] = h(x,y)$$

and the equation of continuity for incompressible fluids becomes:

$$\frac{\partial u}{\partial x} + \frac{\partial v}{\partial y} = 0 \qquad (4.4\text{-}2)$$

Since the velocity components are not functions of z, $\partial p/\partial x$, $\partial p/\partial y$ and $\partial p/\partial z$ are unique functions of x and y. Hence, from Equation (4.4-1):

$$\frac{\partial^2 p}{\partial x\, \partial z} = \frac{\partial f}{\partial z} = \frac{\partial h}{\partial x} = 0$$

$$\frac{\partial^2 p}{\partial y\, \partial z} = \frac{\partial g}{\partial z} = \frac{\partial h}{\partial y} = 0$$

This result implies that $h(x,y)$ is in fact a constant.

The set of Equations (4.4-1) and (4.4-2) can now be uncoupled to obtain two independent sets of equations:

– For the transverse flow:

$$\frac{\partial p}{\partial x} = \eta \left[\frac{\partial^2 u}{\partial x^2} + \frac{\partial^2 u}{\partial y^2} \right]$$

$$\frac{\partial p}{\partial y} = \eta \left[\frac{\partial^2 v}{\partial x^2} + \frac{\partial^2 v}{\partial y^2} \right] \tag{4.4-3}$$

$$\frac{\partial u}{\partial x} + \frac{\partial v}{\partial y} = 0$$

with the three unknown functions $p(x,y)$, $u(x,y)$ and $v(x,y)$.
– For the axial flow:

$$\frac{\partial p}{\partial z} = \text{constant} = \frac{\Delta p}{Z_p} = \eta \left[\frac{\partial^2 w}{\partial x^2} + \frac{\partial^2 w}{\partial y^2} \right] \tag{4.4-4}$$

c) Transverse flow

The solution of the set of Equations (4.4-3) can be obtained by a finite elements method. Figure 4.4-3 shows the streamlines obtained for a channel of square flow section. (BAUDIER and AVENAS, 1973).

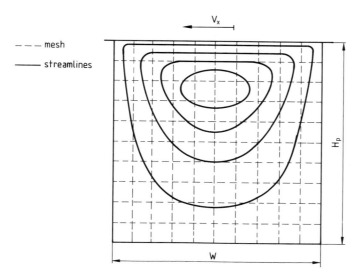

Figure 4.4-3 Transverse flow in a square channel

In the metering zone of an extruder, $H_p \ll W$, the streamlines take the shapes shown in Figure 4.4-4, and the problem can be considerably simplified by assuming that the velocity components u and v are independent of x away from the edges of the channel (near the flight walls) where the flow is necessarily more complex. We note from the equation of continuity that if u and v are not functions of x,

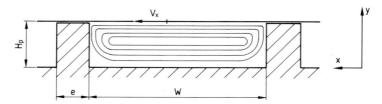

Figure 4.4-4   Transverse flow in the metering zone of an extruder

$\partial v/\partial y = 0$ and $v = $ constant $= 0$ (from the boundary conditions of no normal flow at $y = 0$ or $y = H_p$).

The set of Equations (4.4-3) then reduces to:

$$\frac{\partial p}{\partial x} = \eta \frac{d^2 u}{dy^2}$$

$$\frac{\partial p}{\partial y} = 0 \tag{4.4-5}$$

Since p is not a function of y and u is independent of x, both members of Equation (4.4-5) are constant and we can write:

$$\eta \frac{d^2 u}{dy^2} = \frac{\Delta p}{W} \tag{4.4-6}$$

where Δp is the pressure difference between the front and the back flight. This Equation can be easily integrated and using the boundary conditions:

   B.C. 1), at $y = 0$, $u = 0$ (velocity is zero at the screw)
   B.C. 2), at $y = H_p$, $u = V_x$ (velocity is equal to that of the barrel)

we obtain:

$$u = \frac{1}{2\eta} \frac{\Delta p}{W} y(y - H_p) + V_x \frac{y}{H_p} \tag{4.4-7}$$

Neglecting the leak flow in the clearance between the flight tip and the barrel wall (in Section 4.4-3 we will examine this hypothesis), the transverse flow rate is equal to zero. Hence per unit length:

$$Q = \int_0^{H_p} u(y)\, dy = -\frac{1}{12\eta}\left(\frac{\Delta p}{W}\right) H_p^3 + V_x \frac{H_p}{2} = 0 \tag{4.4-8}$$

The pressure drop across the channel width is then given by:

$$\Delta p = \frac{6\eta V_x W}{H_p{}^2} \tag{4.4-9}$$

and the velocity profile becomes:

$$u = V_x y \frac{3y - 2H_p}{H_p{}^2} \tag{4.4-10}$$

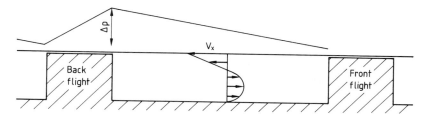

Figure 4.4-5 Velocity and pressure profiles in transverse flow of a screw channel

Figure 4.4-5 illustrates the pressure and velocity profiles across the channel width.

Remark

The pressure drop computed for the reference extruder of Section 4.1-6 is equal to 2.74 MPa using a melt viscosity of 10^3 Pa · s. This is in no way comparable to values attained at the start of plastication (see the solution to Problem 4.3-A).

d) Axial flow

In absence of leakage flow through the clearance between the flight tip and the barrel wall, the flow capacity of an extruder is determined by the axial or longitudinal flow.

The axial flow can be considered as the sum of a drag and pressure flows, as illustrated in Figure 4.4-6.

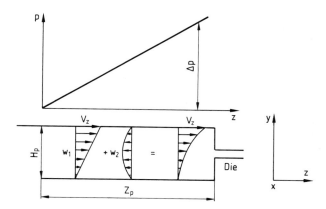

Figure 4.4-6 Axial flow in the metering zone

As for the transverse flow, the condition $H_p \ll W$ implies that the velocity component w is independent of x (except near the flight walls). Equation (4.4-4) reduces to:

$$\frac{\Delta p}{Z_p} = \eta \frac{d^2 w}{dy^2} \tag{4.4-11}$$

Integrating with the following boundary conditions:

B.C. 1), at $y = 0$, $\quad w = 0$ $\quad$ (no slip at the screw surface)
B.C. 2), at $y = H_p$, $w = V_z$ $\quad$ (the melt velocity is equal to the barrel
$\qquad\qquad\qquad\qquad\qquad$ velocity)

we get:

$$w = \frac{1}{2\eta}\frac{\Delta p}{Z_p}\, y(y - H_P) + V_z\frac{y}{H_p} \tag{4.4-12}$$

The flow situation is similar to that of the Rayleigh bearing, analyzed in Section 3.1-3.

- The term $w_1 = V_z y/H_p$ corresponds to the drag or shear flow between two parallel plates separated by a distance H_p from each other; the lower plate is fixed and the upper one is displaced at the velocity V_z.
- The terms $w_2 = \Delta p y(y - H_p)/2\eta Z_p$ corresponds to the Poiseuille flow between two parallel plates at a distance H_p apart, under a pressure gradient $\Delta p/Z_p$. The total flow rate is also the sum of the back (Poiseuille) flow and of the shear flow:

$$Q = W\int_0^{H_p} w(y)\, dy = -\frac{W H_p{}^3}{12\eta}\frac{\Delta p}{Z_p} + W\frac{V_z}{2} H_p \tag{4.4-13}$$

- The term $Q_{1p} = W V_z H_p/2$ is independent of the die dimensions; this is the maximal flow rate that could be obtained from the metering zone, assuming it could be isolated from the rest of the extruder.
- The term $Q_{2p} = W H_p{}^3 \Delta P/(12\eta Z_p)$ is a function of the pressure developed at the end of the screw or inlet of the die; this term then depends on the flow resistance of the extruder die.

The velocity profile can be expressed in terms of the relative flow rate terms Q_{1p} and Q_{2p}:

$$w = V_z\left[\frac{y}{H_p} - 3\frac{Q_{2p}}{Q_{1p}}\frac{y}{Hp}\left(1 - \frac{y}{H_p}\right)\right] \tag{4.4-14}$$

Figure 4.4-7 illustrates four different possible situations.

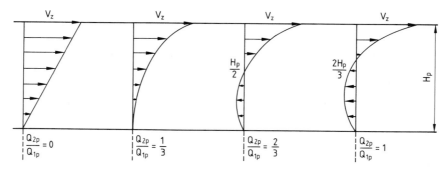

Figure 4.4-7 $\quad$ Axial velocity profiles for different values of the ratio Q_{2p}/Q_{1p}

For $Q_{2p}/Q_{1p} > 1/3$, back flow of the polymer in the metering zone is predicted. To understand the physics of this back flow, it is necessary to consider simultaneously the axial and transverse flows. This is done with the help of Figure 4.4-8, which illustrates a case where back flow is important, i.e. $Q_{2p}/Q_{1p} = 2/3$.

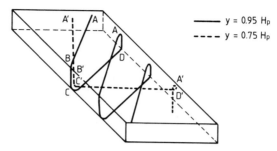

Figure 4.4-8   Pathlines of polymer particles in the metering zone

A particle initially located near the barrel wall (point A, at $y = 0.95\,H_p$ for example), will at high speed travel the distance AB, go down along the flight (BC), then it will come backward slightly at low speed (segment CD) before going up along the front flight and resume a configuration similar to A.

In contrast, a particle located at a greater distance from the barrel wall (point A', at $y = 0.75\,H_p$ in Figure 4.4-8) will at moderate speed travel the distance $A'B'$, go down along the back flight ($B'C'$), then it will continue forward at moderate speed ($C'D'$) to finally reach a configuration of type A'.

We observe that the residence times of particles A and A' are considerably different. Therefore, the distribution of residence times at the outlet of the extruder is quite wide. The higher the die pressure, the more important is the back flow and the larger is the residence time distribution (PINTO and TADMOR, 1970).

When $Q_{2p} = Q_{1p}$, the velocity profile is identical to that of the transverse flow (Figure 4.4-5). The net flow rate of the metering zone is in theory equal to zero. The die pressure is then given by the expression:

$$\Delta p = 6\eta V_z \frac{Z_p}{H_p{}^2} \qquad (4.4\text{-}15)$$

The pressure drop calculated for the reference extruder of Section 4.1-6 is then equal to 250 MPa, a value that considerably exceeds the design pressure for an extruder. In fact, leakage between the flights and the barrel (see Section 4.4-3) results in this case in a non-zero flow rate.

In summary the flow rate in the metering zone may be written as:

$$Q = Q_{1p} - k_p \Delta p \qquad (4.4\text{-}16)$$

where $k_p = WH_p{}^3/(12\eta Z_p)$. The variation of the flow rate as a function of the pressure drop is shown in Figure 4.4-9.

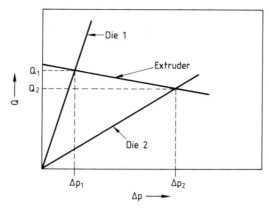

Figure 4.4-9   Relation between flow rate and pressure drop in the metering zone

Assuming Newtonian behaviour, the flow rate in the die is related to the pressure drop by:

$$Q = k_f \Delta p \tag{4.4-17}$$

where k_f is a characteristic of the die. This expression is a straight line on Figure 4.4-9 which intercepts the extruder characteristic curve at the operating point Q and Δp for a given extruder-die system. The metering zone of an extruder acts like a pump and if its characteristic slope k_p is close to zero, the flow rate is controlled by the metering zone and it is not affected by the flow resistance of the die.

Remark

Recent numerical modelling of the metering zones have been proposed using helical coordinates, so avoiding the necessity of unwinding the flow channel (NEBRENSKY et al., 1973; TUNG and LAURENCE, 1975; CHOO et al., 1980; HAMI and PITTMAN, 1980).

4.4-2 Compression zone

In the extruder's compression zone, the diameter of the screw increases with the axial distance, and thus the flow section decreases. Figure 4.4-10 recalls the main characteristics of the compression zone.

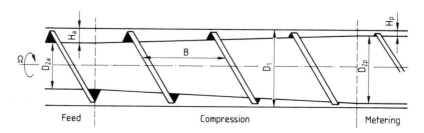

Figure 4.4-10   Compression zone of an extruder

Assuming that the screw is fixed and the barrel rotating, and unwinding the channel, we obtain a flow situation corresponding to the Reynolds bearing, as illustrated in Figure 4.4-11.

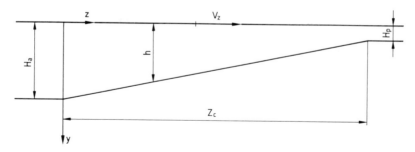

Figure 4.4-11   Unwinding of the compression zone

a) Simplified results

The equation of Reynolds for the axial flow in the compression zone is:

$$\frac{dp}{dz} = 6\eta V_z \frac{h - h^*}{h^3} \qquad (4.4\text{-}18)$$

where h^* is the depth of the channel for which the pressure is maximum.

Assuming that the pressure is equal to zero both at the entrance and outlet of the compression zone, we can directly use the results of Problem 3.1-A:

— the depth for which the pressure is maximum is the harmonic average of H_a and H_p, i.e.:

$$h^* = 2\frac{H_a H_p}{H_a + H_p} \qquad (4.4\text{-}19)$$

— the expression for the polymer flow rate is:

$$Q_{1c} = W V_z \frac{H_a H_p}{H_a + H_p} \qquad (4.4\text{-}20)$$

— and the maximal pressure is given by:

$$p_{\max} = 3\eta V_z Z_c \frac{H_a - H_p}{2 H_a H_p (H_a + H_p)} \qquad (4.4\text{-}21)$$

These expressions applied to the reference extruder of Section 4.1-6 give:

$$h^* = 6.6 \text{ mm}, \ Q_{1c} = 119.6 \text{ mL/s}, \ p_{\max} = 11.7 \text{ MPa}$$

b) Effect of presssure gradient

There is no reason why the pressure should be equal to zero at the two boundaries of the compression zone. Let us consider the following boundary conditions:

B.C. 1), at $z = 0$, $p = p_c$ (inlet of compression zone)
B.C. 2), at $z = Z_c$, $p = p_p$ (outlet of compression zone).

The solution for this problem is analogous to that of Problem 3.1-A. We obtain:

$$h^* = \left[1 - \frac{(p_p - p_c)\, H_p H_a}{6\eta V_z Z_c}\right] \frac{2 H_a H_p}{H_a + H_p} \tag{4.4-22}$$

This result implies that h^* decreases as the pressure drop across the compression zone increases, i.e. the location where the pressure is maximum is pushed towards the end of compression zone. If the value of h^* is less than that of H_p, then the pressure gradient is always positive and the pressure is continuously increasing in the compression zone.

The expression for the flow rate is:

$$Q = Q_{1c} - k_c(p_p - p_c) \tag{4.4-23}$$

where Q_{1c} is given by Equation 4.4-20 and

$$k_c = \frac{W H_a^2 H_p^2}{6\eta Z_c (H_a + H_p)} \tag{4.4-24}$$

As for the metering zone, the flow rate of the compression zone is the sum of the shear flow rate, Q_{1c}, and the back flow rate, $Q_{2c} = k_c(p_p - p_c)$. The expression for the pressure profile is:

$$p(z) = p_c + 6\eta V_z Z_c \frac{(H_a - h)\,(h - H_p)}{h^2 (H_a^2 - H_p^2)} + (p_p - p_c)\left(\frac{H_p}{h}\right)^2 \frac{H_a^2 - h^2}{H_a^2 - H_p^2} \tag{4.4-25}$$

Typical pressure profiles are illustrated in Figure 4.4-12 for various values of the pressure difference, $p_p - p_c$, as a function of the unwound length of the screw channel of our reference extruder. We notice that the pressure build-up in the compression zone depends not only on the compression ratio of the screw, but on the compression factor, $A = (H_a - H_p)/Z_c$, as well. Similar effects were observed for the plastication mechanism in Section 4.3-3.

c) Effect of relative motion

In reality, the screw is rotating inside a fixed barrel, and in the compression zone, the linear velocity of the screw surface is increasing with axial distance. This situation is depicted in Figure 4.4-13.

The equation of Reynolds in the form of (4.4-18) can no longer be used. Instead we make use of the more general Reynolds equation presented in Section 3.1-3. We write Equation (3.1-9) as:

$$Q = -\frac{W h^3}{12\eta} \frac{dp}{dz} + W V(z) \frac{h}{2} \tag{4.4-26}$$

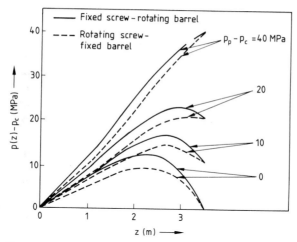

Figure 4.4-12   Pressure profiles in the compression zone

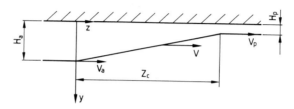

Figure 4.4-13   More realistic sketch of the compression zone

and since $V(z) = \Omega\pi \cos\theta(D_1 - 2h)/60$ where Ω has units of rev/min, Equation (4.4-26) can be written as:

$$\frac{dp}{dz} = \frac{12\eta}{Wh^3}\left[\frac{W\pi\Omega\cos\theta}{120}(D_1 - 2h)\,h - Q\right] \qquad (4.4\text{-}27)$$

Integrating with respect to h and using the same boundary conditions for the pressure as in the previous case, we obtain:

$$Q = Q'_{1c} - k_c(p_p - p_c) \qquad (4.4\text{-}28)$$

The back flow expression, $Q_{2c} = k_c(p_p - p_c)$ is identical to that of the previous case, but the expression for the shear flow is now:

$$Q'_{1c} = Q_{1c} - \frac{2WV_z}{D_1}\frac{H_a^{\,2}H_p^{\,2}}{H_a^{\,2} - H_p^{\,2}}\ln\frac{H_a}{H_p} \qquad (4.4\text{-}29)$$

The flow rate is then somewhat reduced when considering that the screw is rotating. For the reference extruder, we get Q'_{1c} equal to 105.8 mL/s instead of 119.6 mL/s obtained in the previous case. Also, the corrected expression for the pressure profile is:

$$p'(z) = p(z) - \frac{12\eta V_z Z_c}{D_1(H_a - H_p)} \left[\ln \frac{H_a}{h} - \frac{H_a^2 - h^2}{H_a^2 - H_p^2} \left(\frac{H_p}{h} \right)^2 \ln \frac{H_a}{H_p} \right] \quad (4.4\text{-}30)$$

The dashed lines in Figure 4.4-12 refer to the pressure profiles calculated for the reference extruder, compared to profiles obtained under the assumption that the screw is fixed and the barrel is rotating (solid lines). The differences are not negligible.

d) Effect of solid bed

The situation in a plasticating extruder is far more complex than described so far. In the compression zone, there is simultaneous flow of solid and molten polymer. The situation is approximately that as described in Figure 4.4-14.

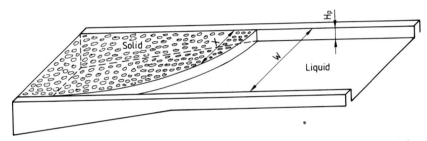

Figure 4.4-14   Simultaneous flow of solid and molten polymer in the compression zone

The following analysis is based on the work of HALMOS et al. (1978). If we assume that the Reynolds equation is still valid locally, we can write:

$$Q_m = \left(-\frac{W'h^3}{12\eta} \frac{dp}{dz} + W'V_z \frac{h}{2} \right) \varrho_l \quad (4.4\text{-}31)$$

where Q_m is the mass flow rate of the molten polymer and ϱ_l its density. The width of the liquid pocket is not constant, but equal to:

$$W' = W - X \quad (4.4\text{-}32)$$

The solid bed width can be calculated as shown in Section 4.3. Moreover, the liquid flow rate increases with the axial distance, according to the following expression:

$$\frac{dQ_m}{dz} = \phi\sqrt{X} \quad (4.4\text{-}33)$$

where X is given by Equation (4.3-14).

Knowing X as a function of z, it is possible to integrate Equations (4.4-33) and (4.4-31) to obtain the pressure profile in the compression zone. Details of the mathematics are given in Appendix E.

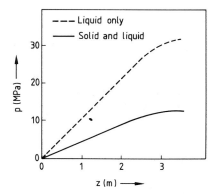

Figure 4.4-15 Effect of solid on the pressure profile in the compression zone

Figure 4.4-15 compares the pressure profile in the compression zone of the reference extruder (see Section 4.1-6), obtained under the assumption that the channel is filled with liquid (dashed line) with that calculated when both liquid and solid are present (solid line). In this last case, it was assumed that plastication started at the beginning of the compression zone.

4.4-3 Clearance flow

We will restrict the following analysis to the metering zone. So far, we have assumed that the screw flights were perfectly in contact with the barrel. In the actual situation, there is a clearance, δ_f, between the tip of the flights and the barrel. The clearance or leakage flow has been analyzed by many authors, mainly MC KELVEY (1962), TADMOR and KLEIN (1970), and MIDDLEMAN (1977).

In presence of clearance flow, the flow rate of the extruder is no longer equal to that of the axial flow of the screw channel. The transverse flow is disturbed: the velocity profile is still given by Equation (4.4-7) i.e.:

$$u = \frac{1}{2\eta} \frac{\Delta p}{W} y(y - H_p) + \frac{V_x y}{H_p}$$

but the transverse flow rate is no longer equal to zero. The axial flow profile is not perturbed by the clearance flow, but the expression for the flow rate must account for the flow through the clearance above the edges of the flights (of thickness, δ_f, and of width e). Details of the mathematics are given in Appendix E.

The total axial flow rate is:

$$Q = Q_{1p}\left(1 - \frac{\delta_f}{H_p}\right) - Q_{2p}(1 + f_L) \tag{4.4-34}$$

where Q_{1p} and Q_{2p} are the flow rate expressions defined below, Equation (4.4-13):

The parameter f_L is:

$$f_L = \frac{\eta e}{\eta_f W}\left(\frac{\delta_f}{H_p}\right)^3 + \left(1+\frac{e}{W}\right)\frac{(1+e/W)/\tan\theta + 6\eta V_z(H_p - \delta_f)/(H_p{}^3(\Delta p/Z_p))}{1+(e/W)\,(\eta_f/\eta)\,(H_p/\delta_f)^3}$$

(4.4-35)

where η_f is the polymer viscosity evaluated at the shear rate of the clearance flow, which for shear-thinning polymers is much less than the viscosity evaluated at the shear rate of the main channel.

The clearance flow reduced the extruder throughput in two ways:

- the shear flow is reduced: $Q'_{1p} = Q_{1p}(1 - \delta/H_p)$
- the back flow is enhanced: $Q'_{2p} = Q_{2p}(1 + f_L)$

The effect of the clearance flow is illustrated in Table 4.1 which reports flow rates for the reference extruder of Section 4.1-6, calculated for three different values for the flight clearance: 0, 50 and 500 μm. The density of the polymer was taken as $\varrho_l = 1000$ kg/m^3.

Table 4.1.   Effect of flight clearance on flow rate

	$\delta_f = 0$			$\delta_f = 50\ \mu m$			$\delta_f = 500\ \mu m$		
Δp MPa	Q_{1p} kg/h	Q_{2p} kg/h	Q kg/h	Q'_{1p} kg/h	Q'_{2p} kg/h	Q' kg/h	Q'_{1p} kg/h	Q'_{2p} kg/h	Q' kg/h
0	258	0	258	255	0	255	232	0	232
1	258	1	257	255	1	254	232	44	188
10	258	10	248	255	10	245	232	58	174
50	258	50	208	255	50	205	232	123	109

For small value of the flight clearance, $\delta_f = 50\ \mu$m, both the shear and pressure flows are weakly affected by leakage. In the case of the large values of δ_f, the shear flow is not much reduced, but the back flow is considerably enhanced and this results in a drastic reduction of the extruder throughput.

4.4-4 Thermal effects

It is important to control the temperature of the polymer in the last section of the extruder for the following reasons:

- to avoid possible thermal degradation;
- a too high temperature, hence a very low melt viscosity may make the post-extrusion processing impossible or very difficult to handle, for example in the case of profile extrusion or wire coating;
- excessive heating is a waste of energy.

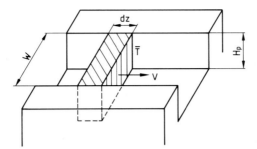

Figure 4.4-16   Differential liquid element in the extruder channel

In this section, we will examine thermal effects in the metering section of the extruder, but the results can be easily extended to the compression zone.

Let us discuss the variation of the average temperature in a liquid element of the channel of length dz as shown in Figure 4.4-16. Assuming that the screw and the barrel walls are at the same temperature, T_w, a heat balance yields:

$$\underbrace{\varrho c_p V \frac{d\bar{T}}{dz} W H_p \, dz}_{\substack{\text{variation of} \\ \text{internal energy}}} = \underbrace{-\frac{2Nu\,k}{H_p}(\bar{T} - T_w)\, W \, dz}_{\substack{\text{heat loss by} \\ \text{conduction}}} + \underbrace{W \, dz \int_0^{H_p} \eta \dot{\gamma}^2 \, dy}_{\substack{\text{heat generated} \\ \text{by viscous forces}}} \qquad (4.4\text{-}36)$$

The expression for the heat generated by viscous forces is, in general, quite complex due to the combined axial and transverse flows. As a first approximation, we consider that the flow is essentially a simple shear flow, for which the rate of heat generated by viscous forces is:

$$\dot{W} = \eta W \frac{V^2}{H_p} \, dz \qquad (4.4\text{-}37)$$

The Nusselt number Nu depends on the balance between shear flow and Poiseuille flow in the metering zone and on the boundary conditions at the screw and barrel surfaces. If shear flow is dominant and if isothermal conditions are considered, the solution of Section 2.3-4 may be applied; the average temperature as a function of the axial distance is expressed by:

$$\bar{T} = T_0 + \eta \frac{V^2}{3k} \left(1 + \frac{3}{\mathrm{Br}}\right) \left(1 - \exp(-12\,\mathrm{Ca}\,z/L)\right) \qquad (4.4\text{-}38)$$

where the Cameron number is:

$$\mathrm{Ca} = \frac{\alpha Z_p}{\bar{V} H_p{}^2} = \frac{\alpha Z_p W}{Q H_p} \qquad (4.4\text{-}39)$$

and the Brinkman number:

$$\mathrm{Br} = \frac{\eta V^2}{k(T_w - T_0)} \qquad (4.4\text{-}40)$$

T_w and T_0 are respectively the wall temperature and the inlet polymer temperature.

If we consider that the screw is adiabatic instead of being maintained at T_w (this hypothesis is closer to reality for most applications), we obtain

$$\bar{T} = T_0 + \eta \frac{4V^2}{3k}\left(1 + \frac{3}{4\,\mathrm{Br}}\right)\left(1 - \exp\{-3\,\mathrm{Ca}\,z/L\}\right) \qquad (4.4\text{-}41)$$

Figure 4.4-17 shows the temperature profile in the metering zone of the reference extruder of Section 4.1-6 for a pressure of 20 MPa at the die inlet. The inlet polymer temperature and barrel temperature are taken to be equal to 180° C. We observe that the temperature rise is important (up to 40° C), but it could be reduced by approximately 10° C by controlling also the screw surface at 180° C.

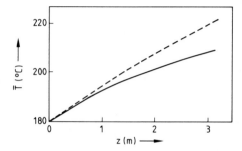

Figure 4.4-17 Temperature profile in the metering zone; ——— Screw and barrel maintained at 180° C, - - - - - Barrel maintained at 180° C and adiabatic screw

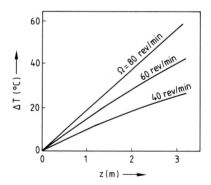

Figure 4.4-18 Shear-heating in the metering zone as a function of the screw rotational speed

Figure 4.4-18 shows the influence of the screw rotational speed on shear-heating in the metering zone. The same die characteristics, $k_f = Q/\Delta P$, as for the calculation of the temperature profile of Figure 4.4-17 has been used. The barrel wall is maintained at 180° C and the screw is adiabatic. We observe that, at high rotational speeds, the temperature rise is a linear function of the axial distance. This is the adiabatic regime discussed in Chapter 2. The influence of the operating conditions is better observed in Figure 4.4-19 which shows the average temperature of the polymer at the end of the metering zone as a function of the screw rotational speed for three different barrel temperatures, 160, 180 and 200° C.

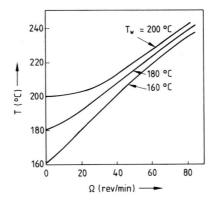

Figure 4.4-19 Influence of the screw rotational speed and barrel temperature on the polymer temperature at the end of the metering zone

At very high rotational speed, the final polymer temperature is not affected much by the barrel temperature. For example, at 80 rev/min, a decrease of 40° C in the barrel temperature results in a decrease of only 6° C of the polymer temperature.

These illustrations allow us to define the concepts of adiabatic and isothermal extruders:

– In general, no extruder can really be operated under isothermal conditions as defined in Section 2.3, i.e. in such conditions that the polymer temperature remains constant along the metering zone.
– In most cases, extruders are rather operated under adiabatic conditions and the temperature rise of the polymer is mainly due to viscous dissipation. Under such conditions, it does not make any sense to try to control the polymer temperature by controlling the barrel temperature.

4.4-5 Pressure profile in the entire extruder

In fact, the metering zone is not isolated from the other zones of the extruder. Considering all the three zones, from the feed to the die entrance, we may observe two types of pressure profiles as illustrated in Figure 4.4-20.
a) If the feed zone is of the standard type (heated with no grooves), the pressure increases monotonously up to the die entrance (curve a), provided the screw compression ratio is not too high.
b) If the feed zone is cooled and has grooves in the barrel surface, the pressure increases very rapidly in the feed zone, then remains relatively constant in the

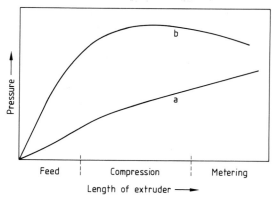

Figure 4.4-20   Pressure profiles in extruders

compression zone and possibly decreases in the metering zone (curve b). With
a negative pressure gradient in the metering zone, the extruder throughput
can be larger than the maximum drag flow. This may seem to be of interest,
but gases (entrapped air or volatiles) can no longer be eliminated through the
feed hooper and the final product will be of poor quality. A solution to the
problem is to use a vent-type extruder as illustrated in Figure 4.4-21.

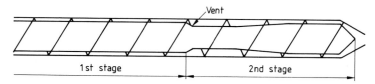

Figure 4.4-21   Two-stage screw with vent

The polymer is plasticated in the first stage of the screw which controls the
flow rate. The second stage is essentially used to pump the molten polymer.

In other designs, the two stages shown in Figure 4.4-21 are completely
separated: there are two extruders in series with a vent station in between.

4.4-6 Conclusion

The analysis of the flow of molten polymer in an extruder presented in Section
4.4 remains highly qualitative. The shear-thinning (and the elastic) properties
of polymers have not been considered and only approximate solutions for the
equations of motion and energy have been presented.

The analysis, nevertheless, reveals several important aspects for the operation
of an extruder:

–   one cannot install dies of any geometrical characteristics to a given extruder
    (see Problem 4.4-A);
–   a too high pressure at the end of the metering zone may cause significant
    increase of the polymer residence time;

- screw wear has catastrophic effects on the flow rate, as shown for the plastication process;
- in most practical situations, the temperature rise of the polymer in the metering zone cannot be controlled by the wall temperature.

Problem 4.4

4.4-A Design criteria of an extruder

Let us consider the reference extruder of Section 4.1-6. We assume that the polymer is entirely molten at the beginning of the compression zone and that its viscosity remains constant and equal to 1 kPa·s during the different stages of extrusion. We assume that the pressure is equal to zero at both the beginning of the compression and outlet of the die. We assume that the screw is fixed and the barrel is rotating at the speed $\Omega' = -\Omega$.

We are only concerned by the axial flow (along the channel) and we wish to study the extruder performances for different die characteristics. The nomenclature for the three zones of the extruder is shown in Figure 4.4-22.

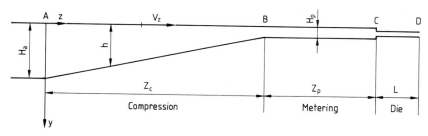

Figure 4.4-22   Nomenclature for the three zones of the extruder

a) Show that for each zone (compression, metering and die) the flow rate can be written in the following form:

$$Q = Q_1 - k\Delta p$$

where Q_1 and k are different for each zone. Obtain the corresponding expressions.

b) Then obtain the expressions for the flow rate and for the pressure at points B and C. Plot the curves of the flow rate and the pressure p_B and p_C as functions of k_f for $10^{-15} < k_f < 10^{-9}\,\mathrm{m}^3/(\mathrm{s}\cdot\mathrm{Pa})$.

c) We consider that extrusion will correctly proceed under the following conditions:
   - the maximum pressure is less than 100 MPa;
   - the pressure is increasing in the metering zone;
   - the pressure is larger than 1 MPa at the extruder die;
   - the distribution of the residence time is narrow.

- Justify briefly these conditions.
- For what value of k_f is the extrusion possible, at $\Omega = 60$ rev/min?
- A circular die of 0.1 m in length is used. For what diameter is the extrusion acceptable?

4.5   Modelling of single screw extrusion

4.5-1   Introduction

Extruders have been used in the plastic industry for some thirty years. The early designs were based on extruders used since a long time in the rubber or food industry. Designs have been considerably improved since the fifties when short L/D screws (12 to 15 diameters) and uniform temperature barrel were common practice. As mentioned in the previous sections, considerable progress has been possible because of a much better understanding of the process.

Progress has been possible so far through a good but often qualitative understanding of the phenomena involved in extrusion. Further progress will be possible only if more complete analyses are provided.

Computer programs have been developed to calculate the performance of a single screw extruder (pressure and temperature profiles, fractions of solid and liquid) from the feed hooper to the die exit. Initial work started with CHUNG (1968), followed by TADMOR and KLEIN (1970) and followed by many: LOSSON (1974), FENNER (1974), AGUR and VLACHOPOULOS (1982), KLEIN (1982), ZAVADSKY and KARNIS (1984), and VIRIYAYUTHAKORN and KASSAHUN (1984). The results presented in this section have been obtained with the computer program CEMEXTRUD of the Centre de mise en forme des matériaux (VERGNES et al., 1983, AGASSANT et al., 1987).

The major difficulties encountered in the modelling of extrusion are on two levels:

- Some of the transport phenomena are highly difficult to model and cannot be described by appropriate equations, mainly transition phenomena between two different zones, viscoelastic properties of polymeric liquids, etc.
- Some of the polymer properties are difficult to measure: friction coefficient of solid granules, compacting coefficient, properties near the melting point or near the glass transition, etc.

4.5-2   Numerical applications for the reference extruder

Figure 4.5-1 reports for the reference extruder of Section 4.1-6 the pressure profile (a), the temperature profile (b) and the ratio of solid width (c) as a function of the position in the extruder, expressed in terms of the number of turns. The main characteristics and the wall temperature are shown on the top of Figure 4.5-1. The following data for high-density polyethylene have been used:

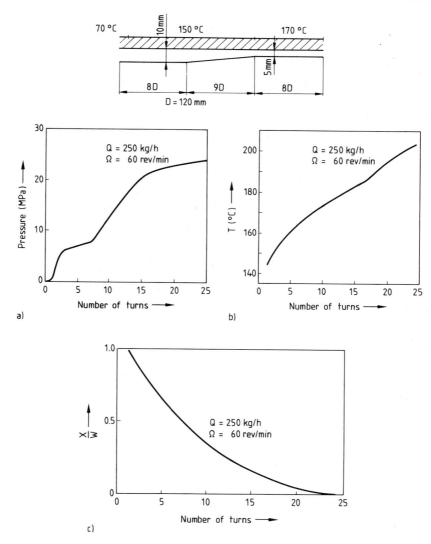

Figure 4.5-1 Simulation results for the reference extruder
a)   Pressure curve
b)   Temperature of the molten polymer
c)   Ratio between the width of the solid bed and the width of the channel

- Compressibility of solid granules:

$$\frac{\varrho_\infty - \varrho}{\varrho_\infty - \varrho_0} = e^{-\alpha' p} \qquad (4.5\text{-}1)$$

where

$\alpha' = 3.7 \times 10^{-7}\,\mathrm{Pa}^{-1}$

$\varrho_0 = 540\ \mathrm{kg/m^3}$ (bulk density of the granules)

$\varrho_\infty = 930\ \mathrm{kg/m^3}$ (density of the solid polymer)

- $\nu = 0.25$ (Poisson's coefficient for the granules)
- $k_s = 0.335\ \mathrm{W/m \cdot {}^\circ C}$ (thermal conductivity of the solid polymer)
- $c_{ps} = 2000\ \mathrm{J/kg}$ (specific heat of the solid polymer)
- $f_1 = 0.5$ (friction coefficient of polymer-barrel surface)
- $f_2 = 0.15$ (friction coefficient of polymer-screw surface)
- $\lambda = 1.3 \cdot 10^5\ \mathrm{J/kg}$ (heat of fusion of the polymer)
- $T_m = 115^\circ\,\mathrm{C}$ (melting temperature of the polymer)
- $k_l = 0.188\ \mathrm{W/m^2 \cdot {}^\circ C}$ (thermal conductivity of the liquid polymer)
- $c_{pl} = 1964\ \mathrm{J/kg \cdot {}^\circ C}$ (specific heat of the liquid polymer)
- $\varrho_l = 870\ \mathrm{kg/m^3}$ (density of the liquid polymer)
- The viscosity of the liquid polymer is given by the following expression:

$$\eta = K_0 \exp\left\{ \frac{E}{R} \left(\frac{1}{T} - \frac{1}{T_0} \right) \right\} |\dot\gamma|^{n-1}$$

where $K_0 = 4 \cdot 10^4\,\mathrm{Pa \cdot s}^{n-1}$, $E/R = 3300\ \mathrm{K}$, $T_0 = 403\ \mathrm{K}$ and $n = 0.3$.

It is possible to incorporate temperature dependent functions for the density, specific heat and thermal conductivity. Also, it is quite easy to replace the power-law expression for the viscosity by experimental data obtained at various temperature values. The data can then be interpolated for shear rate and temperature conditions corresponding at the various positions along the screw channel. Typical experimental values for the viscosity of a polymethyl methacrylate (PMMA) used in the second example are shown in Figure 4.5-2.

Figure 4.5-3 reports the simulation results for the pressure profiles along the extruder barrel for three different values of the rotational speed. In all cases, the pressure is a monotonous function increasing from the feed zone to the end of the metering zone. The corresponding temperature profiles are shown in Figure 4.5-4.

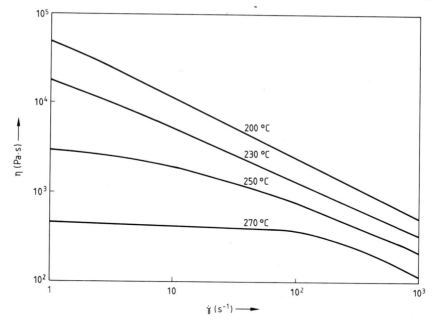

Figure 4.5-2   Viscosity of the PMMA used for the simulation results presented in Figure 4.5-6

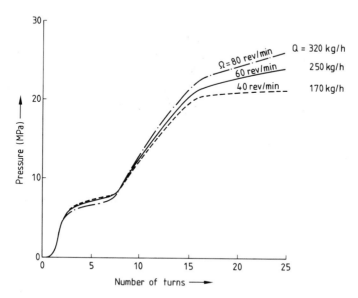

Figure 4.5-3   Pressure profiles for three different values of the rotational speed; same processing parameters as for Figure 4.5-1.

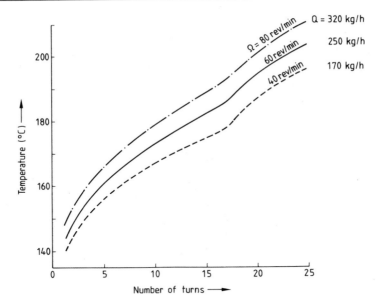

Figure 4.5-4   Temperature profiles for three different values of the rotational speed

4.5-3   A second example

As shown in Figure 4.5-5, a screw of a smaller diameter ($D_1 = 40$ mm) is now analyzed. The compression ratio is now 2.3 instead of 1.9 and the compression slope is much more important ($1.4 \cdot 10^{-3}$ instead of $1.4 \cdot 10^{-5}$).

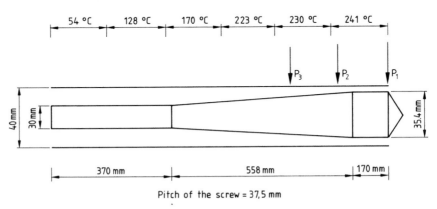

Pitch of the screw = 37,5 mm

Figure 4.5-5   Screw geometry used for the second example

The extrudate is a polymethyl methacrylate (PMMA) for which the viscosity data are read directly from Figure 4.5-2.

Figure 4.5-6 shows three typical pressure profiles calculated for this screw geometry. Due to the much higher compression factor, the pressure is an increasing function up to the middle of the compression zone, then it decreases in the last section of the compression zone and in the metering zone. It is worth noticing that the pressure measurements (at locations shown in Figure 4.4-5) are in a very good agreement with the predicted values.

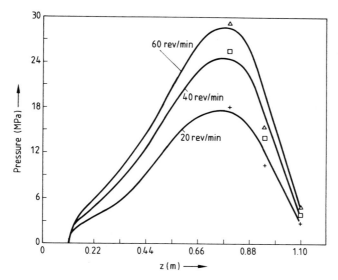

Figure 4.5-6   Pressure profiles for three different values of the rotational screw speed

$\triangle$-pressure measured at 60 rev/min
$\square$-pressure measured at 40 rev/min
+-pressure measured at 20 rev/min

4.5-4 Conclusion

Such computer programs are of considerable interest for the design and the operation of extruders. As of now, however, it is not possible to rely completely on these tools for the design of an extruder simply using the polymer properties (viscosity, thermal conductivity, friction coefficients, etc.) and knowing the operating conditions (temperature profile of the barrel and flow rate). Nevertheless, these computer programs are needed for decision making and trouble-shooting. They can provide useful answers to the following questions:

- How can we scale-up a 100 mm diameter extruder, operating satisfactorily, to a 150 mm one?
- Problems are experienced with a given extruder. Is it useful to use a screw with a deeper feed zone to increase the compression factor or rather is it better to change the temperature profile on the barrel wall? The following two figures, adapted from VERGNES et al. (1983), give partial answers to these questions.

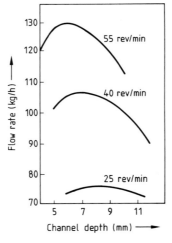

Figure 4.5-7   Flow rate as a function of the channel depth in the metering zone for three different screw speeds

- Figure 4.5-7 shows the influence of the screw channel depth on the flow rate. For each rotational speed, a maximum in the flow rate corresponds to a different channel depth.
- For a given rotational speed, there is an optimum channel depth in the metering zone corresponding to a given die flow characteristic, k_f. This is illustrated in Figure 4.5-8. The flow in the die is given by the expression:

$$Q = k_f (\Delta P)^{1/n}$$

If the flow resistance of the die is very large (small value of k_f), a smaller channel depth has to be used. For a die of low flow resistance, it is worthwhile to use a deeper screw channel.

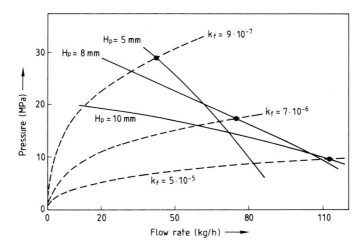

Figure 4.5-8 Relation between flow rate and pressure for various die geometries and screw channel depths – The intercepts represent operating points

In general it is not possible to optimize a screw design for given operating conditions without considering the coupled interactions of the screw and die.

Chapter 5
Dynamics of Film
and Fiber Forming

5.1   Stretching of a liquid

5.1-1   Elongational viscosity

In 1687 Newton defined the shear viscosity as the ratio between the shear stress ($\sigma = F/S$) and the shear rate ($\dot{\gamma} = U/h$), i.e.:

$$\eta = \frac{\sigma}{\dot{\gamma}} \tag{5.1-1}$$

Much later, in 1906, TROUTON defined, for a liquid-like material, the extension viscosity as the ratio between the normal stress ($\sigma_n = F/S$) and the rate of strain ($\dot{\varepsilon} = dL/L\,dt$):

$$\eta_e = \frac{\sigma_n}{\dot{\varepsilon}} \tag{5.1-2}$$

It was shown by TROUTON, for the special case of a Newtonian fluid, that the elongational viscosity is equal to three times the shear viscosity:

$$\eta_e = 3\eta \tag{5.1-3}$$

We can verify Trouton's relation for a Newtonian fluid using the rectangular sample of Figure 5.1-1 a). We pull on the sample in the z-direction assuming an homogeneous deformation and neglecting the weight of the sample compared to the tensile force.

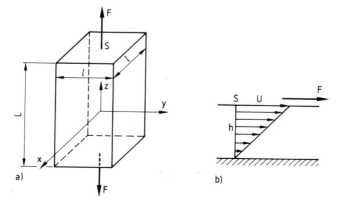

Figure 5.1-1   Elongational (a) and shear flow (b)

- The constant strain rate in the z-direction is:

$$\dot{\varepsilon}_{zz} = \frac{1}{L}\frac{dL}{dt} = \dot{\varepsilon} \qquad (5.1\text{-}4)$$

For incompressible fluid, the continuity equation implies that:

$$\dot{\varepsilon}_{yy} = \dot{\varepsilon}_{xx} = -\frac{1}{2}\dot{\varepsilon} \qquad (5.1\text{-}5)$$

If the shape of the sample remains orthogonal during the deformation, there will be no shear deformation.

- For a Newtonian polymer, the stress tensor is:

$$\boldsymbol{\sigma} = -p\boldsymbol{\delta} + 2\eta\dot{\varepsilon} = \begin{bmatrix} -p - \eta\dot{\varepsilon} & 0 & 0 \\ 0 & -p - \eta\dot{\varepsilon} & 0 \\ 0 & 0 & -p + 2\eta\dot{\varepsilon} \end{bmatrix} \qquad (5.1\text{-}6)$$

and the equation of motion gives:

$$\frac{\partial\sigma_{xx}}{\partial x} = \frac{\partial\sigma_{yy}}{\partial y} = 0$$

As there is no force acting on the sample lateral surfaces, σ_{xx} and σ_{yy} are equal to zero everywhere and consequently:

$$p = -\eta\dot{\varepsilon} \quad \text{and} \quad \sigma_{zz} = 3\eta\dot{\varepsilon} \qquad (5.1\text{-}7)$$

and the definition of Equation (5.1-3) is verified, i.e.:

$$\eta_e = \frac{\sigma_{zz}}{\dot{\varepsilon}} = 3\eta$$

Remarks

Measuring the elongational viscosity is much more difficult than measuring the shear viscosity.

- Indeed, the elongational viscosity cannot be measured under a constant velocity type of flow. To impose a constant strain rate, $\dot{\varepsilon}$, it is necessary to deform exponentially the sample, i.e.:

$$\dot{\varepsilon} = \frac{1}{L}\frac{dL}{dt}$$

and

$$L = L_0 \exp(\dot{\varepsilon}t) \qquad (5.1\text{-}8)$$

- To deform the liquid-like sample at constant temperature, it is necessary to use a temperature controlled bath.

- Finally, to avoid deformation of the sample due to gravitational effects, it is essential to plunge the sample in a bath containing a liquid of specific gravity comparable to that of the sample at the experimental temperature. Moreover, due to instrument physical dimensions, the maximum length or deformation is necessarily restricted.

5.1-2 Fiber spinning

One application of the elongational flow is the fiber spinning operation. All synthetic fibers are manufactured by spinning a polymeric solution or a molten polymeric filament. In this last case, fiber spinning, as illustrated in Figure 5.1-2, includes:

1) Melting of the polymer, using in most cases an extruder.
2) Filtering of the melt through a series of screen packs and feeding to a spinnerette; a high purity polymer melt is required because of the small diameter of the fibers produced (tens of microns).
3) A drawing zone, where the extrudate is simultaneously stretched and cooled. There is two types of drawing:
   - water drawing where the extrudate is first stretched over a short distance (few centimeters) in air, then quenched in a water bath. Stretching in air can be considered isothermal.
   - air drawing where the extrudate is stretched over a long distance (one meter or more), and simultaneously cooled by forced convection. This is the most popular process, for example for producing polyamide and polyester fibers.
4) A variable speed take-up roll. For water drawing, the stretching or drawn-down ratio used is moderate (≈ 20). However, it may be very large in the case of air drawing (> 100).
5) A sophisticated wind-up system.

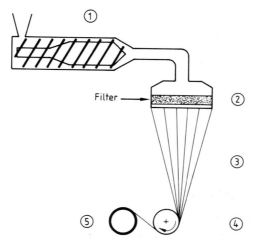

Figure 5.1-2   Sketch of the melt spinning process

5.1-3 Isothermal melt spinning of a Newtonian fluid

Isothermal fiber spinning can only be considered in the case of drawing over a short distance in air and then quenching in a water bath. In the following analysis we will study the variation of velocity and cross section of the filament between the die and the water bath (see Figure 5.1-3). We neglect the swelling of the extrudate at the die exit (see Section 6.1) due to a change in velocity profile from a Poiseuille flow in the die to a uniform velocity over the cross section outside the die.

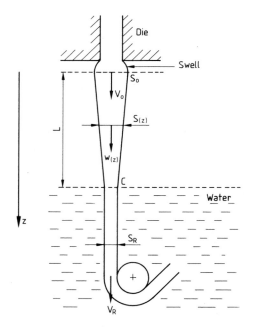

Figure 5.1-3   Spinning of a filament

The cross section of the filament varies from S_0 (which may be assumed equal to the cross-section of the die) to S_R at L (corresponding to the distance between the die and the water bath). The speed of the filament, constant over the cross-section, varies from the average speed in the orifice of the die, V_0, to the wind-up speed V_R. We define the draw-down ratio as:

$$DDR = \frac{V_R}{V_0} \qquad (5.1\text{-}9)$$

and we neglect inertia and gravitational forces and surface tension effects.

a) Dynamic simplifications

Using cylindrical coordinates (r, θ, z), we make the following assumptions:

– The flow is symmetrical, the θ-component of the velocity is zero and the r and z-components, u and w, are not functions of θ.

– We assume that the z-component w is uniform in the r-direction. Consequently, the velocity field is:

$$\boldsymbol{u} \quad \begin{vmatrix} u(r,z) \\ 0 \\ w(z) \end{vmatrix}$$

and the rate-of-deformation tensor is:

$$[\dot{\varepsilon}] = \begin{bmatrix} \frac{\partial u}{\partial r} & 0 & \frac{1}{2}\frac{\partial u}{\partial z} \\ 0 & \frac{u}{r} & 0 \\ \frac{1}{2}\frac{\partial u}{\partial z} & 0 & \frac{dw}{dz} \end{bmatrix} \qquad (5.1\text{-}10)$$

Defining $\dot{\varepsilon} = dw/dz$, we can obtain from the continuity equation an expression for $u(u = -r\dot{\varepsilon}/2)$ so that we can write the rate-of-deformation tensor as:

$$[\dot{\varepsilon}] = \begin{bmatrix} -\frac{\dot{\varepsilon}}{2} & 0 & -\frac{1}{4}r\frac{d\dot{\varepsilon}}{dz} \\ 0 & -\frac{\dot{\varepsilon}}{2} & 0 \\ -\frac{1}{4}r\frac{d\dot{\varepsilon}}{dz} & 0 & \dot{\varepsilon} \end{bmatrix} \qquad (5.1\text{-}11)$$

b) Set of equations

For a Newtonian fluid, the stress tensor is:

$$[\boldsymbol{\sigma}] = \begin{bmatrix} -p - \eta\dot{\varepsilon} & 0 & -\frac{\eta}{2}r\frac{d\dot{\varepsilon}}{dz} \\ 0 & -p - \eta\dot{\varepsilon} & 0 \\ -\frac{\eta}{2}r\frac{d\dot{\varepsilon}}{dz} & 0 & -p + 2\eta\dot{\varepsilon} \end{bmatrix} \qquad (5.1\text{-}12)$$

If we suppose pure elongational flow (neglecting the shear components of the rate-of-deformation tensor), we obtain for the normal stress (Section 5.1-1):

$$\sigma_{zz} = 3\eta\dot{\varepsilon}$$

And, if we neglect gravitational and inertial forces, the force acting on any filament section is expressed by:

$$F = S(z)\,\sigma_{zz} \qquad (5.1\text{-}13)$$

For a constant flow rate:

$$Q = S_0 V_0 = S_R V_R = S(z)\,w(z) \qquad (5.1\text{-}14)$$

The solution for the set of equations is obtained by combining Equations (5.1-7) and (5.1-13):

$$F = 3\eta S(z)\frac{dw}{dz} \qquad (5.1\text{-}15)$$

Combining with Equation (5.1-14) we can write $S(z)$ as a function of $w(z)$ and obtain:

$$F = 3\eta \left(\frac{Q}{w}\right) \left(\frac{dw}{dz}\right) \tag{5.1-16}$$

Applying the boundary condition $w(0) = V_0$, we obtain:

$$w(z) = V_0 \exp\left(\frac{F}{3\eta Q} z\right) \tag{5.1-17}$$

The force F is unknown and can be determined using the second boundary condition $w(L) = V_R$:

$$F = \frac{3\eta Q}{L} \ln \frac{V_R}{V_0} = \frac{3\eta Q}{L} \ln DDR \tag{5.1-18}$$

Comments

In the confined flows studied in Chapters 1, 3 and 4, we have solved the Stokes equations. In this case, we could imagine a solution based on the same approach. Taking into account the kinematic simplifications, the Stokes equations are:

$$\begin{cases} \dfrac{\partial p}{\partial r} = -\eta \dfrac{r}{2} \dfrac{d^2 \dot\varepsilon}{dz^2} \\[2mm] \dfrac{\partial p}{\partial z} = \eta \dfrac{d\dot\varepsilon}{dz} \end{cases} \tag{5.1-19}$$

As the pressure is not defined in a free surface flow of an incompressible fluid, we eliminate it from Equations (5.1-19) by taking the derivative of the first with respect to z and the derivative of the second equation with respect to r. We obtain:

$$\frac{d^3 \dot\varepsilon}{dz^3} = 0 \quad \text{or} \quad \frac{d^4 w}{dz^4} = 0$$

The velocity can be represented by a third-degree polynomial in z, i.e.

$$w(z) = Az^3 + Bz^2 + Cz + D \tag{5.1-20}$$

However, it is not possible to determine the coefficients of the polynomial with only two boundary conditions for the velocity. The difference between Equations (5.1-20) and (5.1-17), comes from the fact that the shear components of the rate-of-deformation tensor is considered in result (5.1-20).

Example 5.1-A Drawing in a water bath

Let us consider the spinning of a polymer filament through a short distance between the die and the water bath. The conditions and properties are the following:

$$
\begin{aligned}
\text{Extrusion speed}: \quad & V_0 = 10 \text{ mm/s} \\
\text{Pulling speed}: \quad & V_R = 100 \text{ mm/s} \\
\text{Pulling length}: \quad & L = 10 \text{ mm} \\
\text{Polymer viscosity}: \quad & \eta = 200 \text{ kPa} \cdot \text{s} \\
\text{Die radius}: \quad & R_0 = 0.5 \text{ mm}
\end{aligned}
$$

From Equation (5.1-17), we obtain $F = 1.085 \cdot 10^{-3} \text{ N}$.

Example 5.1-B Drawing in ambient air

In this case, the isothermal model is not realistic considering the air cooling of the filament along the pulling length with variations of the polymer viscosity of several orders of magnitude. However, we propose for comparison purpose only, the following numerical example:

$$
\begin{aligned}
\text{Extrusion speed}: \quad & V_0 = 120 \text{ mm/s} \\
\text{Pulling speed}: \quad & V_R = 14.6 \text{ m/s} \\
\text{Pulling length}: \quad & L = 1 \text{ m} \\
\text{Polymer viscosity}: \quad & \eta = 200 \text{ kPa} \cdot \text{s} \\
\text{Die radius}: \quad & R_0 = 0.44 \text{ mm}
\end{aligned}
$$

We obtain from Equation (5.1-17) a smaller pulling force, $F = 2.1 \cdot 10^{-4} \text{ N}$.

Since the computed values for the pulling force in both examples are very small, it is not obvious that the gravitational and inertia forces can be neglected. This will be discussed below.

c) Validity of the approximations used

In this first model of fiber spinning of a Newtonian fluid, we have neglected the shear components, gravitational and inertia forces. We will now successively verify these approximations.

Shear components

The shear component of the rate-of-deformation tensor is:

$$
\dot{\varepsilon}_{rz} = -\frac{r}{2}\frac{d\dot{\varepsilon}}{dz} \tag{5.1-21}
$$

Shearing is hence maximum at the fiber surface:

$$
|\dot{\varepsilon}_{rz}| < \dot{\Gamma}(z) = \frac{R}{2}\frac{d\dot{\varepsilon}}{dz} \tag{5.1-22}
$$

Assuming that the previous solution is close to the exact solution, we can calculate the shear component and compare it to the rate of elongation:

$$
\frac{\dot{\Gamma}(z)}{\dot{\varepsilon}(z)} = \frac{R_0 F}{6\eta Q}\, e^{-\frac{F}{6\eta Q}z} = \frac{R(z)}{2L}\ln DDR \tag{5.1-23}
$$

For the two examples, we obtain:

- for water drawing: $0.018 < \dot{\Gamma}/\dot{\varepsilon} < 0.057$
- for air drawing: $10^{-4} < \dot{\Gamma}/\dot{\varepsilon} < 1.05 \cdot 10^{-3}$

In both cases, the shear components are effectively negligible.

Gravitational forces

Considering the small values for the pulling force calculated in the previous examples, the weight of the fiber, even though very small, could be of importance. The pulling force due to gravity is maximum for the entire polymeric filament, i.e.:

$$F_{g\,\text{MAX}} = \varrho g \int_0^L S(z)\, dz \qquad (5.1\text{-}24)$$

Using the following approximate solution for the function $S(z)$ derived from Equation (5.1-17):

$$F_{g\,\text{MAX}} = \varrho g S_0 L \frac{1}{\ln DDR}\left[1 - \frac{1}{DDR}\right] \qquad (5.1\text{-}25)$$

and we obtain:

- For water drawing, $F_{g\,\text{MAX}} = 0.3 \cdot 10^{-4}\,\text{N}$, with $\varrho = 1000\,\text{kg/m}^3$. The weight of the fiber is effectively negligible compared to the drawing force.
- For air drawing, $F_{g\,\text{MAX}} = 1.23 \cdot 10^{-3}\,\text{N}$ and the weight of the fiber is five times greater than the drawing force. It is therefore not possible to neglect the gravitational force in the analysis.

Inertia forces

We can write the following expression for a local Reynolds number:

$$\text{Re} = \frac{\varrho w\left(\dfrac{dw}{dz}\right)}{\eta\left(\dfrac{d\dot{\varepsilon}}{dz}\right)} = \frac{\varrho V_0 L}{\eta \ln DDR} \exp\left(\frac{z}{L}\ln DDR\right) \qquad (5.1\text{-}26)$$

We get:

- For the water drawing case, the Reynolds number varies from 0.022 at the die exit to 0.22 at the entrance of the water bath. Inertia forces can, therefore, be neglected.
- For air drawing, the Reynolds number is 0.125 at the die exit and 15 at the pulling device. Inertia forces are no longer negligible.

The isothermal Newtonian analysis presented above is applicable for drawing in a water bath. However, when the spinning distance in air is large, gravitational and inertia forces have to be included, in conjunction with a temperature-dependent viscosity.

5.1-4 Isothermal melt spinning of a viscoelastic fluid

The viscoelastic nature of polymer melts plays a very important role in melt spinning (see Problem 6.1-A). The following analysis is based on the work of DENN, PETRIE and AVENAS (1975), and DEMAY (1983) for a Maxwell fluid. The constitutive equation is (see Chapter 6):

$$\boldsymbol{\sigma}' + \theta \frac{\delta}{\delta t} \boldsymbol{\sigma}' = 2\eta\dot{\boldsymbol{\varepsilon}} \qquad (5.1\text{-}27)$$

where σ' is the stress tensor at the arbitrary pressure p', $\delta\sigma'/\delta t$ the contravariant (or upper convected) derivative of the stress tensor, and θ a relaxation time.

a) Equations

Neglecting the shear components compared to the elongational ones, the rate-of-deformation tensor is expressed by:

$$[\dot{\varepsilon}] = \begin{bmatrix} -\frac{1}{2}\frac{dw}{dz} & 0 & 0 \\ 0 & -\frac{1}{2}\frac{dw}{dz} & 0 \\ 0 & 0 & \frac{dw}{dz} \end{bmatrix} \qquad (5.1\text{-}28)$$

and the stress tensor is:

$$[\boldsymbol{\sigma}'] = \begin{bmatrix} -p' + \sigma'_{rr} & 0 & 0 \\ 0 & -p' + \sigma'_{\theta\theta} & 0 \\ 0 & 0 & -p' + \sigma'_{zz} \end{bmatrix} \qquad (5.1\text{-}29)$$

Considering only the stresses in the r and z-directions (σ'_{rr} and σ'_{zz}), the Maxwell model for steady state is:

$$\sigma'_{rr} + \theta\left[w\frac{\partial\sigma'_{rr}}{\partial z} + \frac{dw}{dz}\sigma'_{rr}\right] = -\eta\frac{dw}{dz} \qquad (5.1\text{-}30)$$

$$\sigma'_{zz} + \theta\left[w\frac{\partial\sigma'_{zz}}{\partial z} - 2\frac{dw}{dz}\sigma'_{zz}\right] = 2\eta\frac{dw}{dz} \qquad (5.1\text{-}31)$$

Since there is no force acting on the outside surface of the filament (the drag force exerted by ambient air is clearly negligible), we obtain:

$$\sigma_{rr} = -p' + \sigma'_{rr} = 0 \quad \text{or} \quad p' = \sigma'_{rr} \qquad (5.1\text{-}32)$$

implying that:

$$\sigma_{zz} = \sigma'_{zz} - \sigma'_{rr} \qquad (5.1\text{-}33)$$

Substracting Equation (5.1-30) from Equation (5.1-31), we obtain:

$$\sigma_{zz} + \theta\left[w\frac{\partial\sigma_{zz}}{\partial z} + \frac{dw}{dz}\sigma_{zz} - 3\frac{dw}{dz}\sigma'_{zz}\right] = 3\eta\frac{dw}{dz} \qquad (5.1\text{-}34)$$

The other equations are identical to those developed for a Newtonian fluid:
- constant flow rate: $Q = S_0 V_0 = S(z)\, w(z)$
- force balance: $F = \sigma_{zz}(z)\, S(z)$

Viscoelastic melt spinning can be studied by solving a set of five Equations (5.1-31), (5.1-33), (5.1-34), (5.1-13), and (5.1-14) with five unknowns S, w, σ'_{zz}, σ_{zz}, σ'_{rr}. Without lost of generality, for incompressible material, we can set σ'_{rr} equal to zero and ignore Equation (5.1-33). It is interesting to solve the set of equations using dimensionless variables.

b) Dimensionless equations

We define the following dimensionless variables:

$$V = \frac{w}{V_0}; \quad A = \frac{S}{S_0}; \quad \zeta = \frac{z}{L};$$

$$\Phi = \sigma_{zz}\frac{S_0}{F}; \quad \psi = \sigma'_{zz}\frac{S_0}{F} \tag{5.1-35}$$

and we introduce three dimensionless numbers:
- the draw down ratio: $DDR = V_R/V_0$
- the Deborah number: $De = \theta V_0/L$ \hfill (5.1-36)

(The Deborah number is a criterion of the importance of viscoelasticity on the flow properties of the system. It is discussed in Section 6.3).
- a dimensionless force:

$$\frac{1}{E} = \frac{FL}{\eta S_0 V_0} \tag{5.1-37}$$

Equations (5.1-31), (5.1-34), (5.1-14) and (5.1-13) can be rewritten in terms of the dimensionless variables as:

$$\psi + De\left[V\frac{d\psi}{d\zeta} - 2\psi\frac{dV}{d\zeta}\right] = 2E\frac{dV}{d\zeta} \tag{5.1-38}$$

$$\Phi + De\left[2V\frac{d\Phi}{d\zeta} - 3\psi\frac{dV}{d\zeta}\right] = 3E\frac{dV}{d\zeta} \tag{5.1-39}$$

$$AV = 1 \tag{5.1-40}$$

$$A\Phi = 1 \tag{5.1-41}$$

The boundary conditions become:

$$A(0) = 1; \quad V(0) = 1; \quad V(1) = DDR \tag{5.1-42}$$

These boundary conditions are not sufficient to solve the system of partial differential equations. It is necessary to define an extra boundary condition for the stresses, for example, the value of σ'_{zz} at the exit of the die (beginning of the fiber drawing). This extra boundary condition is not easy to determine because of the combined effects of the die geometry, extrudate swell, and operating conditions. We choose here (see DEMAY, 1983) the average value of the stress at the end of a very long capillary (see Problem 6.1-A), i.e.

$$\sigma'_{zz}(0) = 16\eta\theta\left(\frac{V_0}{R_0}\right)^2 \tag{5.1-43}$$

where R_0 is the capillary radius. In dimensionless form:

$$\psi(0) = 16\,\mathrm{De}\,E\left(\frac{L}{R_0}\right)^2 \tag{5.1-44}$$

Other authors have chosen (see DENN, PETRIE and AVENAS, 1975) a zero value of $\sigma'_{zz}(0)$ considering that the stress at the end of the capillary has been relaxed during swelling.

c) Solution

Equations (5.1-40) and (5.1-41) implies that $V = \Phi$, and combining Equations (5.1-38) and (5.1-39) we obtain:

$$\frac{dV}{d\zeta} = \frac{V}{3(E + \mathrm{De}\,\psi) - 2\,\mathrm{De}\,V} \tag{5.1-45}$$

$$\frac{d\psi}{d\zeta} = \frac{2\dfrac{dV}{d\zeta}\,(E + \mathrm{De}\,\psi) - \psi}{\mathrm{De}\,V} \tag{5.1-46}$$

The solution is obtained by selecting an arbitrary value for E (therefore $\psi(0)$), then numerically integrating Equations (5.1-45) and (5.1-46) along the drawing path to calculate $V(\zeta)$. If the calculated value of V at $\zeta = 1$ is different from DDR, E is modified and the process is repeated until convergence is reached. Finally, the velocity along the drawing path and the force are computed.

d) Results

Figure 5.1-4 shows axial velocity profiles along the spinning line for two different Deborah numbers. As elasticity increases, the velocity profile becomes a linear function of the axial distance between the die and the water bath.

Figure 5.1-5 shows that the drawing force increases markedly with the Deborah number (by a factor of 10 to 100).

For a draw ratio above a critical value, at a given Deborah number, it is not possible to solve Equations (5.1-45) and (5.1-46) with the boundary conditions (5.1-42) and (5.1-44). This shows the existence of an unattainable zone, which physically corresponds to conditions under which drawing of the polymer filament is impossible. The result is presented in Figure 5.1-6 for a ratio $L/R_0 = 50$.

The results of Figures 5.1-5 and 5.1-6 show clearly the importance of polymer elasticity on the fiber spinning process.

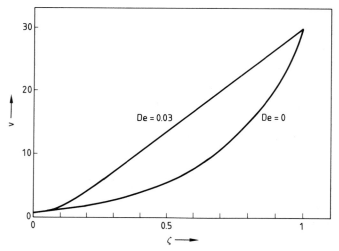

Figure 5.1-4 Velocity profiles along the spinning line for two different Deborah numbers
$DDR = 30$, $L/R_0 = 50$

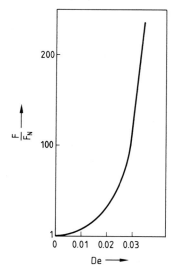

Figure 5.1-5 Ratio of the drawing forces as a function of the Deborah number – F_N is the
corresponding Newtonian drawing force

$$F_N = \frac{3\eta Q}{L} \ln DDR, \quad DDR = 30, \quad \frac{L}{R_0} = 50$$

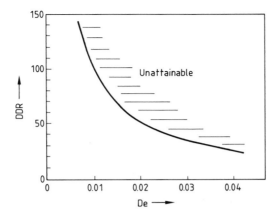

Figure 5.1-6   unattainable zone in fiber spinning
$L/R_0 = 50$

5.1-5 Modelling of the air drawing process

For the modelling of the air drawing process, we must take into account heat transfer, gravitational and inertia forces and air friction. We assume a Newtonian behaviour. Further analyses can be found in KASE   and MATSUO (1965), LAMONTE and HAN (1972), PRASTARO and PARRINI (1975), and DEMAY and AGASSANT (1982).

a) Dynamic equilibrium equation

Following ZIABICKI and KEDZIERSKA (1961) and MATOVICH and PEARSON (1969), we make the following transformations of the equations of motion. Including gravitational and inertia forces, the z-component of the equation of motion is:

$$\frac{\partial \sigma_{zz}}{\partial z} + \frac{1}{r}\frac{\partial}{\partial r}\left(r\sigma_{rz}\right) = \varrho w \frac{dw}{dz} - \varrho g \qquad (5.1\text{-}47)$$

If we integrate this equation over the cross section of the fiber, we obtain a force balance over a length dz of the fiber.

$$2\pi \int_0^R \frac{\partial \sigma_{zz}}{\partial z} r\, dr + 2\pi R \sigma_{rz}(R) = \varrho\left(w\frac{dw}{dz} - g\right)\pi R^2 \qquad (5.1\text{-}48)$$

on the other hand:

$$\frac{d}{dz}\int_0^R \sigma_{zz} r\, dr = \frac{dR}{dz} R\sigma_{zz}(R) + \int_0^R r\frac{\partial \sigma_{zz}}{\partial z}\, dr \qquad (5.1\text{-}49)$$

and the pulling force is:

$$F = 2\pi \int_0^R \sigma_{zz} r \, dr \qquad (5.1\text{-}50)$$

and Equation (5.1-48) becomes:

$$\frac{dF}{dz} + 2\pi R \left[\sigma_{rz}(R) - \frac{dR}{dz} \sigma_{zz}(R) \right] = \varrho \left(w \frac{dw}{dz} - g \right) \pi R^2 \qquad (5.1\text{-}51)$$

b) Force balance at the filament surface

Defining by τ the friction force per unit surface exerted by the surrounding air, a stress balance on the fiber surface yields:

$$\sum_{ij} \sigma_{ij}(R) n_j = \tau t_j \qquad (5.1\text{-}52)$$

where n_j and t_j are respectively the components of the unit normal vector $(\cos \alpha,$ $0, \sin \alpha)$ and of the unit tangential vector $(\sin \alpha, 0, -\cos \alpha)$, as illustrated in Figure 5.1-7.

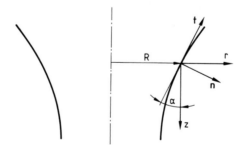

Figure 5.1-7   Stress balance at the fiber surface

Equation (5.1-52) can be written as:

$$\begin{bmatrix} \sigma_{rr}(R) & 0 & \sigma_{rz}(R) \\ 0 & \sigma_{\theta\theta}(R) & 0 \\ \sigma_{rz}(R) & 0 & \sigma_{zz}(R) \end{bmatrix} \begin{bmatrix} \cos \alpha \\ 0 \\ \sin \alpha \end{bmatrix} = \tau \begin{bmatrix} \sin \alpha \\ 0 \\ -\cos \alpha \end{bmatrix} \qquad (5.1\text{-}53)$$

and the two following equations are obtained:

$$\sigma_{rr}(R) \cos \alpha + \sigma_{rz}(R) \sin \alpha = \tau \sin \alpha \qquad (5.1\text{-}54)$$

$$\sigma_{rz}(R) \cos \alpha + \sigma_{zz}(R) \sin \alpha = -\tau \cos \alpha \qquad (5.1\text{-}55)$$

Dividing both sides of Equation (5.1-55) by $\cos\alpha$ and replacing $\tan\alpha$ by $-dR/dz$ (see Figure 5.1-7) we obtain:

$$\sigma_{rz}(R) - \sigma_{zz}(R)\frac{dR}{dz} = -\tau \tag{5.1-56}$$

and we can rewrite Equation (5.1-51) as:

$$\frac{dF}{dz} = \varrho\left(w\frac{dw}{dz} - g\right)\pi R^2 + 2\pi R\tau \tag{5.1-57}$$

This equation expresses the variation of the drawing force as the sum of the inertia, gravitational and air drag forces. The drag force exerted by air was determined experimentally by MATSUI (1976) and GOULD and SMITH (1980). The results are reported in terms of a drag coefficient defined locally by:

$$\tau = \frac{1}{2}C_D\varrho_{\text{air}}w^2 \tag{5.1-58}$$

where C_D is a function of the Reynolds number in the air flow near the fiber, given by:

$$C_D = a\,\mathrm{Re}^{-b} \tag{5.1-59}$$

GOULD and SMITH (1980) has proposed the following values for the two empirical coefficients:

$$a = 0.41, \quad b = 0.61$$

c) Newtonian hypothesis

If now we assume a Newtonian behaviour in a purely elongational flow (see Section 5.1-3), we can write locally for any position z Equations (5.1-7) and (5.1-13) as follows:

$$\sigma_{zz} = 3\eta(T)\frac{dw}{dz} \quad \text{and} \quad F(z) = S(z)\sigma_{zz} = \frac{Q}{\varrho(T)w}\sigma_{zz}$$

By taking the derivative of the force with respect to z and substituting in Equation (5.1-51) we obtain a differential equation of the velocity $w(z)$ and of the temperature $T(z)$:

$$3\nu(T)\left[\frac{d^2w}{dz^2} - \frac{1}{w}\left(\frac{dw}{dz}\right)^2\right] + 3\frac{d\nu}{dT}\frac{dT}{dz}\frac{dw}{dz} = w\frac{dw}{dz} - g + \frac{2\tau}{\varrho(T)R} \tag{5.1-60}$$

where $\nu(T)$ is the kinematic viscosity of the polymer.

Comments

This equation was solved analytically by TROUTON in 1906 for isothermal conditions taking into account only the gravitational and viscous forces.

d) Heat transfer equations

The energy balance over a cross-section of the fiber (illustrated in Figure 5.1-3) is written in terms of the average temperature as (see Chapter 2.2):

$$\varrho c_p w S \frac{d\bar{T}}{dz} = -2\pi R h(\bar{T} - T_a) + \dot{W} \qquad (5.1\text{-}61)$$

where

R :   radius of the fiber at the position z;
h :   heat transfer coefficient;
T_a :   ambient temperature;
$\dot{W}$:   rate of energy generated by viscous dissipation per unit volume.

The heat transfer coefficient is obtained by combining the forced convection coefficient, h_c, and the radiation coefficient, h_r.

Forced convection

The problem is more complicated than the approach presented in Appendix D because of the variable section of the fiber and the perpendicular stream of air coming from a blower.

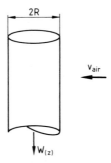

Figure 5.1-8   Heat transfer by forced convection to a cylindrical fiber

Relatively to the fiber, the flow of air may be divided into two components:

– a longitudinal component, $w(z)$, with the corresponding heat transfer correlation for forced convection (Appendix D.2, Figure D.2-2):

$$\mathrm{Nu} = 0.42 \, (\mathrm{Re})^{1/3} \qquad (5.1\text{-}62)$$

with:

$$\mathrm{Nu} = \frac{2h_c R}{k_{\mathrm{air}}}, \quad \text{Nusselt number for longitudinal air flow,}$$

$$\mathrm{Re} = \frac{2wR}{\nu_{\mathrm{air}}}, \quad \text{Reynolds number for longitudinal air flow,}$$

where ν_{air} is the kinematic viscosity of air. Therefore:

$$h_c = 0.21 \, (\text{Re})^{1/3} \, \frac{k_{\text{air}}}{R}$$

— a transverse component V_{air}: the effect of this component was considered by KASE and MATSUO (1965) by introducing an experimentally determined correction factor. The proposed correlation is:

$$h_c = 0.21 (\text{Re})^{1/3} \, \frac{k_{\text{air}}}{R} \left[1 + \left(\frac{8 V_{\text{air}}}{w} \right)^2 \right]^{1/6} \tag{5.1-63}$$

Therefore, the forced convection coefficient varies from the exit of the die to the wind-up system. If we consider the numerical example presented previously for air drawing, with an air velocity $V_{\text{air}} = 2 \, \text{m/s}$, we obtain:

$$39.5 < h_c < 836 \, \text{W/m}^2 \cdot {}^\circ\text{C}$$

from the beginning to the end of the drawing path.

Radiative heat transfer coefficient

As shown in Appendix D.3, the heat transfer coefficient for radiation is defined by:

$$h_r = \sigma \varepsilon \frac{\bar{T}^4 - T_a{}^4}{\bar{T} - T_a} \tag{5.1-64}$$

where σ is the Boltzman constant, equal to $5.66 \, 10^{-8} \, \text{J/m}^2 \cdot \text{K}^4$ and ε is the polymer emissivity (estimated to be equal to $= 0.6$).

For a polyester fiber coming out of a die at 285° C we find:

$$h_r = 11.5 \, \text{W/m}^2 \cdot {}^\circ\text{C}$$

at the die exit (for ambient air at 20° C). On the other hand, at the crystallization point (110° C for polyester under standard fiber spinning cooling conditions),

$$h_r = 5.35 \, \text{W/m}^2 \cdot {}^\circ\text{C}$$

If the radiation heat transfer coefficient is not negligible compared to the forced convection coefficient at the die exit, it will become negligible very rapidly along the spinning line.

Comments

The glass fiber process is approximately described by the equations for non-isothermal Newtonian spinning. However, heat transfer by radiation is more important than heat transfer by convection because of the very high temperature of molten glass (higher than 1000° C).

Viscous dissipation rate during drawing

From the definition:

$$\dot{W} = \pi R(z)^2 \left[2\eta \sum_{ij} \dot{\varepsilon}_{ij}^2 \right] \tag{5.1-65}$$

Neglecting the shear components compared to the elongational ones in the rate-of-deformation tensor, the rate of energy dissipated by viscous forces is simply expressed by:

$$\dot{W} = 3\eta\pi R^2 \dot{\varepsilon}^2 \tag{5.1-66}$$

This term can be estimated and compared to the convection term in Equation (5.1-61) by assuming that the analytical solution obtained for water drawing is valid (obviously, this is a drastic assumption). We obtain:

$$\frac{\text{viscous dissipation rate}}{\text{heat transferred convection}} \cong \frac{3\eta R \dot{\varepsilon}^2}{2h_c(\bar{T} - T_a)} \tag{5.1-67}$$

For the conditions specified in the numerical example presented above, this ratio varies from $8.37 \cdot 10^{-6}$ at the die exit to a value of $1.56 \cdot 10^{-3}$ at the end of the drawing path. Although approximate, these calculations show how negligible is the contribution of the viscous dissipation term to the drawing process. The heat transfer equation reduces then to:

$$\varrho c_p w \frac{d\bar{T}}{dz} = -\frac{2h_c(\bar{T} - T_a)}{R} \tag{5.1-68}$$

e) Solution for the momentum and heat transfer equations

We have a set of three equations:

Equation (5.1-60):

$$3\nu(\bar{T}) \left[\frac{d^2w}{dz^2} - \frac{1}{w}\left(\frac{dw}{dz}\right)^2 \right] + 3\left(\frac{d\nu}{dT}\right)\left(\frac{d\bar{T}}{dz}\right)\left(\frac{dw}{dz}\right) = w\frac{dw}{dz} - g + \frac{2\tau}{\varrho(\bar{T})R}$$

Equation (5.1-14):

$$Q_0 = \varrho_0 V_0 \pi R_0^2 = \varrho w \pi R^2$$

and Equation (5.1-68):

$$\varrho(\bar{T})c_p(\bar{T}) w \left(\frac{d\bar{T}}{dz}\right) = -2h_c\frac{\bar{T} - T_a}{R}$$

and three unknowns, w, R, and $\bar{T}$. The heat transfer coefficient is given by Equation (5.1-63), i.e.:

$$h_c = 0.21(\text{Re})^{1/3}\frac{k_{\text{air}}}{R}\left[1 + \left(\frac{8V_{\text{air}}}{w}\right)^2\right]^{1/6}$$

The viscosity η is a function of temperature following either an Arrhenius, WLF relationship or an empirical relation determined from experimental data as the one used for specific gravity and heat capacity. The boundary conditions are: the initial velocity at the die exit, V_0, the initial temperature, T_0, and the final velocity, V_R. The equations are numerically solved using an iterative method.

The algorithm for the solution is the following:

- the fiber spinning length is divided into n sections;
- from known values of V_0 and T_0, $\nu(T_0)$ and $c_p(T_0)$ are computed;
- for an arbitrary value of $(dw/dz)_0$ we can solve Equation (5.1-60), then Equation (5.1-68) for the first section;
- knowing w_1, T_1 and $(dw/dz)_1$, we can solve the equations for the second section, and so on until the end of the spinning line is reached;
- we compare the computed value for w_n with the boundary condition, V_R. If they are not identical, the initial guess for $(dw/dz)_0$ is modified and the procedure repeated until convergence is attained.

f) Results

Figure 5.1-9 shows two velocity profiles along the spinning line, in stagnant air and for air blown at a velocity equal to 2 m/s. Obviously, heat transfer or cooling affects considerably the velocity profile. This is even more striking if we compare Figure 5.1-9 with Figure 5.1-4 which reports Newtonian and viscoelastic profiles calculated for isothermal conditions.

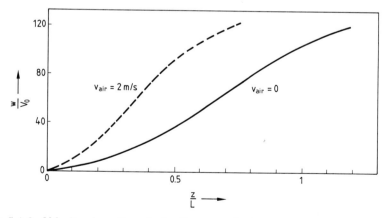

Figure 5.1-9   Velocity along the spinning line – non isothermal conditions

The corresponding temperature profiles are reported in Figure 5.1-10. These two graphs show, as expected, a correspondence between the point of constant velocity and the solidification temperature of the fiber.

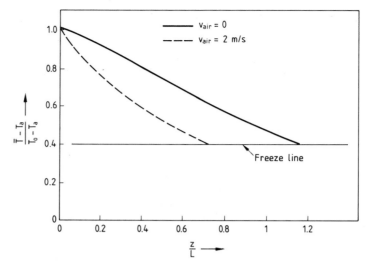

Figure 5.1-10   Temperature profile along the spinning line

Table 5.1-1 reports for the two cases the values of the various forces acting on the fiber from the exit of the die to the solidification point.

Table 5.1-1   Forces acting on the fiber

V_{air}	Initial drawing force	Final drawing force	Inertial force	Gravitational force	Air drag force
m/s	F_0 (mN)	F_1 (mN)	mN	mN	mN
0	0.890	1.96	1.31	0.68	0.43
2	1.60	2.85	1.31	0.36	0.31

We note that these results verify the force balance expressed by Equation (5.1-57), i.e.:

$$F_1 - F_0 = \text{inertial} + \text{friction} - \text{gravitational force.}$$

We also note that the viscous drawing force is an order of magnitude greater than the one calculated for the isothermal case (see Section 5.1-3). The inertial and the viscous drawing forces are of the same order of magnitude. The gravitational and air drag forces are smaller but not negligible.

Comments

Other analyses combining heat transfer and viscoelastic constitutive equations have been proposed in the literature, by FISHER and DENN (1977) and DEMAY (1983) for a Maxwell fluid and GAGON (1980) for the more sophisticated Phan-Thien model (PHAN-THIEN, 1978).

5.2   Biaxial drawing; application to the cast film process

Biaxial drawing or stretching is commonly encountered in many major polymer processing operations, such as blow film production (Chapter 5.3), blow-molding (see Problem 5.2-A), and cast film production, which will be discussed in this section. It is also encountered in pipe extrusion (see Problem 5.2-C).

In every case, biaxial drawing analysis is based on a number of approximations, justified by the small thickness of the stretched material.

5.2-1   Biaxial stretching of a Newtonian liquid

Let us consider the biaxial stretching of a film made of Newtonian material, as illustrated in Figure 5.2-1. The deformation is assumed homogeneous and the thickness uniform. The only non-zero components of the rate-of-deformation tensor are:

$$\dot{\varepsilon}_{xx} = \frac{1}{a}\left(\frac{da}{dt}\right); \quad \dot{\varepsilon}_{yy} = \frac{1}{b}\left(\frac{db}{dt}\right); \quad \dot{\varepsilon}_{zz} = \frac{1}{e}\left(\frac{de}{dt}\right) \tag{5.2-1}$$

with

$$\dot{\varepsilon}_{xx} + \dot{\varepsilon}_{yy} + \dot{\varepsilon}_{zz} = 0 \tag{5.2-2}$$

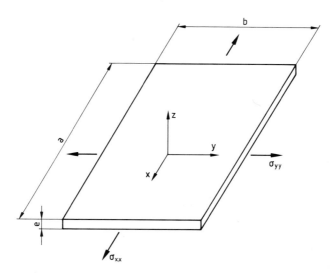

Figure 5.2-1   Biaxial-stretching of a liquid sample

The stress tensor is then for a Newtonian fluid:

$$[\boldsymbol{\sigma}] = \begin{bmatrix} -p + 2\eta\dot{\varepsilon}_{xx} & 0 & 0 \\ 0 & -p + 2\eta\dot{\varepsilon}_{yy} & 0 \\ 0 & 0 & -p + 2\eta\dot{\varepsilon}_{zz} \end{bmatrix} \tag{5.2-3}$$

As the sample is very thin, we may assume that the stress σ_{zz} is constant in the thickness direction (thin shell approximation). Since, at the sample surface, σ_{zz} is equal to the atmospheric pressure (arbitrarily set equal to zero), we may assume then that it is zero everywhere. Therefore, the pressure inside the sample is given by:

$$p = 2\eta \dot{\varepsilon}_{zz} \qquad (5.2\text{-}4)$$

and consequently:

$$[\boldsymbol{\sigma}] = \begin{bmatrix} 2\eta(\dot{\varepsilon}_{xx} - \dot{\varepsilon}_{zz}) & 0 & 0 \\ 0 & 2\eta(\dot{\varepsilon}_{yy} - \dot{\varepsilon}_{zz}) & 0 \\ 0 & 0 & 0 \end{bmatrix} \qquad (5.2\text{-}5)$$

From the continuity Equation (5.2-2), we may write:

$$\begin{cases} \sigma_{xx} = 2\eta(2\dot{\varepsilon}_{xx} + \dot{\varepsilon}_{yy}) \\ \sigma_{yy} = 2\eta(\dot{\varepsilon}_{xx} + 2\dot{\varepsilon}_{yy}) \end{cases} \qquad (5.2\text{-}6)$$

This result reduces for uniform biaxial stretching to:

$$\sigma_{xx} = \sigma_{yy} = 6\eta\, \dot{\varepsilon}_{xx}(= 6\eta\, \dot{\varepsilon}_{yy}) \qquad (5.2\text{-}7)$$

From this result, we define the biaxial elongational viscosity by:

$$\eta_{be} = 6\eta \qquad (5.2\text{-}8)$$

If stretching is not uniform, we can obtain the biaxial longitudinal viscosity by adding the two components of the stress tensor in Equation (5.2-7):

$$\sigma_{xx} + \sigma_{yy} = 6\eta(\dot{\varepsilon}_{xx} + \dot{\varepsilon}_{yy}) \qquad (5.2\text{-}9)$$

The symmetry of biaxial stretching with respect to x and y is now much better perceived.

5.2-2 Extrusion of cast film

The castfilm process has been analyzed by PEARSON (1966), SERGENT (1977) and COTTO (1984).

In the cast film production, a film of molten polymer is extruded from a slit die, stretched in ambient air, and then cooled either on a roller surface (a) or in a water bath (b), as illustrated in Figure 5.2-2.

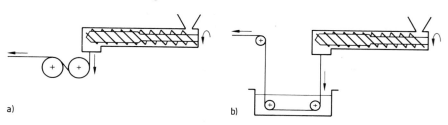

Figure 5.2-2   Cast film extrusion

The distance between the die exit and the contact with the roller or the water is kept to a minimum (a few centimeters) in order to avoid as much as possible lateral shrinkage (neck-in) observed in film stretching.

In the following paragraphs, we will analyze the change of thickness and the lateral shrinkage of a film when stretched vertically in air.

a) Velocity field

It is impossible to solve analytically the motion of a stretched flat film without making further assumptions:

–   We assume that the longitudinal velocity component u depends only on x. This means that any horizontal line traced on the film surface at a certain level will remain horizontal throughout ulterior stretching. This has been experimentally verified.

–   We also assume that the transverse velocity component v does not depend on z and that the component w is independent of y. This implies that the film width and thickness are only dependent on x and that its cross section remains rectangular throughout stretching. This is shown in Figure 5.2-3.

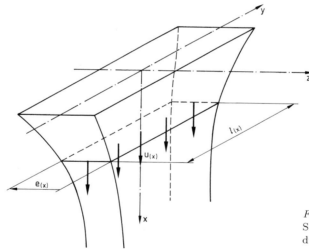

Figure 5.2-3
Sheet deformation
during stretching

This hypothesis is very restrictive; it is observed generally that the film thickness does not remain uniform throughout the width during stretching.

In summary we suppose that the velocity field is of the form:

$$u(x) \qquad v(x,y) \qquad w(x,z)$$

and the rate-of-deformation tensor is given by:

$$[\dot{\varepsilon}] = \begin{bmatrix} \dfrac{du}{dx} & \dfrac{1}{2}\left(\dfrac{\partial v}{\partial x}\right) & \dfrac{1}{2}\left(\dfrac{\partial w}{\partial x}\right) \\[2mm] \dfrac{1}{2}\left(\dfrac{\partial v}{\partial x}\right) & \dfrac{\partial v}{\partial y} & 0 \\[2mm] \dfrac{1}{2}\left(\dfrac{\partial w}{\partial x}\right) & 0 & \dfrac{\partial w}{\partial z} \end{bmatrix} \qquad (5.2\text{-}10)$$

The diagonal terms in $[\dot{\varepsilon}]$ are related through the equation of continuity, i.e.:

$$\frac{du}{dx} + \frac{\partial v}{\partial y} + \frac{\partial w}{\partial z} = 0 \tag{5.2-11}$$

which can be written as:

$$\frac{du}{dx} + \frac{1}{l}\left(\frac{dl}{dt}\right) + \frac{1}{e}\left(\frac{de}{dt}\right) = \frac{du}{dx} + \frac{u}{l}\left(\frac{dl}{dx}\right) + \frac{u}{e}\left(\frac{de}{dx}\right) = 0 \tag{5.2-12}$$

b) Stress boundary conditions

The only forces applied at the film surface are due to the interfacial tension and air friction, which will be systematically neglected in this case. Since the system is symmetrical, we consider only the upper part and the right boundary of the film. The net force per unit surface at the upper surface of the film is zero:

$$T_p = \boldsymbol{\sigma}^P \cdot \boldsymbol{n}^P = 0 \tag{5.2-13}$$

with:

$$\boldsymbol{n}^P = (\sin\alpha,\ 0,\ \cos\alpha), \quad \tan\alpha = -\frac{1}{2}\left(\frac{de}{dx}\right) = -\frac{1}{2}e'$$

and σ^P is the stress tensor on the film upper surface:

$$[\boldsymbol{\sigma}^P] = \begin{bmatrix} \sigma_{xx}^P & \sigma_{xy}^P & \sigma_{xz}^P \\ \sigma_{xy}^P & \sigma_{yy}^P & \sigma_{yz}^P \\ \sigma_{xz}^P & \sigma_{yz}^P & \sigma_{zz}^P \end{bmatrix} \tag{5.2-14}$$

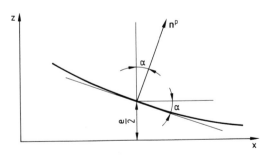

Figure 5.2-4   Boundary conditions at the film surface

The stress boundary conditions (5.2-13) can be written as:

$$\sigma_{xx}^P \sin\alpha + \sigma_{xz}^P \cos\alpha = 0 \tag{5.2-15}$$

$$\sigma_{xy}^P \sin\alpha + \sigma_{yz}^P \cos\alpha = 0 \tag{5.2-16}$$

$$\sigma_{xz}^P \sin\alpha + \sigma_{zz}^P \cos\alpha = 0 \tag{5.2-17}$$

By combining Equations (5.2-15) and (5.2-17) we obtain:

$$\sigma_{xz}{}^P = \frac{1}{2} e' \sigma_{xx}{}^P \tag{5.2-18}$$

$$\sigma_{zz}{}^P = \frac{1}{4} e'^2 \sigma_{xx}{}^P \tag{5.2-19}$$

Defining the stress tensor by $\boldsymbol{\sigma}^{P'}$ and the unit normal vector at the right side surface of the film by $\boldsymbol{n}^{P'} = (\sin\beta,\ \cos\beta,\ 0)$, we obtain for the stress balance:

$$\sigma_{xx}{}^{P'} \sin\beta + \sigma_{xy}{}^{P'} \cos\beta = 0 \tag{5.2-20}$$

$$\sigma_{xy}{}^{P'} \sin\beta + \sigma_{yy}{}^{P'} \cos\beta = 0 \tag{5.2-21}$$

$$\sigma_{xz}{}^{P'} \sin\beta + \sigma_{yz}{}^{P'} \cos\beta = 0 \tag{5.2-22}$$

with

$$\tan\beta = -\frac{1}{2}\frac{dl}{dx} = -\frac{l'}{2}$$

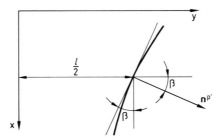

Figure 5.2-5   Boundary conditions on the film side surface

Combining Equations (5.2-20) and (5.2-21), we obtain:

$$\sigma_{xy}{}^{P'} = \frac{1}{2} l' \sigma_{xx}{}^{P'} \tag{5.2-23}$$

$$\sigma_{yy}{}^{P'} = \frac{1}{4} l'^2 \sigma_{xx}{}^{P'} \tag{5.2-24}$$

c) Equation of motion

We neglect the gravitational and inertia forces (this is more reasonable in this case than for fiber spinning since the distance of stretching is short and values for the extrusion and stretching velocities are smaller). The three components of the equation of motion reduce to:

$$\frac{\partial \sigma_{xx}}{\partial x} + \frac{\partial \sigma_{xy}}{\partial y} + \frac{\partial \sigma_{xz}}{\partial z} = 0 \tag{5.2-25}$$

$$\frac{\partial \sigma_{xy}}{\partial x} + \frac{\partial \sigma_{yy}}{\partial y} + \frac{\partial \sigma_{yz}}{\partial z} = 0 \tag{5.2-26}$$

$$\frac{\partial \sigma_{xz}}{\partial x} + \frac{\partial \sigma_{yz}}{\partial y} + \frac{\partial \sigma_{zz}}{\partial z} = 0 \tag{5.2-27}$$

We integrate Equation (5.2-25) over the film cross section:

$$\int_{-l/2}^{l/2} \int_{-e/2}^{e/2} \left[\frac{\partial \sigma_{xx}}{\partial x} + \frac{\partial \sigma_{xy}}{\partial y} + \frac{\partial \sigma_{xz}}{\partial z} \right] dz\, dy = 0 \tag{5.2-28}$$

This result may also be written as:

$$\int_{-l/2}^{l/2} \int_{-e/2}^{e/2} \frac{\partial \sigma_{xx}}{\partial x} dy\, dz + 2 \int_{-e/2}^{e/2} \sigma_{xy}{}^{p'} dz + 2 \int_{-l/2}^{l/2} \sigma_{xz}{}^{p} dy = 0 \tag{5.2-29}$$

The derivative of the expression for the streching force is:

$$\frac{dF}{dx} = \frac{d}{dx} \int_{-l/2}^{l/2} \int_{-e/2}^{e/2} \sigma_{xx} dz\, dy \tag{5.2-30}$$

Taking the derivative of the integral and considering the sign of σ_{xx} with respect to y and z, Equation (5.2-30) can also be written as:

$$\frac{dF}{dx} = l' \int_{-e/2}^{e/2} \sigma_{xx}(x, l/2, z)\, dz + e' \int_{-l/2}^{l/2} \sigma_{xx}(x, y, e/2)\, dy$$

$$+ \int_{-l/2}^{l/2} \int_{-e/2}^{e/2} \frac{\partial \sigma_{xx}}{\partial x} (x, y, z)\, dz\, dy \tag{5.2-31}$$

Noticing that

$$\sigma_{xx}(x, l/2, z) = \sigma_{xx}{}^{p'}$$

and that

$$\sigma_{xx}(x, y, e/2) = \sigma_{xx}{}^{p}$$

Equation (5.2-30) can be written as:

$$\frac{dF}{dx} + \int_{-e/2}^{e/2} (2\sigma_{xy}{p'} - \sigma_{xx}{p'}l') \, dz + \int_{-l/2}^{l/2} (2\sigma_{xz}{p} - \sigma_{xx}{p}e') \, dy = 0 \qquad (5.2\text{-}32)$$

and considering the boundary conditions (5.2-18) and (5.2-23), Equation (5.2-32) reduces to:

$$\frac{dF}{dx} = 0 \qquad (5.2\text{-}33)$$

As in the case of fiber spinning, in absence of gravitational, inertia, air friction drag, and surface tension forces, the drawing force is independent of the axial position.

5.2-3 Isothermal Newtonian stretching

a) Stretching stress

The assumption of isothermal conditions is acceptable since the drawing length is small. Considering the expression for the rate-of-deformation tensor, given in Section 5.2-2, the stress tensor is given by the following expression:

$$[\boldsymbol{\sigma}] = \begin{bmatrix} -p + 2\eta\frac{du}{dx} & \eta\frac{\partial v}{\partial x} & \eta\frac{\partial w}{\partial x} \\ \eta\frac{\partial v}{\partial x} & -p + 2\eta\frac{dv}{dy} & 0 \\ \eta\frac{\partial w}{\partial x} & 0 & -p + 2\eta\frac{dw}{dz} \end{bmatrix} \qquad (5.2\text{-}34)$$

As in Section 5.2-1, we may assume that, for a thin film, the stresses are constant over the thickness and equal to the stresses at the surface. Then, from Equation (5.2-19), we obtain:

$$\sigma_{zz} = \frac{1}{4} e'^2 \sigma_{xx} \qquad (5.2\text{-}35)$$

and if we further assume that the stress component σ_{xx} at any value of x is constant over the film width, we can write σ_{yy} with the help of Equation (5.2-24) as:

$$\sigma_{yy} = \frac{1}{4} l'^2 \sigma_{xx} \qquad (5.2\text{-}36)$$

On the other hand, the hydrostatic pressure is defined by:

$$p = -\frac{1}{3} [\sigma_{xx} + \sigma_{yy} + \sigma_{zz}] \qquad (5.2\text{-}37)$$

or from Equations (5.2-35) and (5.2-36):

$$p = -\frac{\sigma_{xx}}{3} \left[1 + \frac{l'^2}{4} + \frac{e'^2}{4} \right] \cong -\frac{\sigma_{xx}}{3} \left[1 + \frac{l'^2}{4} \right] \qquad (5.2\text{-}38)$$

The approximation on the right hand side follows from the assumption of a thin film, i.e. $e' \ll l'$.

From Equation (5.2-34):

$$\sigma_{yy} = -p + 2\eta \frac{\partial v}{\partial y} \tag{3.2-39}$$

In the one hand, we can write:

$$\frac{\partial v}{\partial y} = \frac{1}{l}\frac{dl}{dt} = \frac{udl}{ldx} \tag{3.2-40}$$

On the other hand, combining Equations (5.2-38, 39 and 40), we obtain:

$$\sigma_{yy} = \frac{\sigma_{xx}}{3}\left[1 + \frac{l'^2}{4}\right] + 2\eta \frac{u}{l} l' \tag{5.2-41}$$

Eliminating the yy-component given by Equation (5.2-36), the final expression for the xx-component of the stress tensor is:

$$\sigma_{xx} = \frac{12\eta u l'}{l(l'^2 - 2)} \tag{5.2-42}$$

This is in agreement with the assumption made above that σ_{xx} depends only on x.

b) Sheet stretching equations

We have the following set of equations:

- from Equation (5.2-33) the drawing force is constant and the component σ_{xx} is independent of y and z:

$$F = \sigma_{xx}el \tag{5.2-43}$$

- the volumetric flow rate is constant given by:

$$Q = uel \tag{5.2-44}$$

- the rheology of the polymer imposes a relationship between the stress and the lateral shrinkage of the film, given by Equation (5.2-42):

$$\sigma_{xx} = \frac{12\eta u l'}{l(l'^2 - 2)}$$

This set of equations has to be solved with the following boundary conditions:
- the die exit conditions: V_0, e_0, l_0;
- the film winding velocity, V_1.

However, these equations and boundary conditions are not sufficient to obtain solutions for F, $l(x)$, $e(x)$ and $u(x)$, but the film shrinkage can be calculated by assuming that the drawing force is known.

Remarks

In fact, it is possible to solve a set of four equations [Eqs. (5.2-42, 43, 44) and Eq. (5.2-53) obtained below] and four unknowns, F, $l(x)$, $e(x)$ and $u(x)$, using an iterative method as in the case of fiber spinning. We propose here a solution for the shrinkage ratio, l_1/l_0 in terms of the drawing force.

c) Film shrinkage

Combining Equations (5.2-42, 43, and 44), the drawing force is expressed by:

$$\frac{F}{12\eta Q} = \frac{l'}{l(l'^2 - 2)} \tag{5.2-45}$$

This result is physically meaningful only if $0 > l' > -\sqrt{2}$ anywhere in the polymer film. This means that the final value of the width, l_1, can only take a value included in the range $l_0 > l_1 > l_0 - \sqrt{2}\,X$, where X is the drawing length. It follows that:

– if $X > l_0/\sqrt{2}$, the inequality is verified and l_1 can be very small;
– if $X < l_0/\sqrt{2}$, l_1 has a lower bound of value increasing as X decreases.

Equation (5.2-45) can be analytically integrated to obtain:

$$-4Ax = (Z - Z_0) + \ln\left[\frac{Z - 1}{Z_0 - 1}\right] \tag{5.2-46}$$

where

$$Z = \sqrt{1 + 8A^2 l^2}, \quad Z_0 = \sqrt{1 + 8A^2 l_0{}^2} \tag{5.2-47}$$

and

$$A = \frac{F}{12\eta Q} \tag{5.2-48}$$

Knowing the pulling force, the variation of the film width, l, as a function of the distance x from the die exit can be calculated with the help of Equations (5.2-46, 47, and 48).

d) Velocity field and film thickness

The drawing force, F, can be measured only with the help of a highly sophisticated instrument. It is much simpler to measure the final width of the film, l_1. Then, we obtain A (therefore F), which is the solution of the following equation:

$$\boxed{-4AX = (Z_1 - Z_0) + \ln\left[\frac{Z_1 - 1}{Z_0 - 1}\right]} \tag{5.2-49}$$

with

$$Z_1 = \sqrt{1 + 8A^2 l_1{}^2} \tag{5.2-50}$$

Considering Equation (5.2-34) for the stress tensor, we may write:

$$\sigma_{xx} - \sigma_{zz} = 2\eta\left[u' - \frac{\partial w}{\partial z}\right] \tag{5.2-51}$$

or, using Equation (5.2-12):

$$\sigma_{xx} - \sigma_{zz} = 2\eta\left[u' - \frac{u}{e}e'\right] \tag{5.2-52}$$

Using Equation (5.2-19), this result may also be written as:

$$\sigma_{xx}\left[1 - \frac{e'^2}{4}\right] = 2\eta u\left[\frac{u'}{u} - \frac{e'}{e}\right] \tag{5.2-53}$$

In general $e' \ll 1$ and:

$$\sigma_{xx}\left[1 - \frac{e'^2}{4}\right] \cong \sigma_{xx} = \frac{F}{el} \tag{5.2-54}$$

On the other hand, the continuity Equation (5.2-12) allows us to write:

$$\frac{u'}{u} = -\left[\frac{e'}{e} + \frac{l'}{l}\right] \tag{5.2-55}$$

Therefore, from Equations (5.2-53 and 54) we write:

$$-\frac{F}{2\eta Q} = -6A = \frac{l'}{l} + 2\frac{e'}{e} \tag{5.2-56}$$

Then, the relationship between the shrinkage ratio, l/l_0, and the thinning ratio, e/e_0, at any point is:

$$\ln\frac{l}{l_0} + 2\ln\frac{e}{e_0} = -6Ax \tag{5.2-57}$$

Knowing $l(x)$ from Equation (5.2-46), we obtain $e(x)$ from (5.2-57) and $u(x)$ from a mass balance.

The relationship (5.2-57) is particularly interesting for $x = X$. The neck-in (final lateral shrinkage) and the total thinning at the vertical distance X are related by:

$$\boxed{\ln\frac{l_1}{l_0} + 2\ln\frac{e_1}{e_0} = -6AX} \tag{5.2-58}$$

e) Comments on shrinkage

Experiments have revealed, as predicted by Equation (5.2-45) that the lateral shrinkage value tends towards a limiting value as the stretching rate is increased; any subsequent increase in velocity will result in a thinning of the film. This is true only if the vertical distance, X, is much smaller than $l_0/\sqrt{2}$. For heights close to or larger than $l_0/\sqrt{2}$, the approximations are no longer valid and the final width of the film may go to zero.

f) Example

Knowing the flow rate ($1.1 \cdot 10^{-6}$ m^3/s), the polymer viscosity ($3 \cdot 10^4$ Pa $\cdot$ s) and the film initial width (0.20 m), it is possible to obtain from any of the following three parameters:

the streching rate, V_1/V_0,
the lateral shrinkage, l_1/l_0,
the thinning, e_1/e_0,

the values for the other two and the stretching force.

We note that in either case the film width tends towards a limiting value corresponding to that obtained from Equation (5.2-46), i.e.:

$$\frac{l_1}{l_0} = 1 - \sqrt{2}\,\frac{X}{l_0} \tag{5.2-59}$$

For $X = 43$ mm,

$$\lim\left(\frac{l_1}{l_0}\right) = 0.7 \tag{Fig. 5.2-6}$$

and for $X = 100$ mm,

$$\lim\left(\frac{l_1}{l_0}\right) = 0.3 \tag{Fig. 5.2-7}$$

The stretching force increases with the stretching rate as in the case of fiber spinning; however, the magnitude of the force for film stretching is much larger.

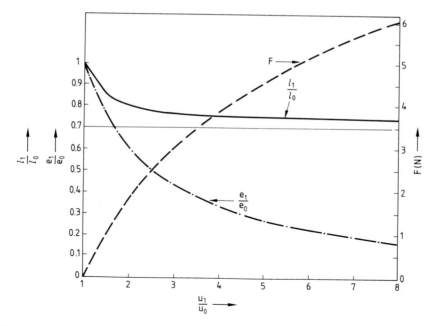

Figure 5.2-6 Calculation of the film stretching: height of stretching, $X = 43$ mm

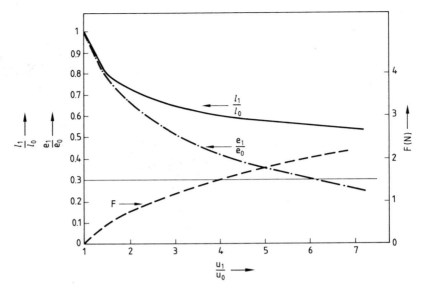

Figure 5.2-7 Calculation of the film stretching: height of stretching, $X = 100$ mm

5.2-4 Conclusion

It is possible to improve the modelling of film stretching, as in the case of fiber spinning, by taking into account heat transfer and viscoelastic effects of the polymer. COTTO et al. (1984) have shown that film cooling is not important on a short distance (a few degrees only). On the other hand, viscoelastic effects are expected to be negligible since the velocities and the stretching rates are much smaller than in the case of fiber spinning.

On the other hand, it is interesting to take into account the film thickness heterogeneities induced by stretching. This phenomenon, currently called "dogbone" defect, has been studied experimentaly by DOBROTH and ERWIN (1986) and analyzed by d'HALEWYN, DEMAY and AGASSANT (1990).

Problems 5.2

5.2-A Blowing of a spherical shell

As a first approach to blow-molding, we consider a liquid spherical shell of radius, r_0, much larger than its shell thickness, e_0. At time $t = 0$, the internal pressure is increased to Δp.

a) Find the expression for the radius, $r(t)$, as a function of time. In terms of geometrical dimensions and Δp, the following quantities must be determined:
  – the stresses acting on the spherical shell,
  – the stretching rate,
  – the relationship between these two quantities for a Newtonian liquid of viscosity equal to η.

b) How long will it take to the spherical bubble to have a radius equal to R? Use the following values:

$$\eta = 10^4 \text{ Pa} \cdot \text{s}, \quad \Delta p = 10^2 \text{ Pa}$$
$$e_0 = 0.1 \text{ mm}, \quad r_0 = 10 \text{ mm}, \quad R = 100 \text{ mm}$$

5.2-B Cast-film process

A Newtonian film of viscosity η is stretched from a velocity of V_0 to V_1. Obviously, it will tend to shrink along its width.

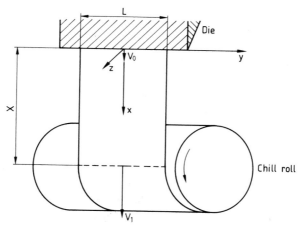

Figure 5.2-8   Cast film process

a) What stress distribution should be applied on the lateral sides in order to keep the film at a constant width, equal to that of the die?
b) Under the conditions determined in a), what are the velocity profile and pulling force?

Use the following values:

$$V_0 = 1 \text{ m/min}, \quad V_1 = 10 \text{ m/min}$$
$$e_0 = 300 \ \mu\text{m}$$
$$X = 100 \text{ mm}, \quad L = 0.50 \text{ m}$$
$$\eta = 10^4 \text{ Pa} \cdot \text{s}$$

5.2-C   Tube extrusion

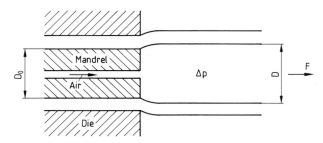

Figure 5.2-9   Tube extrusion

The tube coming out of the die as illustrated in Figure 5.2-9 is pulled with a force F. To avoid its deformation, a pressure difference Δp is applied inside the tube. We wish to make sure that the final diameter D of the tube stays identical to the die diameter D_0.

a)   Show that the results of problem 5.2-B can be directly applied to this flow situation.
b)   Obtain a relationship between F and Δp.
c)   Find the value of F for:

$$\Delta p = 10^4 \text{ Pa}$$
$$D_0 = 400 \text{ mm}$$

5.3   Blown film

5.3-1   Description of the process

The blown film process is illustrated in Figure 5.3-1. The polymer melt is extruded through an annular die of the same type as the one presented in Chapter 3.3 (Figure 3.3-4). The cylindrical polymer tube is inflated by blowing air inside, creating a bubble and drawing with the help of rollers. Under steady-state conditions, the air mass contained between the die and the rollers is maintained constant; the small leak, due to air entrapped between the two sheets of solidified polymers is compensated by injecting air in the die. Cooling is generally achieved by forced convection, blowing air through an air ring situated above the die. In other cases, the film is cooled from inside with the help of a cooling mandrel. Once the film is solidified, its shape does not change; it is then pulled, folded, and wound on

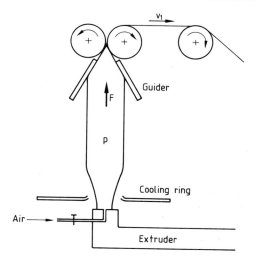

Figure 5.3-1   Schematic view of the blown film process

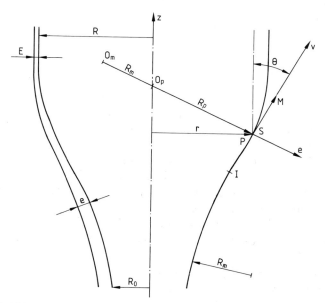

Figure 5.3-2   Bubble geometry

cylindrical cores. Blown film installations are operated in most cases in the vertical position, the film being extruded upward, rarely in the downward direction.

The film, in the molten state, is biaxially stretched by the combined actions of the rollers and the air pressure inside the tube. In the biaxially-oriented zone the linear velocity of the film goes from V_0 to V_1, and the radius from R_0 to R_1, as shown in Figure 5.3-2.

We define:

a draw down ratio,

$$DDR = \frac{V_1}{V_0} \qquad (5.3\text{-}1)$$

and a blowup ratio,

$$BUR = \frac{R_1}{R_0} \qquad (5.3\text{-}2)$$

The blow ratio is usually greater than 1. In some cases (for example, in the downward extrusion of polyamides), it may be smaller than 1.

Here, as in most polymer processing operations, the goal is to obtain a maximum production rate with optimal properties. The properties depend implicitly on the stresses, on the rate of deformation and on the thermal history encountered by the material during biaxial deformation. Therefore, obtaining constant or uniform properties is frequently a difficult task. For most applications, it is desirable to obtain the same mechanical properties (breaking resistance, shrinkage, etc.) in the extrusion as in the circumferential or transverse direction. To obtain such results, it is not sufficient to take the draw ratio equal to the blowup ratio.

The objective of this chapter is to present the basic equations that govern the biaxial stretching of the film.

5.3-2 Film geometry

We assume that the bubble is symmetrical with respect to the z axis and that its thickness is very small. Hence its geometry can be examined as a surface with mechanical properties of a membrane. The surface is defined by the meridional curve $r(z)$, r being the radius from the axis at the distance z (see Figure 5.3-2).

A force balance applied to the membrane requires the knowledge of the shape at every point, completely described by the two main curvature radii and their corresponding orthogonal directions. In the particular cases of surfaces of revolution, the curvature depends only on the shape of the meridian. The principal directions at a point S are the meridian direction M, defined by the tangent to the meridian curve, and the parallel direction P, defined by the tangent to parallel circles (normal to the z-axis).

The center of curvature O_p, corresponding to direction P, is on the z-axis. The radius of curvature is given by:

$$R_p = \frac{r}{\cos\theta} = r\sqrt{1 + r'^2} \qquad (5.3\text{-}3)$$

The curvature following M is that of the meridian. The center of curvature O_m may be inside or outside the film. The radius of curvature is given by:

$$R_m = \frac{-(1 + r'^2)^{3/2}}{r''} \qquad (5.3\text{-}4)$$

Hence R_m is positive when the centers of curvature of the meridian and parallel are on the same side of the surface. The shape of the meridian shows often an inflexion point, I, with R_m being infinite. R_m is negative below and positive above the inflexion point.

In the following sections, we will use a "local" coordinate system M, P, e based on the tangents at the bubble surface, in the M and P directions and normal to the thickness $e(z)$.

5.3-3 Kinematics of the bubble formation

The linear velocity of the film is, in each point, parallel to M and equal to $v(z)$; the volumetric flow rate Q is:

$$Q = 2\pi r(z)\, e(z)\, v(z) \tag{5.3-5}$$

and the rates of deformation in the M, P, and e-directions are:

$$\dot\varepsilon_m = \frac{1}{v}\frac{dv}{dt} = \frac{dv}{dz}\cos\theta \tag{5.3-6}$$

$$\dot\varepsilon_p = \frac{1}{r}\frac{dr}{dt} = \frac{1}{r}\frac{dr}{dz}v\cos\theta \tag{5.3-7}$$

$$\dot\varepsilon_e = \frac{1}{e}\frac{de}{dt} = \frac{1}{e}\frac{de}{dz}v\cos\theta \tag{5.3-8}$$

On the other hand, the continuity equation implies:

$$\dot\varepsilon_m + \dot\varepsilon_p + \dot\varepsilon_e = 0 \tag{5.3-9}$$

We will assume that shearing is negligible so that the rate-of-deformation tensor is therefore a diagonal matrix.

5.3-4 Stresses acting on the bubble

a) Principal stresses

If we assume that the polymer is a purely viscous fluid, the directions of the principal stresses will be the same as the rate of deformation; the rate-of-deformation tensor is then diagonal with respect to the coordinate system $(M$, P, $e)$, and so is the stress tensor. We may say then that there is a meridian principal stress, $\sigma_m(z)$, and a parallel stress, $\sigma_p(z)$. The stress, $\sigma_e(z)$, is negligible since the film is thin (it is of the order of magnitude of the pressure, Δp, inside the bubble (see Problem 5.2-A)).

The inertia and superficial forces are negligible, but the body (gravitational) forces create a problem discussed in the following paragraphs.

b) Force balance in the pulling direction and meridian stresses

Let us take the example of a vertical extrusion. The total pulling force, exerted by the pulling rollers, includes:

- the film weight (solid or liquid material) between the rollers and the die;
- the friction forces of the film on the guiding system;
- the force due to the pressure applied inside the film;
- the force that is responsible of the stretching in the area of biaxial stretching.

The force, measured at position z, decreases from top to bottom. Below the guiding system, the friction forces do not exist anymore. The film weight contribution decreases with decreasing distance from the die to become zero at the die exit.

The force measured at the die level is the pulling force itself, denotes by F_0. It tends to lift the extruder, but it is only of the order of 1 to 10 N (shown below), so it is obvious that direct measurements are impossible to do on industrial installations (such measurements were carried out on the experimental apparatus of HAN and PARK, 1975).

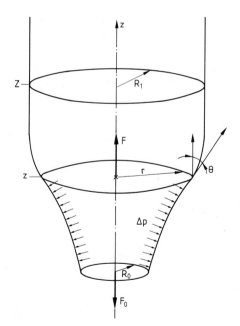

Figure 5.3-3   Vertical force balance at position z

The force F exerted on the film at position z is equal to F_0, the weight of the film, plus the pressure effect on the film, i.e.:

$$F(z) = F_0 + \Delta p \pi (r^2 - R_0{}^2) + \int_0^z \varrho g \, 2\pi r e \frac{dz}{\cos \theta} \qquad (5.3\text{-}10)$$

We see that the pulling force exerted on the film increases with z, partially due to gravity, but mainly because of the internal pressure. Therefore, the internal pressure, which may be believed to be used for blowing only, contributes to the stretching of the film.

The meridian stress σ_m is related to the force by:

$$F(z) = 2\pi r(z) \, e(z) \, \sigma_m(z) \, \cos\theta(z) \tag{5.3-11}$$

This gives the first equilibrium relationship:

$$2\pi r e \sigma_m \, \cos\theta = F_0 + \Delta p\pi(r^2 - R_0^2) + \int_0^z \varrho g \, 2\pi r e \frac{dz}{\cos\theta} \tag{5.3-12}$$

c) Force balance perpendicular to the film

We consider now a small surface element (with dimensions small compared to the sphere radius but large compared to its thickness). We define by S_m and S_p the sides of the surface element that are cut following the meridian and the parallel, as illustrated in Figure 5.3-4, and by e the film thickness. We then write the following expressions for the different forces projected along the normal to the surface, $\boldsymbol{n}$ (see Figure 5.3-5):

a) Force due to the internal pressure Δp: $\Delta p S_m S_p$
b) Force due to the meridian stresses: $2S_p e\sigma_m \sin\alpha_m$
c) Force due to the parallel stresses: $2S_m e\sigma_p \sin\alpha_p$
d) Projection of the element weight on $\boldsymbol{n}$: $\varrho S_m S_p e g \sin\theta$

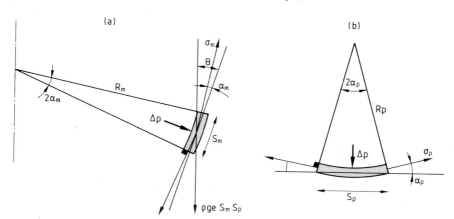

Figure 5.3-4   Stresses acting on a film element
a)   cut view in the meridian plane
b)   cut view in the principal parallel curvature plane

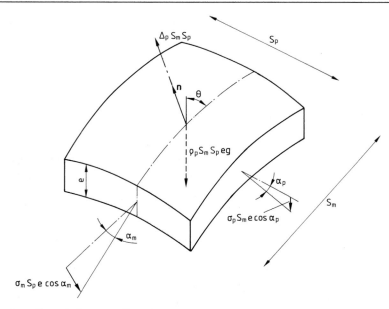

Figure 5.3-5   Force balance on film element

Denoting by R_m and R_p the radius of curvature of the meridian and that of the parallel, we get:

$$\sin\alpha_m \cong \alpha_m \cong \frac{S_m}{2R_m} \quad \text{and} \quad \sin\alpha_p \cong \alpha_p \cong \frac{S_p}{2R_p}$$

A force balance in the *n*-direction, for a small surface element, yields:

$$\Delta p(S_m S_p) = 2 S_p e \sigma_m \left(\frac{S_m}{2R_m}\right) + 2 S_m e \sigma_p \left(\frac{S_p}{2R_p}\right) - \varrho S_m S_p e g \sin\theta \qquad (5.3\text{-}13)$$

which can be written as:

$$\boxed{\frac{\Delta p}{e} = \frac{\sigma_m}{R_m} + \frac{\sigma_p}{R_p} - \varrho g \sin\theta} \qquad (5.3\text{-}14)$$

d) Order of magnitude of the stress components

At the end of the pulling distance, the radius R_m is infinite, the radius R_p is the final radius of the film, R_1, and the angle θ is zero. Hence, the stress is given by:

$$\sigma_p = \Delta p \frac{R_1}{e} \qquad (5.3\text{-}15)$$

For $\Delta p \approx 10^2$ Pa, $R_1 \approx 0.5$ m, $e \approx 0.1$ mm, then $\sigma_p = 5 \cdot 10^5$ Pa. This satisfies the conditions:

$$\sigma_p \gg \sigma_e \approx \Delta p \quad \text{(see Problem 5.2 – A)}$$

5.3-5 The Newtonian approximation

The most difficult point of this analysis is the choice of an appropriate constitutive equation. It is clear that the polymer properties are continually changing along the film line, from a quasi–newtonian behaviour at the exit of the die to a purely elastic behaviour past the freeze line.

The behaviour is in fact viscoelastic from the beginning, with an increasing relaxation time during cooling; then the film becomes partially plastic during solidification. This situation is also complicated by the extensional nature of the flow field and the associated complex extensional or elongational viscosity.

In a first analysis, PEARSON and PETRIE (1970) considered an isothermal Newtonian behaviour. Temperature effects were introduced later by PETRIE (1975).Viscoelasticity was also taken into account by PETRIE (1973) and more recently by SOSKEY and WINTER (1982). HAN and PARK (1975) followed by GUPTA et al. (1982) have analyzed the coupled effects of viscoelasticity and temperature.

We now present the isothermal Newtonian analysis of film blowing. The analysis is qualitative rather than predictive. Biaxial stretching was introduced in Section 5.2-1 and the relationships between stresses and deformations can be written as:

$$\sigma_m = 2\eta(\dot{\varepsilon}_m - \dot{\varepsilon}_e) = 2\eta(2\dot{\varepsilon}_m + \dot{\varepsilon}_p) \tag{5.3-16}$$

$$\sigma_p = 2\eta(\dot{\varepsilon}_p - \dot{\varepsilon}_e) = 2\eta(2\dot{\varepsilon}_p + \dot{\varepsilon}_m) \tag{5.3-17}$$

We have two equations for the stress equilibrium:

Equation (5.3-12):

$$2\pi r e \sigma_m \cos\theta = F_0 + \Delta p \pi (r^2 - R_0{}^2) + \int_0^z \varrho g 2\pi r e \frac{dz}{\cos\theta}$$

and *Equation (5.3-14):*

$$\frac{\Delta p}{e} = \frac{\sigma_m}{R_m} + \frac{\sigma_p}{R_p} - \varrho g \sin\theta$$

The mass balance is given by *Equation (5.3-5):*

$$Q = 2\pi r e v$$

The unknowns for this problem are:

- the geometry and the kinematic of the film r, e, v,
- the stresses σ_m, σ_p and the pulling force F_0,
- the polymer viscosity η, which changes with temperature, therefore with the z-position.

A predictive analysis of the blown film process will require the use of additional equations to take into account the solidification of the film and the temperature-dependent viscosity. Some of these constitutive relations are difficult to define with accuracy and the predictions do not always compare well with the experimental data. For those reasons and to avoid a complex mathematical treatment, we restrict here the analysis of film blowing to the qualitative approach presented by SERGENT (1977) and by PIANA (1984). As we will see, this approach has the advantage of being consistent with the isotherm Newtonian hypothesis presented in the previous section.

The rates of deformation $\dot{\varepsilon}_m$ and $\dot{\varepsilon}_p$ can be determined with the help of periodic tracer marks made on the film surface at the exit of the die (Figure 5.3-6); $r(z)$, $R_p(z)$ and $R_M(z)$ are the results of direct measurement, $e(z)$ is calculated from the flow rate, Q (Equation 5.3-5) and Δp can be measured with the help of a manometer.

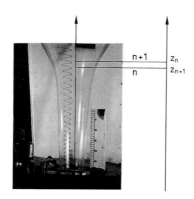

Figure 5.3-6   Film marking technique

Therefore, we obtain a system of 4 equations (5.3-12, 14, 16, 17) and 4 unknowns:

- the stresses $\sigma_m(z)$, $\sigma_p(z)$,
- the pulling force F_0,
- the viscosity $\eta(z)$.

In fact, σ_m, σ_p and η are functions of z, but F_0 is a fixed parameter. It is then by verifying that F_0 is independent of the position z that it is possible to prove the validity of the Newtonian analysis (see Figure 5.3-7).

We note in Figure 5.3-7 that the Newtonian analysis is valid at the exit of the die and up to the zone corresponding more or less to the inflexion point of the

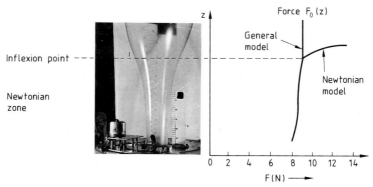

Figure 5.3-7 Variation of the pulling force F_0 with z ($\Delta p = 78$ Pa)

film. Past the inflexion point, it is still possible to calculate the stress components σ_m and σ_p according to the procedure proposed by SERGENT (1977):

– we solve the system of Equations (5.3-12, 14, 16 and 17) in the close neighbourhood of the die where the isothermal Newtonian behaviour is applicable;

– knowing F_0, we apply on the remaining part of the film Equations (5.3-12 and 14) for which the only unknowns are σ_m and σ_p in any position.

Figure 5.3-8, based on the same experiment as used for the results of Figure 5.3-7, shows the deformation rates and stresses. We note that the stresses, σ_m and σ_p, calculated following the last approach, are different from the Newtonian stresses $\sigma_m{}^N$ and $\sigma_p N$ obtained in solving by point the system of Equations (5.3-5, 12, 13, 16, 17) whenever the inflexion point is exceeded.

The stress and the deformation fields can be used to explain and predict orientation and shrinkage effects in film production (PIANA, 1984).

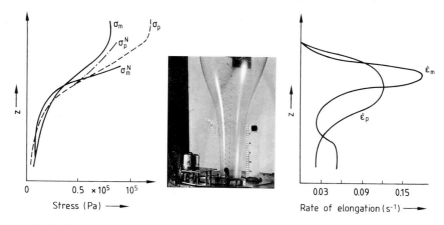

Figure 5.3-8 Measurement of deformation rates and calculation of stresses

In conclusion, the Newtonian analysis and the film marking technique enable us to calculate the pulling force F_0 at the exit of the die; it is therefore possible to calculate the entire stress history in the film, independently of any hypothesis on the material behaviour. This justifies the use of an isothermal Newtonian model in the vicinity of the die exit.

5.3-6 Various bubble shapes

Figure 5.3-7 shows that the Newtonian approximation is only valid near the die exit. Now that we know F_0, it is quite interesting to make use of the five equations presented in Section 5.3-5 to calculate $r(z)$, $e(z)$, $v(z)$, $\sigma_m(z)$ and $\sigma_p(z)$, right at the exit of the die. By neglecting gravity (which is justified near the exit of the die), we obtain a differential equation containing only the radius $r(z)$:

$$
2r^2 \frac{d^2r}{dz^2} \left[F_0 - (R_1^2 - r^2)\,\pi\Delta p\right]
$$

$$
= 6\frac{dr}{dz}\,\eta Q + r\left[1 + \left(\frac{dr}{dz}\right)^2\right]\left[F_0 - (R_1^2 + 3r^2)\,\pi\Delta p\right]
\tag{5.3-18}
$$

This equation can be numerically solved but it doesn't allow us to calculate the shape of the film. First, the viscosity is not known as a function of the position z, and secondly, the boundary conditions in the z-direction are unknown at the level where the polymer ceases to behave as a Newtonian fluid. Nevertheless this equation gives qualitative information on the shape of the film bubble at the die exit.

The solution $r(z) = R_0$ verifies Equation (5.3-18) only if:

$$
F_0 - (R_1^2 + 3R_0{}^2)\,\pi\Delta p = F_0 - \pi\Delta p R_0{}^2(3 + BUR^2) = 0
\tag{5.3-19}
$$

where BUR is the blowup ratio defined by Equation (5.3-2). This result is effectively observed for bubbles of high cylindrical neck. Examination of this equation shows that F_0 is a limit for which the bubble grows larger or shrink at the die level as illustrated in Figure 5.3-9.

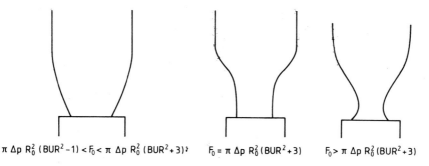

$\pi\,\Delta p\,R_0^2\,(BUR^2-1) < F_0 < \pi\,\Delta p\,R_0^2\,(BUR^2+3)$; $F_0 = \pi\,\Delta p\,R_0^2\,(BUR^2+3)$ $F_0 > \pi\,\Delta p\,R_0^2\,(BUR^2+3)$

Figure 5.3-9 Various shapes of film bubbles

Chapter 6
Viscoelasticity of Polymeric Liquids

6.1　Viscoelasticity of liquids

6.1-1　Phenomena

All the analyses presented so far in the previous chapters are based on the assumption that polymeric liquids obey Newton's law of viscosity or, at most, that they are shear-thinning, inelastic liquids. This assumption is valid in many of the practical applications discussed. However, a number of phenomena encountered in the flow of polymers, directly observed in polymer processing or when measuring rheological properties, cannot be explained from a purely viscous behaviour. These phenomena are first described, and then explained in the following sections.

a)　Extrudate swell

Extrudate swell (wrongly called die swell in the literature) is the increase in diameter of the extrudate as it exits a die, as illustrated in Figure 6.1-1. The diameter of the extrudate in the case of a Newtonian liquid flowing out of a circular capillary is not much different from the diameter of the capillary. At very low Reynolds numbers, the swell is of the order of 12%; it decreases with increasing Reynolds number to become zero at Re $\approx$ 15, and it becomes negative for larger Reynolds numbers.

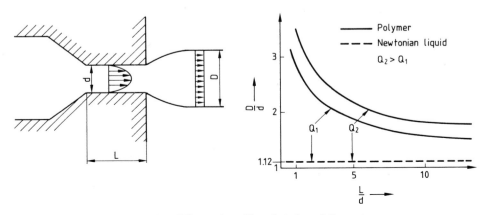

Figure 6.1-1　Extrudate swell – Effects of capillary length and flow rate

The swell observed with polymeric extrudates (melts or solutions) is much larger with a ratio of the extrudate to the die diameter up to 150 or 300%. The swell depends on:

- the length or rather the L/d of the length to the diameter of the die (see Figure 6.1-1);
- the flow rate or the wall shear rate in the die.

This observation suggests that the polymer have a "memory" of the state in which it was in the die reservoir. The fluid memory is better for a short capillary. The extrudate swell for very long capillaries will be explained in Section 6.1-4.

b) Weissenberg effect

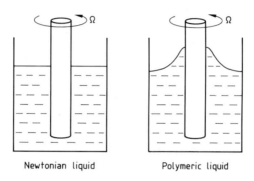

Newtonian liquid   Polymeric liquid

Figure 6.1-2   Weissenberg effect

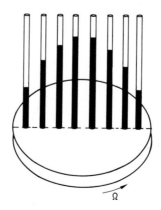

Figure 6.1-3   Pressure generated by shear flow of a polymer between two parallel disks

The Weissenberg effect is described with the help of Figure 6.1-2. Let us consider the tangential flow of a viscous fluid in a Couette geometry (concentric cylinders with one of the cylinders in rotation). In absence of inertial effects, the free surface of the Newtonian fluid remains flat. The free surface of a polymeric liquid would, however rise near the rotating inner cylinder due to extra pressure or normal stresses generated by shear flow. The extra pressure generated by shear flow is better illustrated for the tangential flow of a polymer between two parallel disks, as shown in Figure 6.1-3. The increasing height of the manometric liquid is an indication that the pressure increases with decreasing distance from the disks' axis.

c) Time effects

Let us go back to the Couette flow between two concentric cylinders. The shear rate, $\dot{\gamma}$, in the gap can be calculated from the rotational speed of the inner cylinder, Ω, and geometrical parameters. The shear stress, σ, can be related to the torque required to rotate the inner cylinder. Various relations can be established between the shear stress, σ, and the shear rate, $\dot{\gamma}$, using a Couette viscometer under different transient experiments.

Stress retardation and relaxation

If we apply a sudden and constant velocity, hence a constant shear rate $\dot{\gamma}_0$, and then suddenly stop the rotation of the inner cylinder, we will observe that the shear stress in the case of a purely viscous liquid is a unique function of the shear rate (i.e. $\sigma_0 = f(\dot{\gamma}_0)$). In contrast the response of a polymeric liquid does not follow immediately the applied shear rate. This is a characteristic of viscoelastic behaviour. As illustrated in Figure 6.1-4, the viscoleastic fluid will

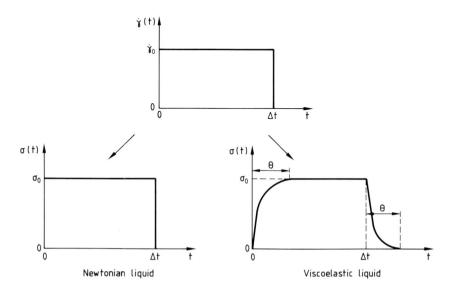

Figure 6.1-4   Stress growth and relaxation

show a retardation or a stress growth phenomenon. If the rotation is suddenly stopped, the stresses do not go back to zero instantaneously as shown for the Newtonian liquid. This is known as the relaxation phenomenon. The retardation time and relaxation time are approximately of the same magnitude and it will be denoted by the symbol θ.

Shear recovery

If instead, we conduct the following experiment: steady-state conditions have been attained and then suddenly, the torque, or the shear stress, is set equal to zero. In absence of inertial effects, the Newtonian liquid will stop immediately with no further deformation as shown by the dashed line of Figure 6.1-5. On the contrary, the viscoelastic polymer will rotate back before coming to a complete stop after a finite time. The phenomenon is known as recoil and the negative deformation is referred to as shear recovery. It is easy to imagine that a purely elastic body in these conditions will exhibit a complete recovery. This is why recoil is also called elastic recovery.

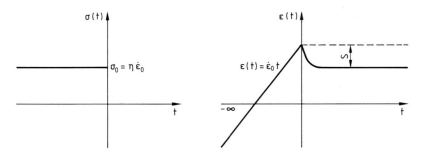

Figure 6.1-5 Shear recovery

Dynamic behaviour

If the inner cylinder of the two concentric cylinder system is submitted to a sinusoidal motion, the torque response of the fluid at very low frequency is in phase with the rate of deformation or at 90° out of phase with the imposed angular displacement of the inner cylinder. On the contrary, at very high frequency, the response of a polymeric liquid is in phase with the imposed angular displacement. The transition from the purely viscous behaviour to the purely elastic behaviour for a viscoelastic polymer is continuous with a characteristic frequency of transition of the order of magnitude equal to $1/\theta$, where θ is the relaxation time defined previously.

These various phenomena illustrate:

- the importance of a characteristic time for polymers;
- the viscous and elastic behaviour of polymers, hence the viscoelastic behaviour.

6.1-2 Linear viscoelasticity

a) General considerations

The framework of linear viscoelasticity (in one dimension) has been presented in various textbooks, mainly in MANDEL (1966), FERRY (1970), BIRD, ARMSTRONG and HASSAGER (1987), TANNER (1985). It is basically based on the analogy between the behaviour of viscoelastic materials and that of mechanical models consisting of various combinations of springs and dashpots in series or in parallel.

The spring shown in Figure 6.1-6 illustrates the purely elastic body for which the stress is proportional to the deformation, i.e. $\sigma = G\gamma$, in shear. The dashpot of Figure 6.1-7, on the other hand, represents the purely viscous behaviour with a constant viscosity η, hence the Newtonian behaviour.

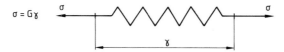

Figure 6.1-6   The elastic body represented by a spring

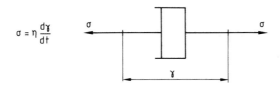

Figure 6.1-7   Viscous behaviour represented by a dashpot

A complex viscoelastic material can be represented by a combination of springs and dashpots placed in parallel or in series. If both ends are connected through a series of springs as shown in Figure 6.1-8a, the behaviour of a solid material is simulated. Otherwise, as illustrated by Figure 6.1-8b, this is the behaviour of a viscoelastic liquid.

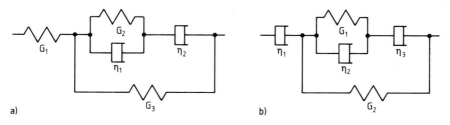

Figure 6.1-8   Typical viscoelastic models: a) viscoelastic solid   b) viscoelastic liquid

The simplest models consist of a spring and a dashpot mounted in parallel, as in the case of the Voigt model shown in Figure 6.1-9, or in series, for the Maxwell model shown in Figure 6.1-10.

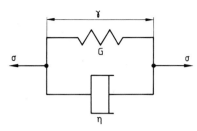

Figure 6.1-9   The Voigt model – solid behaviour

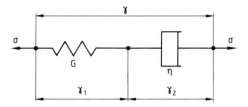

Figure 6.1-10   The Maxwell model – liquid behaviour

b) The Maxwell model

The Maxwell model consists of a spring of constant modulus G in series with a dashpot of constant viscosity η. The total deformation of the mechanical model γ is the sum of the deformation of the spring γ_1 and of the deformation of the dashpot γ_2. The behaviour of the spring is:

$$\sigma = G\gamma_1 \quad \text{or} \quad \frac{d\sigma}{dt} = G\frac{d\gamma_1}{dt}$$

and the dashpot corresponds to:

$$\sigma = \eta\frac{d\gamma_2}{dt}$$

Adding the deformations and defining $\theta = \eta/G$, we get:

$$\boxed{\sigma + \theta\frac{d\sigma}{dt} = \eta\frac{d\gamma}{dt}} \qquad (6.1\text{-}1)$$

We will show that this simple model allows us to qualitatively describe all the time dependent phenomena mentioned above.

Stress relaxation

For the experiment defined in Figure 6.1-4, the deformation is constant for $t > \Delta t$. Hence $d\varepsilon/dt = 0$ and from Equation (6.1-1), we get:

$$\sigma + \theta \frac{d\sigma}{dt} = 0 \quad \Rightarrow \quad \sigma = \sigma_0 \exp\left\{\frac{-(t - \Delta t)}{\theta}\right\} \tag{6.1-2}$$

This result predicts an exponential decay of the stress with a characteristic time equal to θ.

Stress retardation

In the initial part of the experiment illustrated by Figure 6.1-4, a constant shear rate is suddenly applied to the sample. At time zero, the stress is equal to zero and for $t > 0$, the solution of Equation (6.1-1) is:

$$\sigma + \theta \frac{d\sigma}{dt} = \eta\dot{\gamma}_0 \quad \Rightarrow \quad \sigma = \eta\dot{\gamma}_0 \left[1 - \exp\left\{-\frac{t}{\theta}\right\}\right] \tag{6.1-3}$$

We note from this result that:

- if $t \ll \theta$, $\quad \sigma = \eta\dot{\gamma}_0 \dfrac{t}{\theta} = G\gamma$, elastic behaviour
- if $t \gg \theta$, $\quad \sigma = \eta\dot{\gamma}_0$, viscous behaviour.

Shear recovery

For the shear recovery experiment defined in Figure 6.1-5, the stress is set equal to zero at time $t = 0$, i.e.:

$$t < 0, \quad \sigma = \eta\dot{\gamma}_0 = \sigma_0$$
$$t > 0, \quad \sigma = 0$$

and from Equation (6.1-1) we obtain:

$$\frac{d\gamma}{dt} = \frac{1}{\eta}\left(\sigma + \theta\frac{d\sigma}{dt}\right) \tag{6.1-4}$$

This result can be integrated to obtain the relative deformation for $t > t_0$:

$$\gamma(t) - \gamma(0) = \frac{1}{\eta}\int_0^t (\sigma\, dt + \theta\, d\sigma) = \frac{\theta}{\eta}\left[\sigma\right]_0^t = -\frac{\theta}{\eta}\sigma_0 \tag{6.1-5}$$

Hence the relative deformation is independent of time, i.e.:

$$\gamma(t) - \gamma(0) = S = -\theta\dot{\gamma}_0 = -\frac{\sigma_0}{G}$$

The result is illustrated in Figure 6.1-11. The jump in the deformation at $t = 0$ is called the recovery or recoil of the material. This recoil in the particular case of the Maxwell model is instantaneous. This is not the case of real viscoelastic materials, such as polymer melts, for which recoil takes place over a finite time. However, recoil is usually more rapid than stress relaxation.

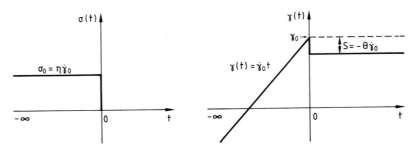

Figure 6.1-11   Recoil for a Maxwell fluid

Sinusoidal deformation

The sinusoidal deformation at the inner cylinder wall can be expressed by:

$$\gamma = \gamma_0 \exp\{i\omega t\} \tag{6.1-6}$$

The stresses exerted on either cylinder walls are out of phase with respect to the deformation and can be expressed by:

$$\sigma = \sigma_0 \exp\{i\omega t + \varphi\} \tag{6.1-7}$$

Replacing the expressions (6.1-6) and (6.1-7) into the Maxwell model (Eq. 6.1-1), we get:

$$\sigma_0 \exp\{i\omega t + \varphi\} + \theta\sigma_0 i\omega \exp\{i\omega t + \varphi\}$$
$$= \eta\gamma_0 i\omega \exp\{i\omega t\} \tag{6.1-8}$$

This result can be expressed as a real and an imaginary part:

$$\sigma_0 \cos\varphi - \theta\sigma_0\omega \sin\varphi = 0 \tag{6.1-9}$$

$$\sigma_0 \sin\varphi - \theta\sigma_0\omega \cos\varphi = \eta\gamma_0\omega \tag{6.1-10}$$

The out-of-phase angle is obtained from Equation (6.1-9):

$$\tan\varphi = \frac{1}{(\omega\theta)} \tag{6.1-11}$$

and knowing φ, the amplitude for the stress can be easily calculated from Equation (6.1-10):

$$\sigma_0 = \gamma_0 \frac{\eta\omega}{\sqrt{1 + \omega^2\theta^2}} \tag{6.1-12}$$

This results shows clearly that for very high frequencies ($\omega \gg 1/\theta$), the fluid behaves as an elastic solid of modulus given by $G = \eta/\theta$. On the other hand, at low frequencies, ($\omega \ll 1/\theta$), the properties of the fluid become those of a Newtonian liquid of viscosity η.

In general, the complex modulus of a viscoelastic liquid is defined by:

$$G^* = G' + iG'' = \frac{\sigma(t)}{\gamma(t)} \tag{6.1-13}$$

where G' is the storage modulus (elastic component) and G'' is the loss modulus (viscous component). One can also define a complex viscosity by:

$$\eta^* = \eta' - i\eta'' = \frac{\sigma(t)}{\dot{\gamma}(t)} \tag{6.1-14}$$

The real and imaginary parts of the complex viscosity are related to the storage or loss modulus by:

$$\eta' = \frac{G''}{\omega} \tag{6.1-15}$$

$$\eta'' = \frac{G'}{\omega} \tag{6.1-16}$$

For the Maxwell fluid defined by Equation (6.1-1), the storage and loss moduli are then given by:

$$\begin{aligned} G' &= \frac{\sigma_0}{\gamma_0} \cos\varphi = \eta\theta \frac{\omega^2}{1 + (\omega\theta)^2} \approx G = \frac{\eta}{\theta}, \quad \text{if} \quad \omega \gg \frac{1}{\theta} \\ G'' &= \frac{\sigma_0}{\gamma_0} \sin\varphi = \eta \frac{\omega}{1 + (\omega\theta)^2} \approx \eta\omega, \qquad \text{if} \quad \omega \ll \frac{1}{\theta} \end{aligned} \tag{6.1-17}$$

Remark

The Voigt model shown in Figure 6.1-9 is mathematically expressed by the following equation:

$$\sigma = G\gamma + \eta\dot{\gamma} \tag{6.1-18}$$

If the Voigt element is submitted to a sinusoidal strain, the out-of-phase angle between the stress response and the imposed strain is given by:

$$\tan\varphi = \omega\theta \tag{6.1-19}$$

The behaviour of the Voigt element is totally different from that of the Maxwell fluid. The stress and the strain are in phase at low frequency; at high frequency, the stress is proportional to the square of ω. This is qualitatively the behaviour of a viscoelastic solid.

6.1-3 Normal stress differences

The linear viscoelasticity defined by the simple Maxwell model in the previous section allows us to explain time effects observed with polymeric materials. This cannot, however, account for the Weissenberg effect (Figure 6.1-2). We will show that normal stress differences are responsible for the rising of the free surface near the rotating inner cylinder.

We know that the stress tensor for a Newtonian fluid is given within an isotropic term by:

$$\boldsymbol{\sigma}' = 2\eta\dot{\boldsymbol{\varepsilon}} \tag{6.1-20}$$

where $\boldsymbol{\sigma}'$ is the extra stress tensor. The total stress tensor for simple shear flow in the 1,2-plane is given by:

$$[\boldsymbol{\sigma}] = \begin{bmatrix} -p & \eta\dot{\gamma} & 0 \\ \eta\dot{\gamma} & -p & 0 \\ 0 & 0 & -p \end{bmatrix} \qquad (6.1\text{-}21)$$

We observe that $\sigma_{11} = \sigma_{22} = \sigma_{33} = -p$, i.e. there is no difference of normal stresses. In fact, the equality of the normal stresses in simple shear results from the use of Newton's law of viscosity or other purely viscous models. We have shown in Chapter 1.3 that the most general form for the total stress tensor in simple shear flow is given by:

$$[\boldsymbol{\sigma}] = \begin{bmatrix} \sigma_{11} & \sigma_{12} & 0 \\ \sigma_{12} & \sigma_{22} & 0 \\ 0 & 0 & \sigma_{33} \end{bmatrix} \qquad (6.1\text{-}22)$$

i.e. that the normal stress components, σ_{11}, σ_{22}, σ_{33}, are not necessarily identical. An example is shown in Appendix C for a Rivlin fluid. A similar behaviour is discussed here.

The first or primary normal stress difference is defined for simple shear flow by the difference between the normal stress component in the flow direction (1) and the normal component in the direction of the shear plane (2):

$$N_1 = \sigma_{11} - \sigma_{22} \qquad (6.1\text{-}24)$$

The secondary normal stress difference is the difference between the normal component in the shear plane and the normal component in the direction normal to the shear plane (3):

$$N_2 = \sigma_{22} - \sigma_{33} \qquad (6.1\text{-}25)$$

The third independent viscometric function is the shear stress, σ_{12}. As an approximation for many shear flow problems, we will assume that these three viscometric functions are unique functions of the shear rate $\dot{\gamma}$.

Whenever a polymeric liquid flows in simple shear, one can measure a shear stress, σ_{12}, and a first normal stress difference, $\sigma_{11} - \sigma_{22}$, that is positive. On the other hand, the second normal stress difference, $\sigma_{22} - \sigma_{33}$, is one order of magnitude smaller than the first difference and it is usually negative (see Chapter 6.2).

This observation allows us to explain the rising of the free surface in the Couette viscometer shown in Figure 6.1-2. The liquid is sheared in the $r\theta$-plane and the shear force is in the θ-direction. The stress tensor may be written as:

$$[\boldsymbol{\sigma}] = \begin{bmatrix} \sigma_{rr} & \sigma_{r\theta} & 0 \\ \sigma_{r\theta} & \sigma_{\theta\theta} & 0 \\ 0 & 0 & \sigma_{zz} \end{bmatrix} \qquad (6.1\text{-}26)$$

and the viscometric functions are:

$$N_1 = \sigma_{\theta\theta} - \sigma_{rr} > 0$$
$$N_2 = \sigma_{rr} - \sigma_{zz} \approx 0 \qquad (6.1\text{-}27)$$

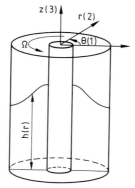

Figure 6.1-12 Liquid rising in a Couette viscometer

and a shear stress, $\sigma = \sigma_{r\theta}$. For creeping flow under steady state, the r-component of the equation of motion reduces to:

$$\frac{\partial \sigma_{rr}}{\partial r} = \frac{\sigma_{\theta\theta} - \sigma_{rr}}{r} > 0 \qquad (6.1\text{-}28)$$

and, if we assume that the secondary normal stress difference is effectively equal to zero (Weissenberg hypothesis), we can write:

$$\frac{\partial \sigma_{zz}}{\partial r} = \frac{\partial \sigma_{rr}}{\partial r} > 0 \qquad (6.1\text{-}29)$$

The normal stress in the z-direction has to be balanced by a liquid column of height $h(r)$, i.e.:

$$\sigma_{zz} = -\varrho g h(r) \qquad (6.1\text{-}30)$$

Hence, as $\partial \sigma_{zz}/\partial r$ is positive, $\partial h/\partial r$ is negative and the free surface rises towards the inner cylinder. From a physical point of view the following reasoning can be used:

- the shear planes in the Couette flow are concentric cylindrical surfaces slicing one over each other;
- due to a tension $(\sigma_{\theta\theta} - \sigma_{rr})$ in the flow direction, the liquid layers are like cylindrical stretched membranes;
- the tension in the cylindrical membrane must necessarily be balanced by an internal extra pressure;
- the equilibrium between each liquid layer must correspond to a pressure gradient that is increasing from the periphery towards the center; this pressure gradient is responsible for the rising of the surface near the inner cylinder.

The flow between parallel disks is treated in Problem 6.1-B:

6.1-4 Extrudate swell

Understanding swelling of a polymer extrudate as it flows out of a die is still a subject of current research activities. As mentioned in Section 6.1-1, the extrudate swell for very short capillaries can be interpreted in terms of recovery of the reservoir dimensions (see Figure 6.1-1). Here we will rather examine the case of swell for very long capillaries.

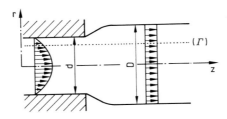

Figure 6.1-13   Swell of a Newtonian extrudate

The swell observed for the creeping flow of a Newtonian fluid can be explained by the rearrangement of the velocity profile as shown in Figure 6.1-13. On the line Γ, the velocity rearrangement usually implies that $\partial w/\partial z \neq 0$, and from the equation of continuity:

$$\frac{\partial u}{\partial r} + \frac{u}{r} = -\frac{\partial w}{\partial z} \neq 0 \qquad (6.1\text{-}31)$$

Hence, there is a radial velocity component, positive or negative that can modify the flow section.

The existence of a radial component of the velocity field and normal stress differences cannot explain in total the large swell observed for the flow of a viscoelastic liquid. Poiseuille flow is a shear flow in the z-direction with shearing surfaces in the rz-plane. In its most general form, the stress tensor is given by:

$$[\boldsymbol{\sigma}] = \begin{bmatrix} \sigma_{rr} & 0 & \sigma_{rz} \\ 0 & \sigma_{\theta\theta} & 0 \\ \sigma_{rz} & 0 & \sigma_{zz} \end{bmatrix} \qquad (6.1\text{-}32)$$

and the three viscometric functions are:

$$N_1 = \sigma_{zz} - \sigma_{rr} > 0$$
$$N_2 = \sigma_{rr} - \sigma_{\theta\theta} \approx 0 \qquad (6.1\text{-}33)$$
$$\sigma = \sigma_{rz}$$

As the polymer flows out of the capillary, the stresses become equal to zero in absence of confinements and the strain recovery components corresponding to the same components of the stress tensor are:

$$\begin{bmatrix} \frac{\varepsilon}{2} & 0 & -\frac{\gamma}{2} \\ 0 & \frac{\varepsilon}{2} & 0 \\ -\frac{\gamma}{2} & 0 & -\varepsilon \end{bmatrix}$$

The shear stress, σ_{rz}, induces a shear strain recovery, $-\gamma$, in the rz-plane whereas the primary normal stress difference, $\sigma_{zz} - \sigma_{rr}$ is responsible for an uniaxial strain, $-\varepsilon$, i.e. a contraction in the flow direction and an expansion in the radial direction. This is illustrated in Figure 6.1-14. The shear recovery does not contribute to the swelling as the liquid flows in planes parallel to the capillary axis. On the other hand, the primary normal stress difference is responsible for the radial expansion and swelling.

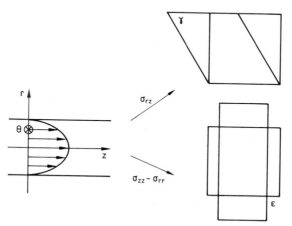

Figure 6.1-14   Interpretation of extrudate swell

The extrudate swell phenomenon results from the combination of two characteristic properties of viscoelastic materials: recoil and the existence of normal stresses in shear flow. Using this approach, TANNER (1970) has derived the following expression for the extrudate swell:

$$\frac{D}{d} = \left[1 + \left(\frac{N_{1w}}{\sigma_w} \right)^2 \right] \tag{6.1-34}$$

where the subscript w means evaluated at the capillary wall shear rate. TANNER has added a correction term to account for the swelling of a Newtonian liquid:

$$\frac{D}{d} = 0.12 + \left[1 + \left(\frac{N_{1w}}{\sigma_w} \right)^2 \right]^{1/6} \tag{6.1-34a}$$

6.1-5 Molecular considerations

a) Time effects

The relaxation time of a liquid is of the order of the time required for a configuration change of the molecules. In simple liquids, consisting of small molecules like pure water, this time is very short, corresponding to the frequency of the Brownian motion, i.e.:

$$\nu_b = \frac{k_b T}{h} = 10^{12}\,\mathrm{s}^{-1} \qquad (6.1\text{-}35)$$

where k_b and h are respectively the Boltzmann and the Planck constant. At room temperature, $\theta \approx 1/\nu_b \approx 10^{-12}\,\mathrm{s}$ and, as such a short relaxation time can not be measured, we may assume that those simple liquids are inelastic and Newtonian.

The time for a configurational change in macromolecular liquids is much longer, typically:

– 10^{-4} to $10^{-2}\,\mathrm{s}$ for dilute or semi-dilute polymer solutions;
– 10^{-2} to $10^2\,\mathrm{s}$ for polymer melts or concentrated polymer solutions.

Such long times are measurable as they are of the order of the inverse of the frequency or shear rate used in conventional rheometers.

b) Normal stress differences

At rest, macromolecular chains take in average an isotropic configuration. In shear flow, the chains are deformed and oriented in the flow direction. This is illustrated in Figure 6.1-15. The sum of all the microscopic tensions from each of the molecules makes up for the macroscopic tension in the 1 (flow)-direction. The difference of tension in the 2 and 3-directions is in contrast very small. This qualitative molecular description explains the values experimentally determined, i.e.:

$$\sigma_{11} - \sigma_{22} > 0$$
$$|\sigma_{22} - \sigma_{33}| \ll \sigma_{11} - \sigma_{22}$$

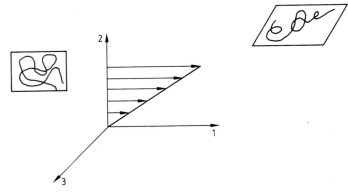

Figure 6.1-15   Macromolecule deformation in shear flow

The statistical mechanics allows us otherwise to predict normal stress differences using, for example, elastic dumbbells as models of macromolecules (see Appendix C). Such a dumbbell is drawn in Figure 6.1-16. The two spheres are

Figure 6.1-16   Elastic dumbbell model

submitted to the drag of the surrounding liquid of viscosity η and the chain elasticity is characterized by a spring of modulus G. In more sophisticate models, the polymeric chains are viewed as a series of dumbbells.

6.1-6   Convected Maxwell model

The study of a number of viscoelastic phenomena has driven us to make use of the mechanical element of Maxwell. The Maxwell model can explain qualitatively in one dimensional configuration the relations between the stress and the deformation in the flow direction. The Maxwell model cannot, however, describe properties observed in the perpendicular directions, and in particular the existence of normal stress differences in simple shear flow.

a)   A first generalization of the Maxwell model

The search for a generalized and tensorially acceptable model leads us to the simplest tensorial form:

$$\boldsymbol{\sigma} + \theta \frac{D\boldsymbol{\sigma}}{Dt} = 2\eta \dot{\boldsymbol{\varepsilon}} \tag{6.1-37}$$

where D/Dt is the material or substantial derivative defined by:

$$\frac{D}{Dt} = \frac{\partial}{\partial t} + u\frac{\partial}{\partial x} + v\frac{\partial}{\partial y} + w\frac{\partial}{\partial z} \tag{6.1-38}$$

The use of the substantial derivative does not yield the desired results. First, Equation (6.1-37) does not predict any normal stress difference in simple shear flow. The substantial derivative of any component of the stress tensor is equal to zero for steady shear flow. For example:

$$\frac{D\sigma_{xx}}{Dt} = \frac{\partial \sigma_{xx}}{\partial t} + u\frac{\partial \sigma_{xx}}{\partial x} + v\frac{\partial \sigma_{xx}}{\partial y} + w\frac{\partial \sigma_{xx}}{\partial z}$$

The only non zero velocity component is u, but the flow is uniform (i.e. independent of x) and in steady conditions. Hence $\partial \sigma_{xx}/\partial x = 0$ and $\partial \sigma_{xx}/\partial t = 0$. It follows that:

$$\frac{D\sigma_{xx}}{Dt} = 0$$

In the same way, the derivatives of the other stress components can be shown to be equal to zero, i.e.:

$$\frac{D\boldsymbol{\sigma}}{Dt} = 0$$

This reasoning shows that the model expressed by Equation (6.1-37) reduces to Newton's law for steady simple shear flow. Therefore it does not predict any normal stress difference. Moreover, we now prove that such a model is not tensorially acceptable, i.e. it does not respect the material objectivity principle.

A quantity respects the objectivity principle if its magnitude is independent of the frame of reference. For example, the rate-of-deformation tensor, $\dot{\boldsymbol{\varepsilon}}$, is objective whereas the velocity gradient, $\nabla\boldsymbol{u}$, is not. The substantial derivative has a physical meaning when operating on a scalar or vectorial quantity, but the substantial derivative of a tensor is no longer objective. Let us consider the following example.

A disk is rotating about its axis at constant speed. Due to the centrifugal force, normal stresses are generated within the disk body. These are axisymmetrical and independent of the time. The situation is illustrated in Figure 6.1-17 and the stress components are:

- radial normal stress: $\sigma_{rr}(r)$ ⎫
- tangential normal stress: $\sigma_{\theta\theta}(r)$ ⎬ not functions of θ
- shear stress: $\sigma_{r\theta} = 0$ ⎭

In a frame of reference (X, Y) rotating with the disk, the stresses are constant with respect to the time (the disk does not "feel" any stress variations with time). Hence the time derivative of the stress tensor must be equal to zero.

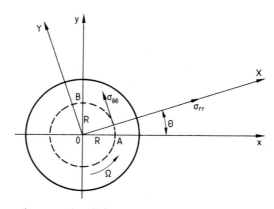

Figure 6.1-17 Stresses in a rotating disk

Let us now use a fixed Cartesian coordinate system. In the xy-plane, the stress components are then:

$$
\begin{bmatrix} \sigma_{xx} & \sigma_{xy} \\ \sigma_{xy} & \sigma_{yy} \end{bmatrix} = \begin{bmatrix} \cos\theta & -\sin\theta \\ \sin\theta & \cos\theta \end{bmatrix} \begin{bmatrix} \sigma_{rr} & 0 \\ 0 & \sigma_{\theta\theta} \end{bmatrix} \begin{bmatrix} \cos\theta & \sin\theta \\ -\sin\theta & \cos\theta \end{bmatrix}
$$

$$
= \begin{bmatrix} \sigma_{rr}\cos^2\theta + \sigma_{\theta\theta}\sin^2\theta & (\sigma_{rr} - \sigma_{\theta\theta})\sin\theta\cos\theta \\ (\sigma_{rr} - \sigma_{\theta\theta})\sin\theta\cos\theta & \sigma_{rr}\sin^2\theta + \sigma_{\theta\theta}\cos^2\theta \end{bmatrix}
$$

$$(6.1\text{-}39)$$

or replacing $\sin \theta$ and $\cos \theta$ by their equivalent:

$$\sin \theta = \frac{y}{\sqrt{x^2 + y^2}}, \quad \cos \theta = \frac{x}{\sqrt{x^2 + y^2}}$$

we get:

$$\begin{bmatrix} \sigma_{xx} & \sigma_{xy} \\ \sigma_{xy} & \sigma_{yy} \end{bmatrix} = \frac{1}{x^2 + y^2} \begin{bmatrix} \sigma_{rr}x^2 + \sigma_{\theta\theta}y^2 & (\sigma_{rr} - \sigma_{\theta\theta})xy \\ (\sigma_{rr} - \sigma_{\theta\theta})xy & \sigma_{rr}y^2 + \sigma_{\theta\theta}x^2 \end{bmatrix} \tag{6.1-40}$$

The substantial derivative of σ_{xx} is for example:

$$\frac{D\sigma_{xx}}{Dt} = u \frac{\partial \sigma_{xx}}{\partial x} + v \frac{\partial \sigma_{xx}}{\partial y}$$

$\partial \sigma_{xx} / \partial t = 0$ in steady state and using the following:

$$\begin{aligned} u &= -r\Omega \sin \theta = -\Omega y \\ v &= r\Omega \cos \theta = \Omega x \end{aligned} \tag{6.1-41}$$

we get:

$$\frac{D\sigma_{xx}}{Dt} = 2 \frac{\Omega xy}{x^2 + y^2} (\sigma_{\theta\theta} - \sigma_{rr}) \tag{6.1-42}$$

For all the stress components, we obtain:

$$\frac{D\sigma}{Dt} = \frac{\Omega}{x^2 + y^2} (\sigma_{rr} - \sigma_{\theta\theta}) \begin{bmatrix} -2xy & x^2 - y^2 \\ x^2 - y^2 & 2xy \end{bmatrix} \tag{6.1-43}$$

All the derivatives are equal to zero at any position only if $\sigma_{rr} = \sigma_{\theta\theta}$. Therefore, the substantial derivative of a stress tensor is not objective since it is equal to zero when written in a rotating frame of reference and different from zero in a fixed Cartesian system.

b) Convected derivative

The convected derivative is defined with respect to a convective or embedded frame of reference. The frame of reference moves with each element of material and it is deformed with it. The convective derivative expresses well the physical variation of the examined property in the material independently of its motion and deformation. The mathematical formalism of this derivative is shown in Appendix B. We report here only the most useful form for the study of viscoelasticity known as the upper convected or Oldroyd derivative:

$$\frac{\delta\sigma}{\delta t} = \frac{D\sigma}{Dt} - \nabla u \cdot \sigma - \sigma \cdot \nabla u^{\dagger} \tag{6.1-44}$$

where ∇u and $\nabla u^{\dagger}$ are respectively the velocity gradient and its transpose.

We could prove, first, that the use of this derivative in the Maxwell model leads to non-zero normal stress differences in simple shear flow. This is the subject of

Problem 6.1-A. Secondly, this derivative is objective. Going back to the previous example of the rotating disk, we show that the convected derivative of the stress tensor is equal to zero. From Equation (6.1-41), the velocity-gradient tensor in the xy-plane is given by:

$$\nabla u = \begin{bmatrix} \frac{\partial u}{\partial x} & \frac{\partial u}{\partial y} \\ \frac{\partial v}{\partial x} & \frac{\partial v}{\partial y} \end{bmatrix} = \begin{bmatrix} 0 & -\Omega \\ \Omega & 0 \end{bmatrix} \tag{6.1-45}$$

and:

$$\nabla u \cdot \sigma = \frac{\Omega}{x^2 + y^2} \begin{bmatrix} -(\sigma_{rr} - \sigma_{\theta\theta})xy & -\sigma_{rr}y^2 - \sigma_{\theta\theta}x^2 \\ \sigma_{rr}x^2 + \sigma_{\theta\theta}y^2 & (\sigma_{rr} - \sigma_{\theta\theta})xy \end{bmatrix} \tag{6.1-46}$$

But $\sigma \cdot \nabla u^\dagger = (\nabla u \cdot \sigma)^\dagger$, and

$$\nabla u \cdot \sigma + \sigma \cdot \nabla u^\dagger = \frac{\Omega}{x^2 + y^2} \begin{bmatrix} -2(\sigma_{rr} - \sigma_{\theta\theta})xy & (\sigma_{rr} - \sigma_{\theta\theta})(x^2 - y^2) \\ (\sigma_{rr} - \sigma_{\theta\theta})(x^2 - y^2) & 2(\sigma_{rr} - \sigma_{\theta\theta})xy \end{bmatrix} \tag{6.1-47}$$

It follows from Equations (6.1-43 and 47) that:

$$\frac{\delta \sigma}{\delta t} = \frac{D\sigma}{Dt} - \nabla u \cdot \sigma - \sigma \cdot \nabla u^\dagger = 0$$

for any value of σ_{rr} and $\sigma_{\theta\theta}$.

c) Convected Maxwell model

The convected Maxwell model can be written:

$$\sigma' + \theta \frac{\delta \sigma'}{\delta t} = 2\eta \dot{\varepsilon} \tag{6.1-48}$$

where the stress tensor, σ', is defined within an arbitrary isotropic (pressure) term, p', such that the total stress tensor is:

$$\sigma = \sigma' - p'\delta \tag{6.1-49}$$

If p' is an arbitrary scalar, the tensor σ' is not necessarily traceless or deviatoric. The thermodynamic pressure may be written as:

$$p = -\frac{1}{3} tr\, \sigma = p' - \frac{1}{3} tr\, \sigma' \tag{6.1-50}$$

The tensorial Maxwell model reproduces all the results obtained for the simple (one dimension) Maxwell model. In particular, for the experiment of stress relaxation, we get:

$$u = 0 \Rightarrow \frac{\delta \sigma'}{\delta t} = \frac{\partial \sigma'}{\partial t}$$

$$\Rightarrow \sigma' + \theta \frac{\partial \sigma'}{\partial t} = 0$$

$$\Rightarrow \sigma'(t) = \sigma'_0 \exp\left\{-\frac{t}{\theta}\right\}$$

Moreover, the convected Maxwell model predicts non-zero primary normal stress differences (Problem 6.1-A). In summary, the model gives for steady simple shear flow the following results:

$$\sigma_{12} = \eta\dot\gamma$$
$$N_1 = \sigma_{11} - \sigma_{22} = 2\eta\theta\dot\gamma^2 \qquad (6.1\text{-}51)$$
$$N_2 = \sigma_{22} - \sigma_{33} = 0$$

d) Dimensionless numbers

The convected Maxwell model leads naturally to the use of the *Weissenberg number* (see WHITE, 1964) defined as the ratio of the elastic forces over the viscous forces, i.e.:

$$\text{We} = \frac{\text{elastic forces}}{\text{viscous forces}} = \frac{\sigma_{11} - \sigma_{22}}{2\sigma_{12}} \qquad (6.1\text{-}52)$$

For the Maxwell model in steady simple shear flow, the Weissenberg is readily obtained from result (6.1-51) as:

$$\text{We} = \theta\dot\gamma \qquad (6.1\text{-}53)$$

This last result stresses that the relaxation time and the normal stress difference are conceptually related. This is through the use of the convected derivative that such a link could be shown. The Weissenberg number can then be defined as the ratio of the fluid characteristic time, θ, and a characteristic time of the rate of deformation, here the inverse of the shear rate, $\dot\gamma$.

Another dimensionless number of interest is the *Deborah number*. We have mentioned in Section 6.1-1 that the elastic stresses, at the end of a capillary for example, depend on the residence time or the time during which the polymer is deformed. This observation leads to the use of the Deborah number defined as the ratio of the fluid characteristic time and a characteristic time of the flow, for example the residence time:

$$\text{De} = \frac{\theta}{t} \qquad (6.1\text{-}54)$$

Obviously, this form appears to be very similar to the result (6.1-53) for the Weissenberg number. The relation between both numbers can be easily seen for a capillary flow for which the shear rate is proportional to V/R (where V is the average velocity) and the average residence time is to L/V. Combining Equations (6.1-53 and 54), we get:

$$\text{We} = \text{De}\,\frac{L}{R} \qquad (6.1\text{-}55)$$

That is, the Weissenberg number is the product of the Deborah number and the geometrical factor of the capillary. The following criteria can be used to establish if elasticity plays an important role or not:

– If $\text{De} \gg 1$, the polymer behaviour is almost that of a purely elastic body;

- if De $\ll 1$, elasticity plays an insignificant role. In the case of capillary flow, one may then assume that the shear and normal stresses are fully established along the capillary axis. Moreover, extrudate swell should be independent of L/R.

6.1-7 Other viscoelastic models

There are almost as many viscoelastic models as there are rheologists working in the area. The reason is that it is impossible to find a model which will explain all the phenomena encountered in various experiments (steady and transient simple shear, sinusoidal, elongational flows and combinations of these flows) and simple enough to be usable for the numerical calculations of complex flow situations.

Extensive reviews of the various types of constitutive relations have been presented by HAN (1976), BIRD, ARMSTRONG and HASSAGER (1977, 1987), WALTERS (1978) and TANNER (1985). Most of the proposed rheological models can be classified in one of three general classes.

a) Generalized Newtonian models

These models have been proposed as generalization of Newton's law of viscosity or extension of purely viscous models (see Appendix C). Derivatives of the rate-of-deformation tensor of various orders are introduced. One of the well known models in this class is the so-called second-order fluid of RIVLIN and ERICKSEN (1955) which may be written as:

$$\boldsymbol{\sigma} = -p'\boldsymbol{\delta} + 2\eta\dot{\boldsymbol{\varepsilon}} + \lambda_1 \frac{\delta\dot{\boldsymbol{\varepsilon}}}{\delta t} + \lambda_2 \dot{\boldsymbol{\varepsilon}} \cdot \dot{\boldsymbol{\varepsilon}} \tag{6.1-56}$$

This second-order model describes primary and secondary normal stress differences, but it cannot predict the elastic effects in transient flow experiments such as stress growth or relaxation or recoil discussed at the beginning of the chapter. The model is then valid only for steady-state or slowly varying flows systems (creeping flows) or for low elasticity fluids for which the relaxation time is much smaller than the residence time (low Deborah number).

b) Integral constitutive equations

The integral models containing a memory or a relaxation appear to be natural candidates to describe viscoelastic properties. For example, one can readily integrate the convected Maxwell equation to obtain:

$$\boldsymbol{\sigma}' = \int_{-\infty}^{t} 2\frac{\eta}{\theta} \exp\left\{ -\frac{t-t'}{\theta} \right\} \dot{\boldsymbol{\varepsilon}}(t,t')\, dt' \tag{6.1-57}$$

On the other hand, Lodge's elastic model is:

$$\boldsymbol{\sigma}' = \int_{-\infty}^{t} m(t,t')\, \boldsymbol{C}^{-1}(t,t')\, dt' \tag{6.1-58}$$

where $m(t, t')$ is a memory function and $\boldsymbol{C}^{-1}$ is the Finger (non-linear) deformation tensor. One can show that the convected Maxwell model can be integrated by parts to obtain the integral model (6.1-58) assuming that the deformation is finite at $t' = -\infty$ and with the memory function defined by (see LODGE, 1968):

$$m(t, t') = \frac{\eta}{\theta^2} \exp \left\{ -\frac{t - t'}{\theta} \right\} \qquad (6.1\text{-}59)$$

If the convected Maxwell model, Equation (6.1-48), and Lodge's model with memory, Equation (6.1-59), explain qualitatively most viscoelastic phenomena, they suffer serious drawbacks for engineering applications. Mostly, they are unable to describe shear-thinning properties and the prediction of a constant viscosity, η, can lead to serious errors in the calculation of flow problems.

Equation (6.1-58) has been the origin of a series of rather successful integral constitutive equations that incorporate shear-thinning effects through the use of invariants of the rate-of-deformation tensor (BIRD and CARREAU, 1968, MEISTER 1971, CARREAU, 1972), or invariants of the deformation tensor (BERNSTEIN et al., 1963, WAGNER, 1976). On the other hands, a few authors, as GODDARD and MILLER (1966), proposed corotational forms (frame reference rotating with the fluid) of Equation (6.1-57). LE ROY and PIERRARD (1973) developed a more realistic model by including the invariants of the rate-of-deformation tensor.

c) Differential constitutive equations

The models in this class contain a derivative of the stress tensor. Their main advantages (as for the integral models) over the generalized Newtonian models is that they can describe time effects as well as normal stresses in steady simple shear flow. Compared to the integral models, the differential models are usually simpler models to use in solving by numerical techniques complex flow problems.

The simplest model of this group is the Maxwell model presented in the previous section. As mentioned above, the major drawback of the Maxwell model is its inability to predict shear-thinning properties. Most of the other forms proposed in the literature aim to remedy this problem.

DE WITT (1955) proposed the following model:

$$\boldsymbol{\sigma}' + \theta \frac{\mathcal{D}\boldsymbol{\sigma}'}{\mathcal{D}t} = 2\eta \dot{\boldsymbol{\varepsilon}} \qquad (6.1\text{-}60)$$

where $\mathcal{D}/\mathcal{D}t$ is the Jaumann derivative, which expresses the rate of variation as one is moving and rotating with a fluid element:

$$\frac{\mathcal{D}\boldsymbol{\sigma}'}{\mathcal{D}t} = \frac{D\boldsymbol{\sigma}'}{Dt} - \boldsymbol{\Omega} \cdot \boldsymbol{\sigma}' - \boldsymbol{\sigma}' \cdot \boldsymbol{\Omega} \qquad (6.1\text{-}61)$$

where $\boldsymbol{\Omega}$ is the vorticity tensor. The De Witt model predicts the following viscometric functions:

$$\sigma_{12}(\dot{\gamma}) = \frac{\eta\dot{\gamma}^2}{1 + (\theta\dot{\gamma})^2} \tag{6.1-62}$$

$$N_1(\dot{\gamma}) = \frac{2\eta\theta\dot{\gamma}^2}{1 + (\theta\dot{\gamma})^2} \tag{6.1-63}$$

$$N_2(\dot{\gamma}) = -\frac{\eta\theta\dot{\gamma}^2}{1 + (\theta\dot{\gamma})^2} \tag{6.1-64}$$

We note that the three functions are now shear-dependent through the relaxation time. The secondary normal stress difference is negative and equal to one half of the primary difference. The Weissenberg number, however, as defined by Equation (6.1-52) will remain constant. Moreover, the shear-rate dependence in these results is not realistic, mainly at large shear rate.

A more realistic, but much more complex, model is the 8-constant model proposed by OLDROYD in 1958: it is written as:

$$\boldsymbol{\sigma}' + \lambda_1 \frac{\mathcal{D}\boldsymbol{\sigma}'}{\mathcal{D}t} - \mu_1\{\boldsymbol{\sigma}' \cdot \dot{\boldsymbol{\varepsilon}} + \dot{\boldsymbol{\varepsilon}} \cdot \boldsymbol{\sigma}'\} + \mu_0\{tr\boldsymbol{\sigma}'\}\dot{\boldsymbol{\varepsilon}} + \nu_1 tr\{\boldsymbol{\sigma}' \cdot \dot{\boldsymbol{\varepsilon}}\}\boldsymbol{\delta}$$

$$= 2\eta\left[\dot{\boldsymbol{\varepsilon}} + \lambda_2\frac{\mathcal{D}\dot{\boldsymbol{\varepsilon}}}{\mathcal{D}t} - 2\mu_2\dot{\boldsymbol{\varepsilon}}^2 + \nu_2 tr\{\dot{\boldsymbol{\varepsilon}}^2\}\boldsymbol{\delta}\right] \tag{6.1-65}$$

WHITE and METZNER (1963) have proposed a generalization of the Maxwell model by taking the viscosity and the relaxation time to be functions of the generalized rate of deformation $\dot{\gamma}$ (see Appendix B1). A particular case of the White-Metzner equation is the model used by PHILIPPE (1981):

$$\boldsymbol{\sigma}' + C\dot{\gamma}^{-p}\frac{\delta\boldsymbol{\sigma}'}{\delta t} = 2K\dot{\gamma}^{(n-1)}\dot{\boldsymbol{\varepsilon}} \tag{6.1-66}$$

where K and n are the power-law parameters for the viscosity and C and p are the corresponding ones for the relaxation function. All these parameters can be easily obtained from measurements made on cone-and-plate rheometers (see Chapter 6.2).

d) Choice of rheological model

The choice of a rheological model depends on the following criteria:
-   Its capacity to describe adequately the rheological behaviour of a given material under flow situations closed to those encountered in the process studied.
-   The possibility to determine from simple rheological measurements the various parameters of the model. In the case of the 8-constant Oldroyd model, only a few parameters can be determined from a simple shear experiment. The other parameters are usually obtained by fitting experimental data with the model predictions for a more complex flow situation. Extrapolation, however, to other flow situations should be carried with caution.

– The possibility or the ease to use the rheological model for numerical simulation. It is worth noting that the predictions can not be more accurate than the initial rheological model. The choice of the model is then necessarily a compromise between accuracy and simplicity.

6.1-8 Viscoelastic fluid flow simulation

– Most simulations of the viscoelastic behaviour carried out up to about ten years ago made use of second order rheological models of the Rivlin-Ericksen type. These models, as discussed in the previous section, do not predict transient effects observed with viscoelastic liquids, but they have the advantage of being explicit with respect to the stress tensor. The mathematical methods developed for Newtonian and inelastic shear-thinning liquids are then usable (see for example RIVLIN, 1978).
– Now most simulations are carried out with integral type constitutive equations or with differential constitutive equations containing the derivative of the stress tensor.
  • The first computations performed with differential constitutive equations overcame the difficulty connected with the implicit character of the stress tensor by using a perturbation technique (see KALONI, 1965; BLACK and DENN, 1976).
  • For integral type constitutive equations, most authors assumed years ago that the flow patterns were identical in a purely viscous as in a viscoelastic case and they integrated the constitutive Equation (6.1-58) along the trajectories (see WINTER and FISHER, 1981).
– The commonly used computation methods are now of two types (see KEUNINGS, 1989).

  • *Decoupled methods* which consist of solving a perturbed viscous problem. For example the Maxwell equation may be written at iteration n:

$$[\boldsymbol{\sigma}']^n = 2\eta[\dot{\varepsilon}]^n - \left(\theta\frac{\delta[\boldsymbol{\sigma}']}{\delta t}\right)^{n-1}$$

  The nonlinear stress terms computed at iteration $(n - 1)$ may be considered as fictitious body forces for iteration n. PERERA and STRAUSS (1978), AGASSANT et al. (1984) and NEYRET (1985) used this method for differential equations as well as VIRIYAYUTHAKORN and CASWELL (1980) for integral equations. They did not take into account the numerical instabilities induced by the convective terms in the stress derivative which limit the convergence.

  Recently, LUO and TANNER (1986, 1989) and MADERS (1990) showed that it was possible to enhance significantly the convergence of these methods for differential as well as integral equations by using streamline upwind techniques to solve the constitutive equation.
  • *Coupled methods* consist in solving directly the whole set of equations (motion, continuity and constitutive equations).

Until recently viscoelastic flows were solved using the same kind of methods as for purely viscous fluids which resulted in severe convergence limitations (MALONE and MIDDLEMAN, 1979; COLEMAN, 1981; KEUNINGS and CROCHET, 1984). Important progress has been achieved recently by using specific integration techniques along the streamlines as in the decoupled methods and by taking into account the compatibility of the finite element velocity and stress fields (MARCHAL and CROCHET, 1987). There is now no real limit point of convergence with increasing Weissenberg number.

– So far, the numerical simulations have been restricted to simple flow geometries.

  • *Contraction flow:* as illustrated in Figure 6.1-19 from the work of DEBBAUT et al. (1988), the size of the corner vortices increases as the Weissenberg number increases.

  • *Extrudate swell at a capillary exit:* Figure 6.1-20 compares the extrudate swell calculated by CROCHET et al. (1984), using a numerical method and the one estimated by the Tanner expression (TANNER, 1970, 1985), given by Equation (6.1-35). Both results show an increase of the swell ratio with the Weissenberg number. The agreement between the two approaches is quite good up to a value of We between 3 and 4. However these results are not necessarily in agreement with experimental observations, even at a low Weissenberg number.

These methods have now to be applied to more complex die or mold geometries, as illustrated in Chapter 3.

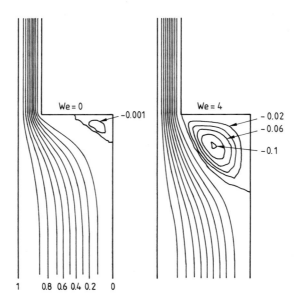

Figure 6.1-18   Flow in a planar 4 to 1 contraction. Influence of the Weissenberg number on the corner vortices (DEBBAUT et al., 1988).

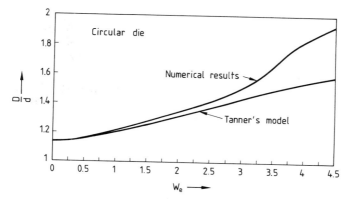

Figure 6.1-19 Extrudate swell ratio as a function of the Weissenberg number (from CROCHET et al., 1984).

Problems 6.1

6.1-A Simple shear of a Maxwell fluid

Let us consider a simple steady shear flow in the xy-plane defined by:

$$u = \dot{\gamma} y$$

$$v = w = 0$$

The components of the stress tensor are for any rheological behaviour (see Problem 1.3-A):

$$\begin{bmatrix} \sigma_{11} & \sigma_{12} & 0 \\ \sigma_{12} & \sigma_{22} & 0 \\ 0 & 0 & \sigma_{33} \end{bmatrix}$$

or

$$\begin{bmatrix} \sigma'_{11} & \sigma'_{12} & 0 \\ \sigma'_{12} & \sigma'_{22} & 0 \\ 0 & 0 & \sigma'_{33} \end{bmatrix} - p' \begin{bmatrix} 1 & 0 & 0 \\ 0 & 1 & 0 \\ 0 & 0 & 1 \end{bmatrix}$$

where p' is an arbitrary pressure. We assume that the liquid obeys the convected Maxwell model (with a constant viscosity η and a constant relaxation time θ).

a) Obtain the expression for ∇u and calculate $\delta \boldsymbol{\sigma}'/\delta t$.

b) From the Maxwell model, obtain four equations for σ'_{11}, σ'_{22}, σ'_{33}, σ'_{12}.

c) Obtain then the expression for σ'_{ij} and for the three viscometric functions: $\sigma_{12} = \sigma(\dot{\gamma})$, $\sigma_{11} - \sigma_{22} = N_1(\dot{\gamma})$, $\sigma_{22} - \sigma_{33} = N_2(\dot{\gamma})$.

d) What is the expression for the pressure, $p(\dot{\gamma}, p')$?

6.1-B Rotating disk extruder

The principle of the rotating disk extruder is illustrated in Figure 6.1-20. The polymer is fed in the gap of two concentric disks, one rotating while the other is fixed. The gap is z_0 and the radius is R. We consider the following assumptions:

I)   the flow is axisymmetrical and steady,
II)   the liquid is isothermal and follows the convected Maxwell model.
III) the shearing surfaces are planes parallel to the disks, rotating at a speed $\Omega(z)$ about the same axis as the rotating disk.

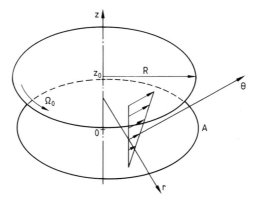

Figure 6.1-20   Shear flow between rotating disks

a)   Obtain the expressions for the velocity vector and for the rate-of-strain tensor as functions of r, z, $\Omega(z)$. What is the direction of the shear force and the magnitude of the shear rate?

b)   What is the general form of the stress tensor?

c)   Obtain the convected derivative of the extra stress tensor $\boldsymbol{\sigma}'$.

d)   What are the components of the stress tensor $\boldsymbol{\sigma}$?
Write down the components of the equation of motion for negligible inertial and gravitational forces. Show then that:

–   the pressure p' is not a function of z;
– $\Omega(z)$ is a linear function; write down the expression for $\Omega(z)$ as a function of Ω_0, z_0 and z;
–   the pressure p' is a parabolic function of r; express $p'_0 - p_a$ as a function of η, θ, Ω_0, z_0 and R (where p'_0 is the pressure at the center and p_a is the atmospheric pressure at the rim of the disks).

e)   Obtain the expression for the normal force F that tends to separate the disks and that for the torque exerted on any of the disks (the normal force is a direct result of the Weissenberg effect). Use the following values for the numerical application:

$$R = 75 \text{ mm}, \quad \Omega_0 = 100 \text{ rev/s}, \quad \theta = 0.1 \text{ s}$$
$$z_0 = 5 \text{ mm}, \quad \eta = 1 \text{ kPa} \cdot \text{s}$$

f) Show how this set-up can be used in extrusion (this is why it is called rotating disk extruder). Discuss the pros and cons.

6.1-C Couette flow of a Maxwell fluid

Referring to Figure 6.1-2, consider the flow of a Maxwell fluid in the gap of concentric cylinders (Couette viscometer) and make use of the same assumptions as for a Newtonian fluid.

a) Obtain successively the expressions for the rate-of-deformation tensor $\dot{\varepsilon}$, the tensor $\nabla \boldsymbol{u}$, and the convective derivative of the extra stress tensor $\boldsymbol{\sigma'}$.

b) Obtain the three viscometric functions, $\sigma(\dot{\gamma})$, $N_1(\dot{\gamma})$, and $N_2(\dot{\gamma})$.

c) Write down a stress balance and show that the velocity field is the same as for the Newtonian case. Obtain the zz-component of the stress tensor $\boldsymbol{\sigma}$.

d) Assuming that the zz-component is balanced by the gravitational force, obtain the expression for the free surface, $h(r)$.

6.1-D Drawing of a Maxwell liquid

This problem is taken from DENN and MARRUCCI (1971). We consider the uniform drawing of an incompressible liquid sample in the x direction. For coordinates axes located at the center of the sample, the velocity field is:

$$u = \dot{\varepsilon}x$$

$$v = -\frac{1}{2}\dot{\varepsilon}y$$

$$w = -\frac{1}{2}\dot{\varepsilon}z$$

The liquid rheology is described by the convected Maxwell model and the atmospheric pressure is arbitrarily taken to be equal to zero.

a) For a constant elongational rate $\dot{\varepsilon}$ under steady state conditions, obtain the expression for the elongational viscosity, η_e. Compare to the expression for a Newtonian liquid. Show that the elongational viscosity of a Maxwell liquid goes to infinity as the elongational rate approaches a critical value related to the relaxation time θ. Discuss the practical implications of this mathematical result.

b) Let us consider now the transient response:
   –   all the stresses (σ_{xx}, σ_{yy}, σ_{zz}) are equal to zero at $t = 0$,
   –   a constant elongational rate $\dot{\varepsilon}$ is applied at time $t = 0$.
   Obtain the expressions for the transient stress $\sigma(\dot{\varepsilon}, t)$ and the transient elongational viscosity $\eta_e(\dot{\varepsilon}, t)$. Illustrate the curves of $\eta_e(\dot{\varepsilon}, t)$ for various values of $\dot{\varepsilon}$. Find the particular expression of $\eta_e(t)$ for the critical elongational rate obtained in part a).

6.2   Measurement of viscoelastic properties

Various methods for measuring the viscosity of molten polymers have been presented in Section 1.5-5. In this section, two methods for measuring viscoelastic properties are presented and the interpretation of the results is done with the help of the convected Maxwell model. We first show how normal stress differences can be determined with a cone-and-plane rheometer. Then, we discuss of the measurement of the dynamic properties with an orthogonal or balance rheometer.

6.2-1   Cone-and-plate rheometer

We can observe from the results of Problem 6.1-B that a viscoelastic liquid in simple shear flow between two parallel disks exerts a force than tends to separate the disks. However, as shown in Problem 1.1-A, the shear rate is not homogeneous in that geometry, varying from zero at the center to a maximum at the periphery. If the normal stresses depend on the shear rate, the resulting normal force is an integral of a variable quantity from the center to the periphery, hence more difficult to use. As suggested by WEISSENBERG (1948), it is preferable to use the cone-and-plate geometry for which the shear rate is constant in the gap (see Figure 6.2-1).

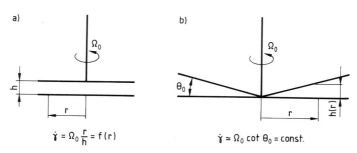

Figure 6.2-1   Shear between parallel disks (a) and in a cone-and-plate (b)
a) $\dot{\gamma} = \Omega_0(r/h) = f(r)$
b) $\dot{\gamma} = \Omega_0 \cot\theta_0 = \text{const.}$

First, we will analyze the steady-state flow of a viscoelastic liquid in the cone-and-plate geometry, considering that the cone is rotating at constant speed and the lower platen is fixed. The analysis will proceed along the following steps:

–   simplification of the kinematics of the flow,
–   definition of general viscometric functions at any points in the flow,
–   solution of the equations of motion.

We will see that the solution requires hypotheses which will impose restrictions on the design of the cone-and-plate rheometer.

In the second part, we will examine transient flow problems (sudden start-up and sudden cessation of flow, sinusoidal motion of the cone).

a) Kinematics approximation

We consider the same hypotheses as in Problem 1.1-A:

- the shearing surfaces are cones rotating about the same axis and with the same apex as the cone-and-plate system;
- these shearing surfaces rotate at the angular speed $\Omega(\theta)$;
- the angular direction θ is such that:

$$\Omega\left(\frac{\pi}{2}\right) = 0, \quad \Omega\left(\frac{\pi}{2} - \theta_0\right) = \Omega_0 \tag{6.2-1}$$

where θ_0 is the angle between the cone and the platen and Ω_0 is the rotational speed of the cone.

The velocity profile can be written in spherical coordinates as illustrated in Figure 6.2-2:

$$\boldsymbol{u} \begin{vmatrix} u = 0 \\ v = 0 \\ w = r\sin\theta\,\Omega(\theta) \end{vmatrix} \tag{6.2-2}$$

The velocity gradient and the rate-of-strain tensor are then:

$$\nabla\boldsymbol{u} = \begin{bmatrix} 0 & 0 & -\sin\theta\,\Omega(\theta) \\ 0 & 0 & -\cos\theta\,\Omega(\theta) \\ \sin\theta\,\Omega(\theta) & \cos\theta\,\Omega(\theta) + \sin\theta\frac{d\Omega}{d\theta} & 0 \end{bmatrix} \tag{6.2-3}$$

and:

$$\dot{\varepsilon} = \begin{bmatrix} 0 & 0 & 0 \\ 0 & 0 & \frac{1}{2}\sin\theta\frac{d\Omega}{d\theta} \\ 0 & \frac{1}{2}\sin\theta\frac{d\Omega}{d\theta} & 0 \end{bmatrix} \tag{6.2-4}$$

This is simple shear flow, the velocity being in the θ-direction in the $\theta\varphi$ plane and the velocity gradient is $\dot{\gamma}(\theta) = (d\Omega/d\theta)\sin\theta$, called the shear rate.

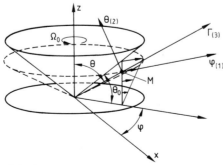

Figure 6.2-2   Cone-and-plate flow

b) Viscometric functions

The stress tensor for any rheological behaviour can be written as (see Section 6.1-3):

$$\boldsymbol{\sigma} = \begin{bmatrix} \sigma_{rr} & 0 & 0 \\ 0 & \sigma_{\theta\theta} & \sigma_{\theta\varphi} \\ 0 & \sigma_{\theta\varphi} & \sigma_{\varphi\varphi} \end{bmatrix} \tag{6.2-5}$$

We defined the following three material functions which depend on $\nabla \boldsymbol{u}$ and $\dot{\varepsilon}$, hence of θ:

$$\sigma_{\theta\varphi} = \sigma(\theta) = \eta\dot{\gamma}(\theta) \tag{6.2-6}$$

$$\sigma_{\varphi\varphi} - \sigma_{\theta\theta} = N_1(\theta) = \psi_1\dot{\gamma}^2 \tag{6.2-7}$$

$$\sigma_{\theta\theta} - \sigma_{rr} = N_2(\theta) = \psi_2\dot{\gamma}^2 \tag{6.2-8}$$

where η is the viscosity and ψ_1 and ψ_2 are respectively the primary and secondary normal stress coefficients.

c) Equation of motion

Under the assumption that the inertial and gravitational forces are negligible, the three components of the equation of motion reduce to:

$$\frac{\partial \sigma_{rr}}{\partial r} + \frac{1}{r}(2\sigma_{rr} - \sigma_{\theta\theta} - \sigma_{\varphi\varphi}) = 0 \tag{6.2-9}$$

$$\frac{1}{r}\frac{\partial \sigma_{\theta\theta}}{\partial \theta} + \frac{1}{r}(\sigma_{\theta\theta} - \sigma_{\varphi\varphi})\cot\theta = 0 \tag{6.2-10}$$

$$\frac{1}{r}\frac{\partial \sigma_{\theta\varphi}}{\partial \theta} + \frac{2}{r}\sigma_{\theta\varphi}\cot\theta = 0 \tag{6.2-11}$$

Equation (6.2-11) can be solved without any difficulty to obtain:

$$\sigma_{\theta\varphi} = \frac{\sigma_{\theta\varphi}(\pi/2)}{\sin^2\theta} \tag{6.2-12}$$

This shows that the shear stress increases continuously from the plate ($\theta = \pi/2$) to the cone ($\theta = \pi/2 - \theta_0$). On the other hand, Equations (6.2-9 and 10) are not compatible as shown by taking the derivative of the first with respect to θ and changing the order:

$$\frac{\partial^2 \sigma_{rr}}{\partial r\,\partial \theta} = \frac{1}{r}\frac{\partial}{\partial \theta}(\sigma_{\varphi\varphi} + \sigma_{\theta\theta} - 2\sigma_{rr}) \tag{6.2-13}$$

and the derivative of the second one with respect to r is:

$$\frac{\partial^2 \sigma_{\theta\theta}}{\partial r\,\partial \theta} = \frac{\partial}{\partial r}(\sigma_{\varphi\varphi} - \sigma_{\theta\theta})\cot\theta \tag{6.2-14}$$

This is equal to zero since $(\sigma_{\varphi\varphi} - \sigma_{\theta\theta})$ is a unique function of the shear rate, hence of θ. We can write:

$$\frac{\partial^2 \sigma_{\theta\theta}}{\partial r\, \partial\theta} - \frac{\partial^2 \sigma_{rr}}{\partial r\, \partial\theta} = \frac{\partial}{\partial r}\left[\frac{\partial}{\partial\theta}(\sigma_{\theta\theta} - \sigma_{rr})\right] \qquad (6.2\text{-}15)$$

This is again equal to zero since $(\sigma_{\theta\theta} - \sigma_{rr})$ is a unique function of θ. These three results imply that:

$$\frac{\partial}{\partial\theta}(\sigma_{\varphi\varphi} + \sigma_{\theta\theta} - 2\sigma_{rr}) = 0$$

or:
$$(6.2\text{-}16)$$

$$\frac{d}{d\dot\gamma}(N_1 + 2N_2)\frac{d\dot\gamma}{d\theta} = 0$$

Obviously, there is no reason why this last relation between normal stress differences should be verified for any viscoelastic liquid. In general, the second normal stress difference is small compared to the first and Equation (6.2-16) would then imply that N_1 is independent of θ or of the shear rate, $\dot\gamma$. This is physically inadmissible. We therefore consider that result (6.2-16) is not verified and that Equations (6.2-9 and 10) are incompatible.

It means that the flow profile assumed a priori is possible for a Newtonian liquid (since $N_1 = N_2 = 0$), but not for a viscoelastic liquid. A similar incompatibility could be shown for Newtonian liquids if inertial forces are not negligible. In fact, we observe with viscoelastic liquids (as well as in Newtonian liquids at high Reynolds numbers) the formation of secondary flows, as illustrated in Figure 6.2-3. The previous analysis is not valid as the flow is no longer simple shear but contains also elongational components. The cone-and-plate geometry does not allow for the measurement of rheological functions except in the limit of small cone angles as shown in the next section.

Figure 6.2-3   Secondary flows in a cone and plate

d) Small cone angle limit

It is obvious that the analysis outlined above cannot be carried further under the same hydrodynamics assumptions except in the limit of small cone angles, θ_0. If θ_0 (in radiants) is much smaller than 1, then the shear rate becomes independent of θ since $d\Omega/d\theta \approx \Omega_0/\theta_0$ and $\sin\theta \approx 1$. Hence:

$$\dot\gamma(\theta) = \frac{\Omega_0}{\theta_0} = \dot\gamma = \text{constant} \qquad (6.2\text{-}17)$$

This result is qualitatively described in Figure 6.2-1(b). It follows that ∇u becomes independent of Ω_0 and the three rheological functions σ, N_1 and N_2 become unique functions of $\dot\gamma$.

The shear stress, $\sigma(\dot\gamma) \approx \sigma_{\theta\varphi}(\pi/2) \approx \sigma_{\theta\varphi}(\theta)$, is obtained directly from the torque measurement:

$$\boxed{C(\dot\gamma) = \sigma(\dot\gamma)\frac{2}{3}\pi R^3}$$

(6.2-18)

and since $\cot\theta \ll 1$, Equation (6.2-10) implies:

$$\frac{\partial\sigma_{\theta\theta}}{\partial\theta} \ll \sigma_{\theta\theta} - \sigma_{\varphi\varphi}$$

(6.1-29)

This is the usual approximation, neglecting the variation of the stresses in the direction of the thickness for thin geometries. Equation (6.2-10) is automatically verified and the incompatibility with Equation (6.2-9) is eliminated. The kinematic hypothesis is now admissible. This is why the cone angle, θ_0, of cone-and-plate rheometers is very small, typically between 1 and 5 degrees.

Noting that:

$$\frac{\partial}{\partial r}(\sigma_{\theta\theta} - \sigma_{rr}) = 0$$

Equation (6.2-9) can be re-arranged in terms of the normal stress exerted on the cone or on the plate, $\sigma_{\theta\theta}$, i.e.:

$$\frac{\partial\sigma_{\theta\theta}}{\partial r} = \frac{1}{r}[N_1(\dot\gamma) + 2N_2(\dot\gamma)]$$

(6.2-20)

and integrating:

$$\sigma_{\theta\theta}(r) = \sigma_{\theta\theta}(R) + [N_1(\dot\gamma) + 2N_2(\dot\gamma)]\ln\frac{r}{R}$$

(6.2-21)

The unknown $\sigma_{\theta\theta}(R)$ can be evaluated from:

$$\sigma_{\theta\theta}(R) - \sigma_{rr}(R) = N_2(\dot\gamma)$$

and noting that at the free surface, $\sigma_{rr}(R) = -p_a$, the atmospheric pressure, this can be written as:

$$\sigma_{\theta\theta}(R) = N_2(\dot\gamma) + p_a$$

Hence, it follows that the normal stress exerted on the cone or on the plate is given by the following expression:

$$\sigma_{\theta\theta}(r) - p_a = N_2(\dot\gamma) + [N_1(\dot\gamma) + 2N_2(\dot\gamma)]\ln\frac{r}{R}$$

(6.2-21)

The magnitude of the stress is increasing as the radius decreases to become infinite at the center. In most cases, the cone is truncated to avoid solid friction between the cone and the plate and the stress remains finite (see Figure 6.2-4).

From measurements of normal stresses at the wall using pressure transducers, one can in principle determine the primary and secondary normal stress differences, N_1 and N_2. However, the size of the commercially available pressure transducers

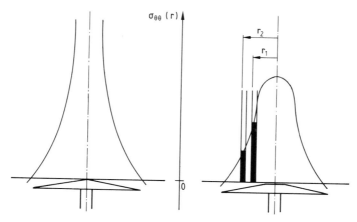

Figure 6.2-4   Normal stress profiles in cone-and-plate geometry

is not small enough compared to the cone and plate dimensions to make accurate measurements. A notable exception is the work of CHRISTIANSEN (see for example GAO et al., 1981), who used specially designed pressure transducers. He obtained for various polymer solutions negative values for N_2 with magnitude in the order of 10 to 20% of N_1.

Cone-and-plate rheometers are usually designed to measure the total normal thrust. This follows from the simple expression obtained by integrating the normal stress over the plate (or cone) surface, assuming that the surrounding is at the atmospheric pressure, p_a:

$$F(\dot{\gamma}) = \int\limits_{0}^{R} (-\sigma_{\theta\theta}(r) + p_a)\, 2\pi r\, dr = N_1(\dot{\gamma})\pi\frac{R^2}{2} \qquad (6.2\text{-}22)$$

This is remarkable that in the cone-and-plate geometry the normal thrust is a unique function of the primary normal stress difference, $N_1(\dot{\gamma})$.

In summary, the cone-and-plate rheometer can be used to measure the three viscometric functions without assuming any rheological behaviour of the material:

- the shear stress, σ (hence the viscosity, η) is directly related to the torque;
- the primary normal stress differences, N_1 (hence ψ_1), is readily determined from the normal thrust or force that tends to separate the cone from the plate;
- the secondary normal stress difference, N_2 (and ψ_2), can be determined from the more difficult measurement of the radial pressure profile.

e) Example

Figure 6.2-5 illustrates typical shear stress and primary normal stress difference – shear rate data obtained on a cone-and-plate rheometer for a polypropylene melt at 200° C. We note here that the normal stresses are of the same magnitude and even larger than the shear stresses. The measurements in a cone and plate are restricted to rather low shear rates, of order of 10 to 50 s⁻¹, due to flow instabilities and ejection of the sample from the gap. This is considerably lower than values encountered in most processing conditions.

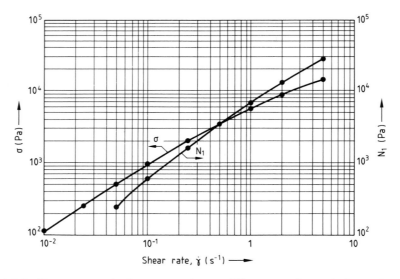

Figure 6.2-5   Shear stress and primary normal stress difference vs. shear rate for a polypropylene at 200° C

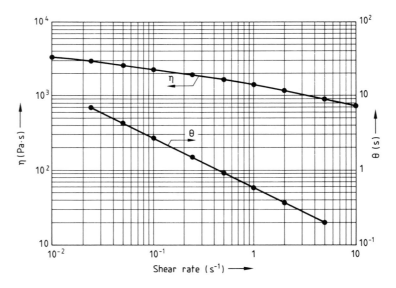

Figure 6.2-6   Viscosity and relaxation time as functions of shear rate (polypropylene at 200° C)

If one assumes that the rheology of this polypropylene is correctly described by a convected Maxwell model, the rheological data can be described in terms of the viscosity, $\eta = \sigma/\dot{\gamma}$ and a time constant defined by:

$$\theta = \frac{N_1}{2\sigma\dot{\gamma}} \qquad (6.2\text{-}23)$$

The polypropylene data are re-plotted in Figure 6.2-6 in terms of η and θ. It is clear that these two rheological functions are shear-thinning (decrease with increasing shear rate). Generalized Maxwell models as proposed by WHITE and METZNER (1963) or by PHILIPPE (1981) can correlate correctly these data.

f) Unsteady-state experiments

The experiment and rheological functions described above were restricted to steady-state viscometric flow. We briefly discuss two types of unsteady-state experiments.

Oscillatory shear flow

The cone-and-plate rheometer can be used in the oscillatory mode as discussed for the Couette rheometer in Sections 6.1-1 and 6.1-2. A sinusoidal rotation is imposed either to the cone or to the plate and the torque response is recorded. For small deformations, the response is of sinusoidal shape but out-phase with the imposed deformation. Following the analysis presented in Section 6.1-2, the complex modulus, G^*, the complex viscosity, η^*, (or equivalently the real and imaginary parts of the complex viscosity, η'', η') can be determined. Such typical dynamic rheological properties are illustrated in Figure 6.2-7.

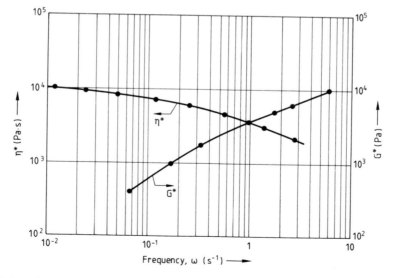

Figure 6.2-7   Complex modulus and complex viscosity vs. frequency

Transient experiments

A first classical transient experiment consists of imposing a sudden and constant rotational speed (hence a shear rate): This is known as the stress growth experiment. When the torque has attained a steady-state value, the cone (or the plate) can be stopped suddenly to record the relaxation of the stresses (i.e. torque and normal thrust). This is referred to as stress relaxation experiment following steady shear flow. Typical recordings of the normal thrust and of the torque are shown in Figure 6.2-8 for three different imposed shear rates. At low shear rate, the torque (hence viscosity) increases in an exponential way to reach a plateau, and then decreases in a similar way under the relaxation experiment. The normal thrust response (hence first normal stress difference) is quite similar but the time constant is much longer. At higher shear rates, the time responses become much shorter and one observs stress overshoots in stress growth experiments. With highly elastic materials, these overshoots can be quite large compared to the steady-state values.

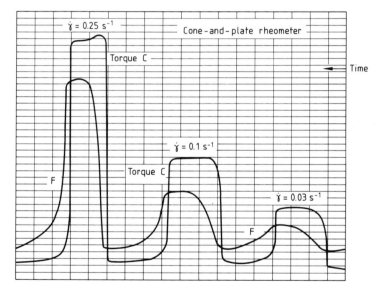

Figure 6.2-8   Stress growth and stress relaxation experiments

It is interesting to note that if the material could be described by a convected Maxwell model, both the stress growth and relaxation curves would be simple exponential curves independent of the applied shear rate (see Section 6.1-2). This is clearly not the case here and the transient curves can be used to determine shear-rate and time dependent relaxation time. Using the relaxation curves, the time constants can be obtained from:

$$\theta = \frac{C}{dC/dt} \quad \text{or} \quad \theta = \frac{N_1}{dN_1/dt} = \frac{F}{dF/dt}$$

The figure shows that the relaxation time for the torque response is much shorter than that for the normal thrust.

6.2-2 Orthogonal rheometer

The principle of the orthogonal or Maxwell rheometer (MAXWELL and CHARTOFF 1965) is illustrated in Figure 6.2-9.

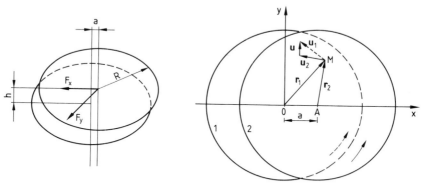

Figure 6.2-9 Orthogonal rheometer

a) Kinematics

The liquid sample is placed between two parallel disks of radius R, but off center, with eccentricity a $(a \ll R)$. The disk ① is under imposed rotation at the angular speed Ω whereas this rotational speed is transmitted to the disk ② (rotating freely) so exerting forces in the plane of rotation:

- F_x,   force along the plane of the disk axes;
- F_y,   force normal to that plane.

The relative velocity at any point M in the gap is constant:

- velocity of the lower disk: $\boldsymbol{u}_1 = \Omega \wedge \boldsymbol{r}_1$
- velocity of the upper disk: $\boldsymbol{u}_2 = \Omega \wedge \boldsymbol{r}_2$
- relative velocity: $\boldsymbol{u} = \boldsymbol{u}_1 - \boldsymbol{u}_2 = \Omega \wedge (\boldsymbol{r}_1 - \boldsymbol{r}_2) = \Omega \wedge \boldsymbol{a}$

Hence, at any point, the relative velocity is in the y-direction with magnitude $u = \Omega a$. The shear rate and the deformation are then given respectively by:

$$\dot{\gamma} = \Omega \left(\frac{a}{h} \right) \tag{6.2-24}$$

$$\gamma = \frac{a}{h} \tag{6.2-25}$$

The deformation is along the x-direction.

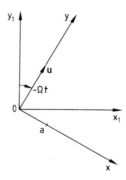

Figure 6.2-10   Coordinates in the orthogonal rheometer

b) General viscoelastic behaviour

Let us choose the axes x_1, y_1 fixed with respect to disk ① and axes x and y, fixed in space and rotating at the angular velocity $-\Omega$ with respect to the axes x_1 and y_1. This is depicted in Figure 6.2-10.
The deformation in the x_1y_1-plane is:

$$x_1\text{-component:} \quad \left(\frac{a}{h}\right) \cos \Omega t = \left(\frac{a}{h}\right) \mathrm{Re}\{e^{i\Omega t}\}$$

$$y_1\text{-component:} \quad -\left(\frac{a}{h}\right) \sin \Omega t = \left(\frac{a}{h}\right) \mathrm{Re}\{e^{i(\Omega t + \pi/2)}\}$$

where Re means the real part of $\{\ \}$: The shear stress is then expressed in terms of the complex modulus $G^* = G' + iG''$ by:

x_1-component:

$$\left(\frac{a}{h}\right)(G' \cos \Omega t - G'' \sin \Omega t) = \left(\frac{a}{h}\right) \mathrm{Re}(G' + iG'')\,e^{i\Omega t}$$

y_1-component:

$$\left(\frac{a}{h}\right)(-G' \sin \Omega t - G'' \cos \Omega t) = \left(\frac{a}{h}\right) \mathrm{Re}(G' + iG'')\,e^{i(\Omega + \pi/2)}$$

These expressions become in the x, y coordinates:

x-component:

$$\left(\frac{a}{h}\right)[(G' \cos \Omega t - G'' \sin \Omega t) \cos \Omega t - (-G' \sin \Omega t - G'' \cos \Omega t) \sin \Omega t] = G'\left(\frac{a}{h}\right)$$

y-component:

$$\left(\frac{a}{h}\right)[(G' \cos \Omega t - G'' \sin \Omega t) \sin \Omega t + (-G' \sin \Omega t - G'' \cos \Omega t) \cos \Omega t] = -G''\left(\frac{a}{h}\right)$$

Therefore the x and y-components of the force acting on disk ② are:

$$F_x = -G' \left(\frac{a}{h} \right) \pi R^2 \tag{6.2-26}$$

$$F_y = G'' \left(\frac{a}{h} \right) \pi R^2 \tag{6.2-27}$$

Figure 6.2-11 reports the values of G' and of G'' for a polypropylene determined from the force measurements on the orthogonal rheometer.

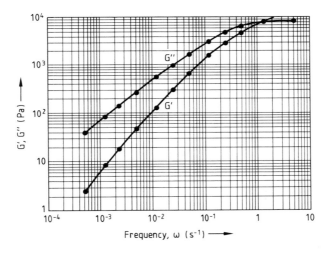

Figure 6.2-11   Real and imaginary parts of the complex modulus for a polypropylene at 200° C

c) Special cases

– If the liquid is *purely viscous*, $G' = 0$ and $G'' = \eta\Omega$ and Equations (6.2-26, 27) reduce to:

$$F_x = 0$$

$$F_y = \eta\Omega \left(\frac{a}{h} \right) \pi R^2 = \eta\dot\gamma\pi R^2 \tag{6.2-28}$$

– If the material placed in the gap is *purely elastic* (elastic rubber assumed isotropic when the disks are concentric), then $G' = G$ (shear modulus) and $G'' = 0$. The force components are then:

$$F_x = -G\gamma\pi R^2 = -G \left(\frac{a}{h} \right) \pi R^2 \tag{6.2-29}$$

$$F_y = 0$$

where $\gamma (= a/h)$ is the shear.

–   The stress tensor for a *Maxwell fluid* can be determined completely from the kinematics. In the x, y, z-coordinates system, the velocity components in the orthogonal rheometer are:

- on disk ①:
$$u = -\Omega y$$
$$v = \Omega x$$
$$w = 0$$

- on disk ②:
$$u = -\Omega y$$
$$v = \Omega x - \Omega a$$
$$w = 0$$

- in the polymer:
$$u = -\Omega y$$
$$v = \Omega x - \Omega a(z/h)$$
$$w = 0$$

Then the velocity gradient tensor is (see Appendix B):

$$\nabla \boldsymbol{u} = \begin{bmatrix} 0 & -\Omega & 0 \\ \Omega & 0 & -\Omega(a/h) \\ 0 & 0 & 0 \end{bmatrix} \tag{6.2-30}$$

As this tensor is uniform and steady, it follows that the stress tensor is also uniform and steady. The use of the Maxwell model leads to the following expression for the two non-zero stress components:

$$\sigma'_{xz} = \frac{\eta\theta\Omega^2}{1+(\Omega\theta)^2}\left(\frac{a}{h}\right) = G'(\Omega)\left(\frac{a}{h}\right) = \sigma_{xz} \tag{6.2-31}$$

$$\sigma'_{yz} = \frac{-\eta\Omega}{1+(\Omega\theta)^2}\left(\frac{a}{h}\right) = -G''(\Omega)\left(\frac{a}{h}\right) = \sigma_{yz} \tag{6.2-32}$$

We note that Equations (6.2-31 and 32) lead to the same results for G' and G'' as given by Equation (6.1-17).

The viscosity η and the relaxation time θ can then be determined from the values of G' and G''. Figure 6.2-12 compares the values of η obtained from G'' as a function of frequency with that determined from cone-and-plate measurements as a function of shear rate (see Figure 6.2-6). The agreement between both sets of data and capillary data (see Figure 1.5-8) is remarkable.

On the other hand, the relaxation time deduced from the steady-shear measurements is considerably different from that calculated from the orthogonal rheometer data. Figure 6.2-13 illustrates the differences between the two sets of data for the same polymer melt. This is a clear indication that a single relaxation time is not sufficient to describe polymer melts elasticity.

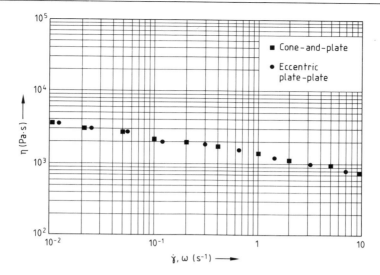

Figure 6.2-12 Comparison of viscosities obtained from the orthogonal and cone-plate rheometers (polypropylene at 200° C)

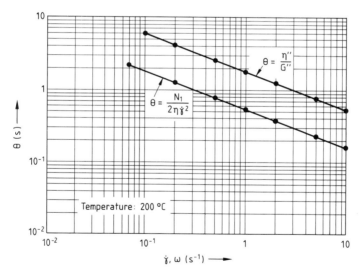

Figure 6.2-13 Relaxation times from steady-shear measurements compared to values obtained from the orthogonal rheometer (polypropylene at 200° C)

The other non-zero components of the stress tensor do not exert any force on the disks. The expressions for these components are:

$$\sigma'_{xy} = -3\eta\Omega(\Omega\theta)^2 \left(\frac{a}{h}\right)^2 \left[\frac{1}{1+(\Omega\theta)^2}\right] \left[\frac{1}{1+4(\Omega\theta)^2}\right]$$

$$\sigma'_{xx} = 6\eta\Omega(\Omega\theta)^3 \left(\frac{a}{h}\right)^2 \left[\frac{1}{1+(\Omega\theta)^2}\right] \left[\frac{1}{1+4(\Omega\theta)^2}\right] \qquad (6.2\text{-}33)$$

$$\sigma'_{yy} = 2\eta\Omega(\Omega\theta)^2 \left(\frac{a}{h}\right)^2 \left[\frac{1}{1+4(\Omega\theta)^2}\right]$$

We note that these components are of the order of $(a/h)^2$ whereas σ_{xz} and σ_{yz} are of the order of (a/h). Obviously, for small (a/h), σ'_{xy}, σ'_{xx} and σ'_{yy} are negligible. This is a requirement for valid measurements. In the orthogonal rheometer, the fluid is then submitted to small strain as in the case of the sinusoidal (small strain) shear experiment described in the previous section. As for the shear rate in the cone-and-plate geometry, the frequency is restricted to values less than 10 to 50 s^{-1}.

6.2-3 Balance rheometer

The balance rheometer designed by KEPES (see WALTERS, 1975) is based on the same principles as the orthogonal rheometer with the following differences (Figure 6.2-14):

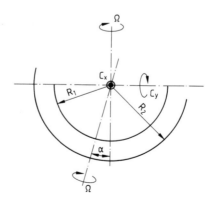

Figure 6.2-14   Balance rheometer of KEPES

— The disks are replaced by two concentric half-spheres of radii R_1 and R_2 and rotating about axes forming a small angle α in the plane P;
— the forces are replaced by torques:
   C_x, about an axis normal to the plane P and resulting from the elastic component of the material,

C_y, about an axis normal in the plane P, due to the viscous contribution of the sample.

The main advantage of this geometry over the orthogonal rheometer comes from the fact that the material being tested is contained inside the wall of the outer sphere; hence the effects of the centrifugal forces are less important and higher frequencies can be attained.

6.3 Flow instabilities in viscoelastic liquids

6.3-1 Main phenomena

Many flow instabilities are observed with viscoelastic liquids, but two main distinct phenomena are frequently encountered. draw resonance and melt fracture.

Draw resonance phenomenon is encountered in the spinning of a liquid polymeric filament. As illustrated in Figure 6.3-1, the liquid filament (polymer melt or solution) extruded from a die is stretched in ambient air by the pulling action of a roller. The final diameter of the filament is controlled by the ratio of the roller velocity to that of the average polymer velocity in the die. This is called the draw down ratio, DDR, (in cases, as shown in the figure, the filament is plunged in a water bath which freezes the dimensions). One observes under certain conditions that the surface of the stretched filament is no longer smooth. For a draw down ratio smaller than a critical value, DDR_c, the diameter is uniform; for a ratio larger than DDR_c, the diameter of the filament may vary drastically in a periodic manner as shown in the figure. This phenomenon has been encountered with purely viscous as well as with viscoelastic liquids. The pulsations decrease as the drawing distance in air increases.

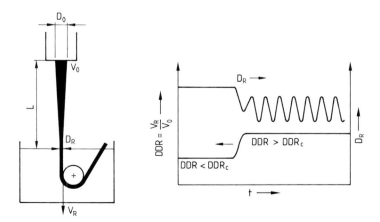

Figure 6.3-1   Draw resonance in spinning

Melt fracture is observed as a polymer is extruded freely from a die at a rate exceeding a critical value. The diameter of the extrudate is no longer uniform and may exhibit various distortions, all being referred to as melt fracture. Figure 6.3-2 illustrates various shapes of melt fracture encountered in different flow situations.

– Defects known as sharkskin are shown in Figure 6.3-2 (a) and (b). This is a surface instability, often periodic which depends on the flow rate, temperature and properties of the polymer. In (a), the extrudate was a linear low density polyethylene whereas in (b), it was a high density polyethylene.

– In some cases, one observes smooth surface followed by sharkskin zones. This is referred to as bamboo (attributed to the stick-slip phenomenom (photo c)).

– As the extrusion rate is increased, the sharkskin may disappear and the surface of the extrudate may become again smooth as shown in the photo (d).

– Helical or screw shapes are frequently encountered in the flow of polystyrene (photo (e)) or in the flow of polypropylene (f). The amplitude of the distortions increases with increasing flow rate. As the flow rate is further increased, polyethylene, polystyrene an polypropylene exhibit chaotic distortions (photo (g)).

The other polymers may show one or many of the distortions shown in Figure 6.3-2.

Distortions similar to melt fracture have been observed in polymer processing, for example, in calendering of PVC as discussed in Section 3.2. The film of polyvinylchloride, usually transparent, becomes partly opaque and the surface which is not in contact with the roller shows surface defects as the rollers' velocity increases or if the nip between the rollers becomes too narrow. Such distortions are shown in Figure 6.3-3.

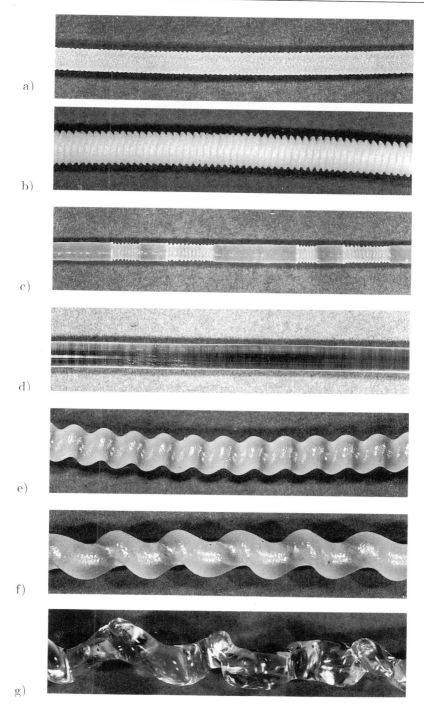

Figure 6.3-2   Various shapes of extrudates under melt fracture

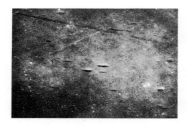

Figure 6.3-3 Surface distortions in calendering
left: surface in contact with roller
right: surface not in contact with roller

6.3-2 Analysis of draw resonance

Draw resonance has been mathematically analyzed as a Hopf bifurcation (see IOOSS and JOSEPH, 1980). If the draw ratio is larger than a critical value, the different steady-state solutions presented in Section 5.1 are no longer stable. A small perturbation about the steady-state value leads to transient and periodic solutions for the diameter and the velocity of the filament.

a) Isothermal Newtonian fluid

PEARSON and MATOVICH (1969) were probably the first authors to theoretically predict the existence of a critical draw ratio for the isothermal spinning of a Newtonian liquid. ISHIHARA and KASE (1975), FISHER and DENN (1975) followed by DEMAY (1983) have reported calculated values for the periodic diameter of the liquid filament. Their method was based on the following. Using the time effect in the basic equations for the spinning process (Equations (5.1-12, 13 and 14)) we can write:

Flow rate:

$$\frac{\partial S}{\partial t} + \frac{\partial}{\partial z}(Sw) = 0 \tag{6.3-1}$$

Force:

$$\frac{\partial}{\partial z}\left(\eta S \frac{\partial w}{\partial z}\right) = 0 \tag{6.3-2}$$

The filament velocity and section are now time-dependent about average values that are the steady solutions obtained previously, i.e.:

$$S_{ss}(z) = S_0 \exp\left(-\frac{z}{L}\ln(DDR)\right) \tag{6.3-3}$$

$$w_{ss}(z) = V_0 \exp\left(\frac{z}{L}\ln(DDR)\right) \tag{6.3-4}$$

and we look for time-dependent solutions of the form:

$$S(z,t) = S_{ss}(z) + S^*(z,t) \quad \text{with} \quad S^*(z,t) = S_1(z)e^{\lambda t} \qquad (6.3\text{-}5)$$

$$w(z,t) = w_{ss}(z) + w^*(z,t) \quad \text{with} \quad w^*(z,t) = w_1(z)e^{\lambda t} \qquad (6.3\text{-}6)$$

where S^* and w^* are the perturbations with respect to the steady-state solution; λ is a complex eigenvalue determined by introducing Equations (6.3-5) and (6.3-6) into Equations (6.3-1) and (6.3-2) and keeping only the terms in the first order with respect to $e^{\lambda t}$.

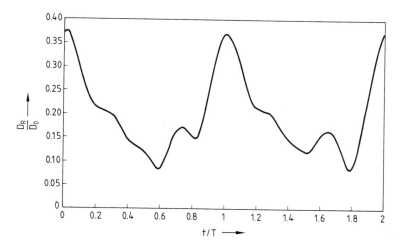

Figure 6.3-4   Diameter variations with time ($DDR = 21.5$). From DEMAY (1983)

If the draw down ratio, DDR, is smaller than 20, the real part of λ is found to be negative. This implies that any perturbations are rapidly dampened and the steady-state solution is stable. On the contrary, for $DDR > 20$, the real part of λ is positive. The solution is of periodic nature with a frequency related to the imaginary part of λ. A typical periodic solution for the diameter is shown in Figure 6.3-4.

b) Non isothermal Newtonian fluid

The effect of heat transfer for the spinning of Newtonian fluids have been analyzed by SHAH and PEARSON (1972) and by DEMAY (1983). To Equations (6.3-1 and 6.3-2), one must add the equation of thermal energy. Equation (5.1-57) has to be rewritten as:

$$\varrho c_p \left(\frac{\partial \bar{T}}{\partial t} + w \frac{\partial \bar{T}}{\partial z} \right) = -2h_c \frac{\bar{T} - T_a}{R} \qquad (6.3\text{-}7)$$

It is also important to include temperature dependent properties for η, ϱ, and c_p.

Using a similar method as for the isothermal case, one can show that there exists a critical draw down ratio, depending on the heat transfer conditions, below which the solution is stable. This is reported in Figure 6.3-5 in term of the Stanton number. The Stanton number for heat transfer to a cylindrical filament is defined with respect to the initial conditions by:

$$\text{St} = 0.26 \frac{k_{\text{air}}}{\nu_{\text{air}}^{1/3}} \frac{L}{\varrho c_p V_0^{2/3} R_0^{5/3}} \tag{6.3-8}$$

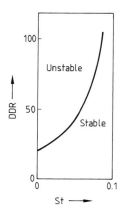

Figure 6.3-5 Critical draw down ratio as a function of the Stanton number, for free convection in ambient air (from DEMAY et al., 1982)

It is interesting to note that when increasing the heat transfer rate, for example by increasing the spinning length in air, larger critical draw down values can be attained. If the Stanton number is large (long enough distance), spinning is stable as experimentally observed for polymer spinning in ambient air.

c) Viscoelastic fluids

To analyze the stability of a Maxwell fluid in spinning, one must consider time effects in Equations (5.1-38 to 5.1-41) (see ZEICHNER, 1973, FISHER and DENN, 1976, and DEMAY, 1983). The dimensionless form of the equations shows another dimensionless number, the Deborah number, a characteristic of the fluid elasticity. The role of this number was discussed in Section 6.1-6. For low Deborah numbers, spinning is stable up to a draw down ratio slightly larger than 20 (Newtonian isothermal case). Above, the solution is unstable and then becomes again stable for $DDR > DDR_2$ (stable drawing for values larger than DDR_2 have never been observed).

For a Deborah number larger than De_c, the periodic instabilities are no longer predicted. The unattainable zone mentioned in Section 5.1 corresponds to an unstable region which is not a bifurcation to a periodic solution. The results are reported in Figure 6.3-6.

6.3-3 Melt fracture

Distortions of the polymer extrudate are observed in conjunction with other phenomena encountered in capillary or flat dies.

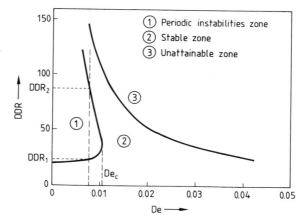

Figure 6.3-6 Critical draw down ratio as a function of the Deborah number (from DEMAY, 1983)

a) Entry flow vortices

Flow visualization of the entry flow in the case of a sudden contraction reveals two main patterns depending on the liquid rheology and flow conditions. These patterns are shown in Figure 6.3-7.

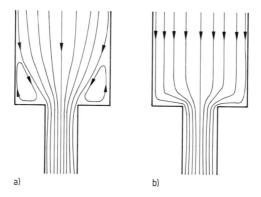

Figure 6.3-7 Main flow patterns in sudden contraction flow
a)   Branched polymer melts such as low density polyethylene, polystyrene and polymethyl methacrylate;
b)   Linear polymer melts as high density polyethylene and linear low density polyethylene

For a large class of polymer melts (usually with branched chains) such as low density polyethylene, polystyrene, polymethylmethacrylate, polyamide, ..., one observes vortices formed in the corners of the entry reservoir (Figure 6.3-7a). The size of these vortices increases with flow rate and then becomes unstable. The unstable flow in the reservoir appears at the same moment as the helical distortion illustrated in Figure 6.3-2. This has been reported by quite a few authors (DEN OTTER, 1970, 1971; BALLENGER et al., 1971; BOGER and RAMAMURTHY, 1972).

For other polymer melts, usually linear polymers, for examples high density polyethylene, linear low density polyethylene, polyisobutadiene, the streamlines in the entry reservoir are all converging as illustrated in Figure 6.3-7b) and no vortices are observed. As reported by VINOGRADOV et al. (1972) and SORNBERGER et al. (1987), this flow pattern remains stable even when "sharkskin" defects are observed.

Other polymers could show a different behaviour. BALLENGER and WHITE (1971) observed helical distortions for the high flow rate of a polypropylene, but no vortices could be detected in the entry reservoir. EL KISSI (1989) observed vortices of different sizes for linear and branched polydimethylsiloxane. These vortices remained stable even when bamboo and chaotic defects were observed, but particles located in the main flow were shown to be suddenly accelerated and decelerated.

b) Flow visualization in capillaries

Flow visualization inside capillaries would shed light on the boundary condition at the polymer die wall interface. This is, however, impractical since

- the shear-thinning behaviour of most polymer melts results in very high velocity gradients at the wall; a small inaccuracy in the measurement of the velocity profile would lead to a non conclusive determination of slip or non slip conditions;

- the technique requires the use of tracer particles of dimensions non negligible with respect to the wall roughness.

The existence of the so-called slip-stick phenomenon may be or not related to the other types of distortion discussed above. BARTOS and HOLOMEK (1971), followed by VINOGRADOV et al. (1972), have reported measurements of pulsating velocities near the capillary wall. However, DEN OTTER (1970, 1971) has shown that the flow pulsations are not related to slip at the wall for several types of polymer melts exhibiting the distortions mentioned above.

c) Pressure-flow rate relationship in a capillary rheometer

A typical pressure profile along the flow axis is sketched in Figure 6.3-8.

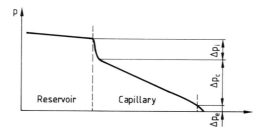

Figure 6.3-8   Pressure profile in a capillary rheometer

There is a small pressure drop in the entry reservoir which is of a large diameter compared to the capillary. The pressure drop is of importance in the capillary and depends on the shear stress evaluated at the capillary wall, i.e.:

$$\Delta p_c = 2\sigma_w \left(\frac{L}{R} \right) \tag{6.3-9}$$

where L and R are respectively the length and radius of the capillary. There is also an excess pressure drop, Δp_i at the entry of the capillary due to the converging flow in the reservoir and due to the flow rearrangement inside the capillary. Finally, one may expect an exit pressure drop, Δp_e, related to the extrudate swell phenomenon and flow rearrangement at the exit.

The work of DEN OTTER (1970, 1971) has shown that melt fracture occurs when either Δp_i or σ_w reaches a critical value. The first case is typical of the flow of low density polyethylene (LDPE) whereas the second has been encountered for high density polyethylene (HDPE). Figure 6.3-9 illustrates the differences observed for the two different polymers (Q is the flow rate).

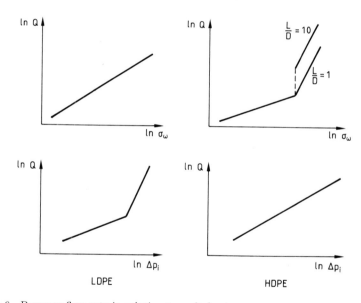

Figure 6.3-9   Pressure-flow rate in relation to melt fracture

For *LDPE*, the occurrence of melt fracture is not related to the L/D ratio (D is the capillary diameter). The helical distortions occur as the slope of the $\ln Q$ vs. $\ln \Delta p_i$ curve is suddenly increased. The curve $\ln Q$ vs. $\ln \sigma_w$ is not affected by the extrudate distortions. The magnitude of the distortions, however, decreases with the increasing length of the capillary.

For HDPE, melt fracture is directly related to the L/D ratio of the capillary. The occurrence of sharkskin does not coincide with discontinuities in either the $\ln Q$ vs. $\ln \Delta p_i$ or $\ln Q$ vs. $\ln \sigma_w$ curve. Chaotic distortions are observed as the slope of the $\ln Q - \ln \sigma_w$ curve is suddenly increased. The $\ln Q$ vs. $\ln \Delta p_i$ curve remains, however, unchanged. The bamboo or slip-stick distortions are observed for long capillaries and characterized by a discontinuity in the $\ln Q - \ln \sigma_w$ curve.

d) Flow birefringence

Flow birefringence is an application of photo-elasticimetry in transparent flow medium (JANESCHITZ-KRIEGL, 1969). The interference patterns, obtained from a transmitted polarized light wave across the medium in motion, can be translated in terms of stresses under the following assumptions:

– the axes of the ellipsoid of indices of propagation are parallel to the stress tensor principal axes;
– the optical sensibility coefficient of the material, C', is independent of temperature and shear.

In polarized circular light, the extinction lines are isochromes related to the difference of the principal stresses by:

$$\sigma_{11} - \sigma_{22} = \frac{K\lambda}{C'e} \qquad (6.3\text{-}10)$$

where λ is the wave length of the light source used, K is the order number of the ray and e is the material thickness. The principal stresses are obtained in writing the stress tensor in the principle axes coordinates. A diagonal matrix is obtained and there is only one difference of the principal stresses for planar flow.

From a photograph as that of Figure 6.3-10, it is possible to obtain a map of the difference of the principal stresses. Such birefringence patterns are characteristics of a given polymer and sometimes extrusion distortions can be detected.

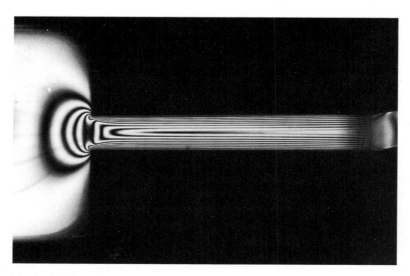

Figure 6.3-10 Birefringence patterns for the flow of LLDPE in a rectangular die (from SORNBERGER et al., 1986)

TORDELLA (1963) observed maximum birefringence at the capillary entrance for the flow of LDPE. The birefringence patterns showed fluctuations as soon as helical distortions occurred. For HDPE, no maximum was observed at the entrance in stable flow conditions. The maximum in birefringence appeared at the same time as chaotic distortions. VINOGRADOV et al. (1972) have reported that the first stage of chaotic distortions in the case of polybutadiene corresponded to perturbations of the extinction lines near the capillary wall. When the flow rate was increased, the perturbations were propagated to the whole flow section. Bamboo distortions corresponded to an alternating thickening and thinning of the interference rings. For linear low density polyethylene, BEAUFILS et al. (1989) quantify the sharkskin defect, using roughness measurements. They link the amplitude of the defect to the maximum value of the flow birefringence observed along the flow axis.

6.3-4   Analysis of extrudate distortions

From the observation of extrudate distortions, flow visualization in the entry reservoir, relationships between flow rate and pressure, birefringence patterns, one may suggest two main types of distortions:
-   helical followed by chaotic distortion (LDPE type);
-   sharkskin, bamboo and then chaotic (HDPE type).

It is not possible to imagine a single physical interpretation for all the distortions observed. Various theories have been proposed to interpret the various distortions. We will not discuss here the details, but present a brief summary. These analyses come sometimes to contradictory results, but more often they generate complementary information.

a)   Poiseuille flow stability

Using the bifurcation method as for the draw resonance study, ROTHENBERGER et al. (1973) have shown that the Poiseuille flow of a Maxwell fluid will degenerate into a periodic pulsating flow when the shear at the wall exceeds a critical value. MCINTIRE (1972) used the same method, but for an integral type viscoelastic constitutive equation.

b)   Highly elastic state

VINOGRADOV and MANIN (1965) has postulated that a polymer melt will attain a state of high elasticity when submitted to a high enough shear stress. The entanglements become almost permanent and the deformation or strain in the quasi-solid material is restricted. The flow behaviour is no longer that of a fluid, hence bamboo distortions (or slip-stick phenomenon) are observed.

c)   Melt fracture

From the work of TORDELLA (1956) and that of DEN OTTER (1971), melt fracture can be interpreted in terms of elastic energy storage. The energy stored in a viscoelastic liquid is related to the deformation of the polymer chains. This energy is necessarily limited and critical deformation rates can not be attained without affecting the structure of the polymer:

–   The stress under extensional flow rapidly increases with the extensional rate. Above a critical rate, melt fracture occurs.

–   Melt fracture is also possible in simple shear flow as shown by the basic experiment of HUTTON (1963). In a cone-and-plate geometry, the torque (hence the shear stress) remains constant with time provided the shear rate does not exceed a critical value. For $\dot\gamma > \dot\gamma_c$, one can observe a decreasing torque with time which finally attains a stable but much lower value than the initial pseudo steady-state value (see Figure 6.3-11). Melt fracture is then detected along shear surfaces with a part of the liquid attached to the cone and the rest attached to the plate. The low shear stress then measured corresponds to the shear force acting on the fracture surface.

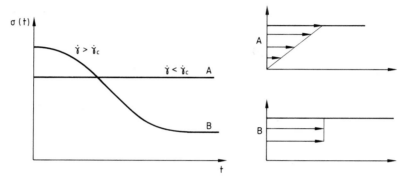

Figure 6.3-11   Melt fracture in simple shear

If the cone (or plate) is stopped in the situation B illustrated in Figure 6.3-11, both parts of the liquid will weld and a subsequent experiment will yield the same torque response as the first one. The phenomenon is reversible and possibly due to disentanglement followed by re-entanglement of the polymer chains along the fracture surface. As there are two possible values of $\dot\gamma$ for the same value of the torque, this is an unstable system.

The distortion of the HDPE type can be interpreted as a "liquid state fracture". The oscillatory flow regime corresponds to the following series of mechanisms:

increasing stresses $\rightarrow$ fracture $\rightarrow$ increasing flow rate

$\uparrow$ $\downarrow$

decreasing flow rate $\leftarrow$ re-welding $\leftarrow$ stress relaxation

This, however, has not yet been quantified adequately. In particular, the following questions have not been satisfactorily answered:

–   Is melt facture occurring at the polymer-capillary wall interface or inside the polymer flow?

–   Is the fracture occurring at the entrance of the capillary (stresses in excess) along the capillary or at the exit in the extrudate swell?

d) Slip-stick phenomenon

BENBOW et al. (1961) have postulated that the polymer melt looses adhesion at the metallic wall and slips when the wall shear stress exceeds a critical value. The wall shear stress then decreases to a value lower than the critical one and adhesion is promoted. The shear stress increases again as observed by the cyclic flow of HDPE. This hypothesis was reexamined by UHLAND (1976) who divided the die into two zones:

– a zone without slippage close to the entrance;
– a downstream zone with progressive increasing polymer slip velocity.

e) Flow re-arrangement at the die exit

BERGEM (1976) has related the occurrence of the sharkskin defects to flow re-arrangement between the core and the surface of the extrudate at the exit of the die.

f) Non-monotonous stress response

HUSEBY (1966) has calculated the stress response, $\sigma(\dot{\gamma})$, of a HDPE from its relaxation spectrum. The result was a non monotonous curve illustrated in Figure 6.3-12. This was experimentally confirmed by TORRE–GROSSA (1984) using a Couette viscometer.

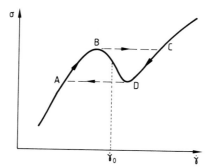

Figure 6.3-12   Unstable flow of HDPE

The decreasing part of the curve corresponds to an unstable regime in capillary flow. To a given flow rate corresponding to a shear rate $\dot{\gamma}_0$, the stress response will follow an hysteresis loop described by the parallelogram ABCD shown on the figure. This hysteresis explains the cyclic distortions observed for HDPE.

g) Cyclic relaxation

WEILL (1978, 1980) has associated the capillary rheometer to a relaxation oscillator made of a capacitance (the entry reservoir in which the polymer compressibility can not be neglected) and of a resistance (the capillary). Under given conditions, the flow rate in the capillary is smaller than that imposed in the reservoir. The polymer is then compressed in the reservoir, and suddenly extruded through the capillary in a cyclic manner exhibiting cork and bamboo type distortions.

The sharkskin phenomenon may be interpreted in a similar fashion by noticing that the pressure drop could be larger at the capillary entrance (see Figure 6.3-8) than across the capillary length. For an intermediate flow rate for which the die is capable of extruding the polymer at constant velocity, the cyclic oscillations could originate between the reservoir and the entrance of the capillary. The amplitude of the oscillations could then be dampened along the capillary length. The distortions are of small amplitude as they correspond to the small volume of the converging flow at the capillary entrance.

h) Secondary flow instability

The extrudate distortions for the flow of low density polyethylene (LDPE) appear to be due to the periodic discharge in the capillary of the vortices of the secondary flow observed in the entry reservoir (see Figure 6.3-13). How can we explain that?

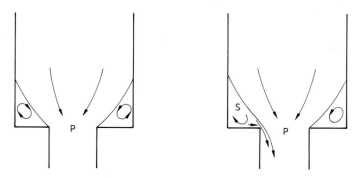

Figure 6.3-13   Secondary flow and extrudate distortions

- Let us consider that the main flow is under steady-state regime. The stress in the z-direction at the contraction is σ_{zp}, corresponding to the extension rate $\dot{\varepsilon}_p$.
- Let us assume now that a small perturbation at time $t = 0$ initiates the discharge of the vortex into the capillary. The extensional rate in the secondary flow, $\dot{\varepsilon}_s$ is larger that $\dot{\varepsilon}_p$ due to the relative dimensions of both flows. The stress $\sigma_{zs}(t)$ increases from zero to attain with time a theoretical value larger than σ_{zp} since $\dot{\varepsilon}_s$ is larger than $\dot{\varepsilon}_p$ (see Problem 6.1-D).
- As far as the stress $\sigma_{zs}(t)$ does not exceed the value of σ_{zp}, the perturbation will tend to grow since the contribution to the pressure drop due to the secondary flow is less than the contribution due to the main flow. However, when $\sigma_{zs}(t)$ exceeds σ_{zp}, the energy in the main flow becomes less than that coming from the vortices which then slow down. This alternating process generates a periodic partial discharge of the vortices and periodic distortions of the flow.
- The influence of the flow rate on the distortions frequency can be qualitatively explained by the increasing size of the vortices with flow rate. Hence $\dot{\varepsilon}_s$ is

more and more important with respect to $\dot{\varepsilon}_p$ and $\sigma_{zs}(t)$ attains more rapidly the value of σ_{zp}.

– As the phenomenon occurs at the entry of the capillary, it is then expected that the magnitude of the extrudate distortions will decrease as longer capillaries are used.

– Finally, we notice that the propagation of the perturbations along the circumference of the contraction as described above could lead to helocoidal distortions, as frequently observed.

6.3-5 Instability criteria

Due to the complexity of the various phenomena encountered and the impossibility to find satisfactory explanations, many authors have proposed various empirical criteria for the occurrence of distortions.

a) Critical shear stress

The interpretation of the various distortions presented above suggest that there exists a critical shear above which instabilities or distortions will appear.

AGASSANT (1984) has shown that the surface distortions in the calendering of PVC occur at a critical shear stress equal to 0.52 MPa. His results are reported in Figure 6.3-14.

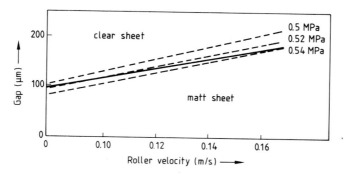

Figure 6.3-14 Critical shear stress in calendering of PVC (from AGASSANT, 1984)
——— experimental,   - - - - - - - computed

A similar analysis has been used by HAMMOND (1960) and by KERTSCHER (1973) to predict the instabilities in wire coating with low density polyethylene. These authors have shown, however, that the critical shear stress varied with entry angle of the die. This is then in line with the physical interpretation of extrudate distortions presented above for LDPE.

VLACHOPOULOS and ALAM (1972) have proposed a simple correlation to relate the critical shear stress to the weight-average molecular weight, M_w, and the absolute temperature, T. Their result restricted to linear polymers is:

$$\frac{\sigma_c}{T} = 1717 + \frac{2.7 \times 10^8}{M_w} \tag{6.3-11}$$

b) Weissenberg number

BALLENGER et al. (1971) and McINTIRE (1972) have used the Weissenberg number as a criterion of distortions occurrence, i.e.:

$$We = \frac{N_1}{2\sigma_w} = \text{constant} \qquad (6.3\text{-}12)$$

and VLACHOPOULOS and ALAM (1972) have suggested the following correlation of the critical Weissenberg number for polystyrene:

$$We = \frac{2.65}{M_z M_{z+1}/M_w{}^2} \qquad (6.3\text{-}13)$$

where M_z and M_{z+1} are respectively the second and third moment weight-average molecular weights (see for example RUDIN, 1982). These higher average molecular weights give a better representation of the long chains of polydisperse polymers.

Appendices

Appendix A
Basic Formulae

A.1 Cylindrical coordinates

Gradient of a scalar:

$$\nabla f = \left(\frac{\partial f}{\partial r}, \; \frac{1}{r}\frac{\partial f}{\partial \theta}, \; \frac{\partial f}{\partial z} \right) \tag{A.1-1}$$

Material derivation or substantial derivative of a scalar:

$$\frac{Df}{Dt} = \frac{\partial f}{\partial t} + u\frac{\partial f}{\partial r} + \frac{v}{r}\frac{\partial f}{\partial \theta} + w\frac{\partial f}{\partial z} \tag{A.1-2}$$

Laplacian of a scalar:

$$\nabla^2 f = \frac{1}{r}\frac{\partial}{\partial r}\left(r\frac{\partial f}{\partial r} \right) + \frac{1}{r^2}\frac{\partial^2 f}{\partial \theta^2} + \frac{\partial^2 f}{\partial z^2} \tag{A.1-3}$$

Velocity gradient tensor:

$$\nabla \boldsymbol{u} = \begin{bmatrix} \dfrac{\partial u}{\partial r} & \dfrac{1}{r}\left(\dfrac{\partial u}{\partial \theta} - v \right) & \dfrac{\partial u}{\partial z} \\[2mm] \dfrac{\partial v}{\partial r} & \dfrac{1}{r}\left(\dfrac{\partial v}{\partial \theta} + u \right) & \dfrac{\partial v}{\partial z} \\[2mm] \dfrac{\partial w}{\partial r} & \dfrac{1}{r}\left(\dfrac{\partial w}{\partial \theta} \right) & \dfrac{\partial w}{\partial z} \end{bmatrix} \tag{A.1-4}$$

Rate-of-strain tensor:

$$\dot{\varepsilon} = \begin{bmatrix} \dfrac{\partial u}{\partial r} & \dfrac{1}{2}\left(\dfrac{\partial v}{\partial r} - \dfrac{v}{r} + \dfrac{1}{r}\dfrac{\partial u}{\partial \theta} \right) & \dfrac{1}{2}\left(\dfrac{\partial u}{\partial z} + \dfrac{\partial w}{\partial r} \right) \\[2mm] \dfrac{1}{2}\left(\dfrac{\partial v}{\partial r} - \dfrac{v}{r} + \dfrac{1}{r}\dfrac{\partial u}{\partial \theta} \right) & \dfrac{1}{r}\dfrac{\partial v}{\partial \theta} + \dfrac{u}{r} & \dfrac{1}{2}\left(\dfrac{1}{r}\dfrac{\partial w}{\partial \theta} + \dfrac{\partial v}{\partial z} \right) \\[2mm] \dfrac{1}{2}\left(\dfrac{\partial u}{\partial z} + \dfrac{\partial w}{\partial r} \right) & \dfrac{1}{2}\left(\dfrac{1}{r}\dfrac{\partial w}{\partial \theta} + \dfrac{\partial v}{\partial z} \right) & \dfrac{\partial w}{\partial z} \end{bmatrix} \tag{A.1-5}$$

Dynamic equilibrium:

r-component:

$$\frac{\partial \sigma_{rr}}{\partial r} + \frac{1}{r}\frac{\partial \sigma_{r\theta}}{\partial \theta} + \frac{\partial \sigma_{rz}}{\partial z} + \frac{\sigma_{rr} - \sigma_{\theta\theta}}{r} + F_r - \varrho\gamma_r = 0 \tag{A.1-6}$$

θ-component:

$$\frac{\partial \sigma_{r\theta}}{\partial r} + \frac{1}{r}\frac{\partial \sigma_{\theta\theta}}{\partial \theta} + \frac{\partial \sigma_{\theta z}}{\partial z} + 2\frac{\sigma_{r\theta}}{r} + F_\theta - \varrho\gamma_\theta = 0 \tag{A.1-7}$$

z-component:

$$\frac{\partial \sigma_{rz}}{\partial r} + \frac{1}{r}\frac{\partial \sigma_{\theta z}}{\partial \theta} + \frac{\partial \sigma_{zz}}{\partial z} + \frac{\sigma_{rz}}{r} + F_z - \varrho \gamma_z = 0 \qquad \text{(A.1-8)}$$

Navier-Stokes equations:

r-component:

$$-\frac{\partial p}{\partial r} + \eta\left(\frac{\partial}{\partial r}\left(\frac{1}{r}\frac{\partial}{\partial r}(ru)\right) + \frac{1}{r^2}\frac{\partial^2 u}{\partial \theta^2} - \frac{2}{r^2}\frac{\partial v}{\partial \theta} + \frac{\partial^2 u}{\partial z^2}\right) + F_r$$

$$= \varrho\left(\frac{du}{dt} - \frac{v^2}{r}\right) \qquad \text{(A.1-9)}$$

θ-component:

$$-\frac{1}{r}\frac{\partial p}{\partial \theta} + \eta\left(\frac{\partial}{\partial r}\left(\frac{1}{r}\frac{\partial}{\partial r}(rv)\right) + \frac{1}{r^2}\frac{\partial^2 v}{\partial \theta^2} + \frac{2}{r^2}\frac{\partial u}{\partial \theta} + \frac{\partial^2 v}{\partial z^2}\right) + F_\theta$$

$$= \varrho\left(\frac{dv}{dt} + \frac{uv}{r}\right) \qquad \text{(A.1-10)}$$

r-component:

$$-\frac{\partial p}{\partial z} + \eta\left(\frac{1}{r}\frac{\partial}{\partial r}\left(r\frac{\partial w}{\partial r}\right) + \frac{1}{r^2}\frac{\partial^2 w}{\partial \theta^2} + \frac{\partial^2 w}{\partial z^2}\right) + F_z = \varrho\frac{dw}{dt} \qquad \text{(A.1-11)}$$

Material or substantial derivative of a tensor:

$$\frac{D\mathbf{a}}{Dr} = \begin{bmatrix} \frac{Da_{rr}}{Dt} - 2\frac{v}{r}a_{r\theta} & \frac{Da_{r\theta}}{Dt} + \frac{v}{r}(a_{rr} - a_{\theta\theta}) & \frac{Da_{rz}}{Dt} - \frac{v}{r}a_{\theta z} \\ \frac{Da_{r\theta}}{Dt} + \frac{v}{r}(a_{rr} - a_{\theta\theta}) & \frac{Da_{\theta\theta}}{Dt} + 2\frac{v}{r}a_{r\theta} & \frac{Da_{\theta z}}{Dt} + \frac{v}{r}a_{rz} \\ \frac{Da_{rz}}{Dt} - \frac{v}{r}a_{\theta z} & \frac{Da_{\theta z}}{Dt} + \frac{v}{r}a_{rz} & \frac{Da_{zz}}{Dt} \end{bmatrix} \qquad \text{(A.1-12)}$$

A.2   Spherical coordinates

Gradient of a scalar:

$$\nabla f = \left(\frac{\partial f}{\partial r}, \frac{1}{r}\frac{\partial f}{\partial \theta}, \frac{1}{r\sin\theta}\frac{\partial f}{\partial \varphi}\right) \qquad \text{(A.2-1)}$$

Material or substantial derivative of a scalar:

$$\frac{Df}{Dt} = \frac{\partial f}{\partial t} + u\frac{\partial f}{\partial r} + \frac{v}{r}\frac{\partial f}{\partial \theta} + \frac{w}{r\sin\theta}\frac{\partial f}{\partial \varphi} \qquad \text{(A.2-2)}$$

Laplacian of a scalar:

$$\Delta f = \frac{1}{r^2} \frac{\partial}{\partial r}\left(r^2 \frac{\partial f}{\partial r}\right) + \frac{1}{r^2 \sin\theta} \frac{\partial}{\partial\theta}\left(\sin\theta \frac{\partial f}{\partial\theta}\right) + \frac{1}{r^2 \sin^2\theta} \frac{\partial^2 f}{\partial\varphi^2} \qquad \text{(A.2-3)}$$

Velocity gradient tensor:

$$\nabla \boldsymbol{u} = \begin{bmatrix} \dfrac{\partial u}{\partial r} & \dfrac{1}{r}\dfrac{\partial u}{\partial\theta} - \dfrac{v}{r} & \dfrac{1}{r\sin\theta}\dfrac{\partial u}{\partial\varphi} - \dfrac{w}{r} \\[2mm] \dfrac{\partial v}{\partial r} & \dfrac{1}{r}\dfrac{\partial v}{\partial\theta} + \dfrac{u}{r} & \dfrac{1}{r\sin\theta}\dfrac{\partial v}{\partial\varphi} - \dfrac{\cot\theta}{r} w \\[2mm] \dfrac{\partial w}{\partial r} & \dfrac{1}{r}\dfrac{\partial w}{\partial\theta} & \dfrac{1}{r\sin\theta}\dfrac{\partial w}{\partial\varphi} + \dfrac{v}{r}\cot\theta + \dfrac{u}{r} \end{bmatrix} \qquad \text{(A.2-4)}$$

Rate-of-deformation tensor:

$$\dot{\varepsilon} =$$

$$\begin{bmatrix} \dfrac{\partial u}{\partial r} & \dfrac{1}{2}\left(\dfrac{1}{r}\dfrac{\partial u}{\partial\theta} + r\dfrac{\partial(v/r)}{\partial r}\right) & \dfrac{1}{2}\left(\dfrac{1}{r\sin\theta}\dfrac{\partial u}{\partial\varphi} + r\dfrac{\partial(w/r)}{\partial r}\right) \\[2mm] \dfrac{1}{2}\left(\dfrac{1}{r}\dfrac{\partial u}{\partial\theta} + r\dfrac{\partial(v/r)}{\partial r}\right) & \dfrac{1}{r}\dfrac{\partial v}{\partial\theta} + \dfrac{u}{r} & \dfrac{1}{2}\left(\dfrac{1}{r}\dfrac{\partial w}{\partial\theta} + \dfrac{1}{r\sin\theta}\dfrac{\partial v}{\partial\varphi} - \dfrac{\cot\theta}{r} w\right) \\[2mm] \dfrac{1}{2}\left(\dfrac{1}{r\sin\theta}\dfrac{\partial u}{\partial r} + r\dfrac{\partial(w/r)}{\partial r}\right) & \dfrac{1}{2}\left(\dfrac{1}{r}\dfrac{\partial w}{\partial\theta} + \dfrac{1}{r\sin\theta}\dfrac{\partial v}{\partial\varphi} - \dfrac{\cot\theta}{r} w\right) & \dfrac{1}{r\sin\theta}\dfrac{\partial w}{\partial\varphi} + \dfrac{v}{r}\cot\theta + \dfrac{u}{r} \end{bmatrix}$$

$$\text{(A.2-5)}$$

Dynamic equilibrium:

r-component:

$$\frac{\partial\sigma_{rr}}{\partial r} + \frac{1}{r}\frac{\partial\sigma_{r\theta}}{\partial\theta} + \frac{1}{r\sin\theta}\frac{\partial\sigma_{r\varphi}}{\partial\varphi} + \frac{1}{r}\left(2\sigma_{rr} - \sigma_{\theta\theta} - \sigma_{\varphi\varphi} + \sigma_{r\theta}\cot\theta\right)$$

$$+ F_r - \varrho\gamma_r = 0 \qquad \text{(A.2-6)}$$

θ-component:

$$\frac{\partial\sigma_{r\theta}}{\partial r} + \frac{1}{r}\frac{\partial\sigma_{\theta\theta}}{\partial\theta} + \frac{1}{r\sin\theta}\frac{\partial\sigma_{\theta\varphi}}{\partial\varphi} + \frac{1}{r}\left(3\sigma_{r\theta} + (\sigma_{\theta\theta} - \sigma_{\varphi\varphi})\cot\theta\right)$$

$$+ F_\theta - \varrho\gamma_\theta = 0 \qquad \text{(A.2-7)}$$

φ-component:

$$\frac{\partial\sigma_{r\varphi}}{\partial r} + \frac{1}{r}\frac{\partial\sigma_{\theta\varphi}}{\partial\theta} + \frac{1}{r\sin\theta}\frac{\partial\sigma_{\varphi\varphi}}{\partial\varphi} + \frac{1}{r}\left(3\sigma_{r\varphi} + 2\sigma_{\theta\varphi}\cot\theta\right)$$

$$+ F_\varphi - \varrho\gamma_\varphi = 0 \qquad \text{(A.2-8)}$$

Navier-Stokes equations:

r-component:

$$-\frac{\partial p}{\partial r} + \eta\left(\nabla^2 u - \frac{2u}{r^2} - \frac{2}{r^2}\frac{\partial v}{\partial\theta} - \frac{2}{r^2} v\cot\theta - \frac{2}{r^2\sin\theta}\frac{\partial w}{\partial\varphi}\right)$$

$$+ F_r - \varrho\gamma_r = 0 \qquad \text{(A.2-9)}$$

θ-component:

$$-\frac{1}{r}\frac{\partial p}{\partial \theta} + \eta\left(\nabla^2 v + \frac{2}{r^2}\frac{\partial u}{\partial \theta} - \frac{v}{r^2 \sin^2 \theta} - \frac{2\cos\theta}{r^2 \sin^2 \theta}\frac{\partial w}{\partial \varphi}\right)$$
$$+F_\theta - \varrho\gamma_\theta = 0 \qquad\qquad (A.2\text{-}10)$$

φ-component:

$$-\frac{1}{r\sin\theta}\frac{\partial p}{\partial \varphi} + \eta\left(\nabla^2 w - \frac{w}{r^2 \sin^2 \theta} + \frac{2}{r^2 \sin^2 \theta}\frac{\partial u}{\partial \varphi} + \frac{2\cos\theta}{r^2 \sin^2 \theta}\frac{\partial v}{\partial \varphi}\right)$$
$$+F_\varphi - \varrho\gamma_\varphi = 0 \qquad\qquad (A.2\text{-}11)$$

where:

$$\gamma_r = \frac{du}{dt} - \frac{v^2 + w^2}{r} \qquad\qquad (A.2\text{-}12)$$

$$\gamma_\theta = \frac{dv}{dt} + \frac{uv}{r} - \frac{w^2 \cot\theta}{r} \qquad\qquad (A.2\text{-}13)$$

$$\gamma_\varphi = \frac{dw}{dt} + \frac{uw}{r} + \frac{vw}{r}\cot\theta \qquad\qquad (A.2\text{-}14)$$

A.3   Table of the error function (erf x)

x	erf x	x	erf x	x	erf x
0	0	0.70	0.677801	1.8	0.989091
0.05	0.056372	0.75	0.711156	1.9	0.992790
0.10	0.112463	0.80	0.742101	2.0	0.995322
0.15	0.167996	0.85	0.770668	2.1	0.997021
0.20	0.222703	0.90	0.796908	2.2	0.998137
0.25	0.276326	0.95	0.820891	2.3	0.998857
0.30	0.328627	1.0	0.842701	2.4	0.999311
0.35	0.379382	1.1	0.880205	2.5	0.999593
0.40	0.428392	1.2	0.910314	2.6	0.999764
0.45	0.475482	1.3	0.934008	2.7	0.999866
0.50	0.520500	1.4	0.952285	2.8	0.999925
0.55	0.563323	1.5	0.966105	2.9	0.999959
0.60	0.603856	1.6	0.976348	3.0	0.999978
0.65	0.642029	1.7	0.983790		

Appendix B
Elements of Tensor Analysis

We review in this appendix basic notions of tensor analysis useful for the understanding of principles on which are based constitutive equations.

B.1 Invariants of a tensor

B.1-1 Definitions

A second-order tensor can be represented as a matrix $\boldsymbol{M}$ in a frame of reference R. Defining $\boldsymbol{P}$ as the matrix of transformation from the frame of reference R to R', the tensor in the frame R' is then given by:

$$\boldsymbol{M'} = \boldsymbol{P}^{-1}\boldsymbol{M}\boldsymbol{P} \tag{B.1-1}$$

For many tensors, for example all symmetric tensors, there is a unique frame of reference in which $\boldsymbol{M}$ is a diagonal matrix:

$$\boldsymbol{M} = \begin{bmatrix} \lambda_1 & 0 & 0 \\ 0 & \lambda_2 & 0 \\ 0 & 0 & \lambda_3 \end{bmatrix} \tag{B.1-2}$$

The λ are the eigenvalues or principal values and the frame of reference is defined by the directions of the eigenvectors, usually called in fluid mechanics principal directions. If $\boldsymbol{M} = \boldsymbol{\varepsilon}$, then the λ are principal extensions; if $\boldsymbol{M} = \boldsymbol{\sigma}$, the λ are principal stresses. By definition the eigenvalues are characteristics of the tensor at a given point and are independent of the frame of reference (or choice of coordinate system). They are solutions of the characteristic equation:

$$\lambda^3 - I_1\lambda^2 + I_2\lambda - I_3 = 0 \tag{B.1-3}$$

where:

$$I_1 = \lambda_1 + \lambda_2 + \lambda_3 \tag{B.1-4}$$

$$I_2 = \lambda_1\lambda_2 + \lambda_2\lambda_3 + \lambda_3\lambda_1 \tag{B.1-5}$$

$$I_3 = \lambda_1\lambda_2\lambda_3 \tag{B.1-6}$$

I_1, I_2, I_3 are the invariants of the tensor $\boldsymbol{M}$. We can define other invariants:

$$J_1 = m_{11} + m_{22} + m_{33} = \operatorname{tr} \boldsymbol{M} \tag{B.1-7}$$

$$J_2 = \frac{1}{2} \sum_{ij} m_{ij} m_{ji} = \frac{1}{2} \operatorname{tr} \boldsymbol{M}^2 \tag{B.1-8}$$

$$J_3 = \frac{1}{3} \sum_{ijk} m_{ij} m_{jk} m_{ki} = \frac{1}{3} \operatorname{tr} \boldsymbol{M}^3 \tag{B.1-9}$$

These invariants are in fact combinations of the first three:

$$J_1 = I_1 \tag{B.1-10}$$

$$2J_2 = I_1{}^2 - 2I_2 \tag{B.1-11}$$

$$3J_3 = 3I_3 - 3I_1 I_2 + I_1{}^3 \tag{B.1-12}$$

B.1-2 Invariants used in fluid mechanics

The most frequently used invariants in fluid mechanics are:

The trace

-   hydrostatic pressure: $p = -\dfrac{1}{3} \operatorname{tr} \boldsymbol{\sigma}$
-   compressibility: $\dfrac{1}{V} \dfrac{dV}{dt} = \operatorname{tr} \dot{\boldsymbol{\varepsilon}}$

For incompressible fluids, $\operatorname{tr} \dot{\boldsymbol{\varepsilon}} = 0$, and both sets of invariants (I and J) of the rate-of-strain tensor, $\dot{\boldsymbol{\varepsilon}}$, have the same absolute value (the tensor $\dot{\boldsymbol{\varepsilon}}$ is symmetrical).

The second invariant J_2

If one considers $\boldsymbol{M}$ as a vector of 9 coordinates m_{ij} in a 9 dimensions space, J_2 is proportional to the square of the length of the vector. $\sqrt{J_2}$ is a measure of the length of the vector, i.e. of the scalar intensity of the tensor. Following the theory of plasticity, one can use the Von Misès criterion:

$$\sqrt{J_2(\boldsymbol{\sigma})} = \frac{1}{\sqrt{3}} \sigma_0 \tag{B.1-13}$$

A generalized rate of deformation can be defined by:

$$\dot{\bar{\varepsilon}} = \frac{2}{\sqrt{3}} \sqrt{J_2(\dot{\boldsymbol{\varepsilon}})} = \sqrt{\frac{2}{3} \sum_{i,j} \dot{\varepsilon}_{ij}^2} \tag{B.1-14}$$

However, as shear flows play an important role in fluid mechanics, it is more appropriate to define the following generalized deformation rate:

$$\dot{\bar{\gamma}} = \sqrt{2 \sum_{ij} \dot{\varepsilon}_{ij}^2} = 2\sqrt{J_2(\dot{\boldsymbol{\varepsilon}})} \tag{B.1-15}$$

This definition reduces in simple shear flow to:

$$\dot{\varepsilon} = \begin{bmatrix} 0 & \frac{\dot{\gamma}}{2} & 0 \\ \frac{\dot{\gamma}}{2} & 0 & 0 \\ 0 & 0 & 0 \end{bmatrix}, \quad \dot{\bar{\gamma}} = \dot{\gamma} \tag{B.1-16}$$

and in extensional flow:

$$\dot{\varepsilon} = \begin{bmatrix} \dot{\varepsilon} & 0 & 0 \\ 0 & -\frac{\dot{\varepsilon}}{2} & 0 \\ 0 & 0 & -\frac{\dot{\varepsilon}}{2} \end{bmatrix}, \quad \dot{\bar{\gamma}} = \sqrt{3}\dot{\varepsilon} \tag{B.1-17}$$

The third invariant

For a large class of flow situations, the third invariant is equal to zero: for example in planar and viscometric flows. For flow situations where it is not equal to zero, it is difficult to account for as there are no simple experiments that can test its influence.

B.2   Convected derivative

It has been shown in Chapter 6.1 that the partial or substantial derivative has no longer physical meaning when applied to a tensor. OLDROYD (1950) has introduced an appropriate derivative that is independent of the frame of reference. The basic idea was to look after the variation of the tensor in a frame embedded in the fluid and deformed with it. This frame is called convected coordinate system and the derivative is the convected derivative. At a given time, the convected coordinates may be orthonormal, but they will not be at another time. So tensorial notions in curvilinear coordinates have to be introduced.

The objective of the appendix is to present the general formalism that leads to the development of the convected derivative and to obtain useful and meaningful expressions for the convected derivatives of the tensors arising in fluid mechanics and rheology.

B.2-1 Curvilinear coordinates

Let us define an orthogonal coordinate system, e_1, e_2, e_3. Any vector connecting the origin and a point M is given, in terms of contravariant components, by:

$$\boldsymbol{OM} = x^1\boldsymbol{e}_1 + x^2\boldsymbol{e}_2 + x^3\boldsymbol{e}_3 = x^i\boldsymbol{e}_i \tag{B.2-1}$$

Curvilinear coordinates may be defined by the following changes of variables:

$$\xi^1 = \psi^1(x^1, x^2, x^3) \tag{B.2-2}$$

$$\xi^2 = \psi^2(x^1, x^2, x^3) \tag{B.2-3}$$

$$\xi^3 = \psi^3(x^1, x^2, x^3) \tag{B.2-4}$$

where the ψ^i are continuously differentiable functions with Jacobian $\partial(\xi^1, \xi^2, \xi^3)/\partial(x^1, x^2, x^3)$ different from zero or infinity at any point. This implies that there exists a unique inverse relation:

$$\xi^i = \psi^i(x^j) \quad \Leftrightarrow \quad x^i = f^i(\xi^j) \tag{B.2-5}$$

If all the points (x^1, x^2, x^3) from a material domain D are uniquely related by the f^i and the ψ^i with the points (ξ^1, ξ^2, ξ^3) of the domain Δ in R^3, D is then by definition a differentiable (class C_1) three-dimension manifold.

Let us consider the infinitesimal vector about point M:

$$d\boldsymbol{M} = dx^i \, \boldsymbol{e}_i \tag{B.2-6}$$

From the chain rule we can write:

$$dx^i = \frac{\partial f^i}{\partial \xi^j} \, d\xi^j, \quad \text{hence} \quad d\boldsymbol{M} = d\xi^j \varepsilon_j \quad \text{with} \quad \varepsilon_j = \frac{\partial f^i}{\partial \xi^j} \boldsymbol{e}_i \tag{B.2-7}$$

The base vectors, ε_1, ε_2, ε_3, satisfying (B.2-6) are thus defined as the vectors tangent to the curvilinear axes at point M. From the inverse relation:

$$\varepsilon_j = \frac{\partial f^i}{\partial \xi^j} \boldsymbol{e}_i \quad \Leftrightarrow \quad \boldsymbol{e}_i = \frac{\partial \psi^j}{\partial x^i} \varepsilon_j \tag{B.2-8}$$

From the definition of the quantities $\partial x^i / \partial \xi^j$, transformations (B.2-8) may be also written as:

$$\varepsilon^j = \frac{\partial x^i}{\partial \xi^j} \boldsymbol{e}^i \quad \Leftrightarrow \quad \boldsymbol{e}^i = \frac{\partial \xi^j}{\partial x^i} \varepsilon^j \tag{B.2-9}$$

These are the transformation rules of the base vectors in curvilinear coordinates.

B.2-2 Expression for a tensor

The transformation from the frame of reference $\boldsymbol{e}_i$ to $\boldsymbol{e}'_i$ can be achieved with the help of a transformation matrix $a_i{}^j$ such that:

$$\boldsymbol{e}'_i = a_i{}^j \boldsymbol{e}_j \quad \Leftrightarrow \quad x^i = a_j{}^i x'^j \tag{B.2-10}$$

In general, any quantity $t^{...n}_{...m}$ of order n in contravariant components and of order m in covariant components is a tensorial quantity if it obeys the following transformation rule:

$$\prod_j a_j^n t'^{...j}_{...m...}{}^{...} = \prod_i a_m^i t^{...n...}_{...i...} \tag{B.2-11}$$

where a_i^j are the transformation matrices.

Metric tensor

In the coordinate system defined by e_1, e_2, e_3, any infinitesimal distance about point M is, using (B.2-6), given by:

$$dM \cdot dM = dM^2 = dx^i \, dx^j \, e_i \cdot e_j \qquad (\text{B.2-12})$$

We define the tensor metric by:

$$g_{ij} = e_i \cdot e_j \qquad (\text{B.2-13})$$

where g_{ij} are the covariant components of the metric tensor. If we consider a change of coordinates with the transformation matrix $a_i{}^j$:

$$g'_{kl} = e'_k \cdot e'_l = a^i_k a^j_l e_i \cdot e_j = a^i_k a^j_l \, g_{ij} \quad \text{hence}: \quad a^k_i g'_{kl} = a^j_l g_{ij}$$

and the quantity g_{ij} follows the rule (B.2-11) for the covariant components of a second order tensor.

The metric tensor in a Cartesian coordinate system (orthonormal base vectors) becomes:

$$g_{ij} = \delta_{ij} \qquad (\text{B.2-14})$$

where δ_{ij} is the Kronecker delta.

B.2-3 Transformation of a tensor

In the following section, coordinates x^1, x^2, x^3, ξ^1, ξ^2, ξ^3 will be denoted x and ξ respectively.
It is not possible to specify a tensorial field based on coordinates ξ^i since, in general, there are no linear relations betweens x^i and ξ^i. This requires the specification of a tangent vector space.

Let us consider the components of a tensor $t^{\cdots n \cdots}_{\cdots m \cdots}(x)$ in a fixed coordinate system e_j. At each point in the tangent vector space ε_i, we can write:

$$\varepsilon_i = \frac{\partial x^j}{\partial \xi^i} e_j \quad \Leftrightarrow \quad dx^i = \frac{\partial x^i}{\partial \xi^j} d\xi^j \qquad (\text{B.2-15})$$

hence the tensor $\tau^{\cdots n \cdots}_{\cdots m \cdots}(\xi)$ in the space ε_i obeys to:

$$\prod_j \left(\frac{\partial x^n}{\partial \xi^j} \right) \tau^{\cdots j \cdots}_{\cdots m \cdots}(\xi) = \prod_i \left(\frac{\partial x^i}{\partial \xi^m} \right) t^{\cdots n \cdots}_{\cdots i \cdots}(x) \qquad (\text{B.2-16})$$

This result can be written as:

$$\tau^{\cdots n \cdots}_{\cdots m \cdots}(\xi) = \prod_i \left(\frac{\partial x^i}{\partial \xi^m} \right) \prod_j \left(\frac{\partial \xi^n}{\partial x^j} \right) t^{\cdots j \cdots}_{\cdots i \cdots}(x) \qquad (\text{B.2-17})$$

This is the transformation rule of a tensor from a fixed coordinates system to a nonorthogonal curvilinear system.

Example

The metric tensor defined above in the space e_1, e_2, e_3 can be written in curvilinear coordinates using the transformation rule (B.2-17). The covariant components are then:

$$\gamma_{ij}(\xi) = \frac{\partial x^k}{\partial \xi^i} \frac{\partial x^l}{\partial \xi^j} g_{kl}(x) = \frac{\partial x^k}{\partial \xi^i} \frac{\partial x^l}{\partial \xi^j} \delta_{kl} \qquad \text{(B.2-18)}$$

B.2-4 Weight of a tensor

Physical tensorial quantities have units. For example, the stress tensor $\boldsymbol{\sigma}$ has units of N/m^2 or Pa as clearly seen from the expression of a force acting on a surface element, i.e.

$$dF^i = \sigma^{ij} n_j \, dS \qquad \text{(B.2-19)}$$

where n_j are the components of the unit vector normal to the surface element dS.

Any scalar quantity (zero-order tensor) will not change dimensions when transformed from a coordinate system to another, for example energy or work done expressed by:

$$dW = dF^i \, dl_i = \sigma^{ij} n_j \, dS \, dl_i \qquad \text{(B.2-20)}$$

The unit vector has no dimension and σ^{ij} varies as the inverse of a volume, i.e. under a change of coordinates:

$$a_i{}^k a_j{}^l \sigma'^{ij} = |a_i{}^j| \sigma^{kl} \qquad \text{(B.2-21)}$$

where $|a_i{}^j|$ is the determinant of the transformation matrix $a_i{}^j$ or the Jacobian of the transformation. By definition the stress tensor is a tensorial density.

In curvilinear coordinates, the components of the stress tensor $P^{ij}(\xi)$ are given by:

$$\frac{\partial x^k}{\partial \xi^i} \frac{\partial x^l}{\partial \xi^j} P^{ij}(\xi) = \left| \frac{\partial x}{\partial \xi} \right| \sigma^{kl}(x) \qquad \text{(B.2-22)}$$

where $|\partial x / \partial \xi|$ is the determinant of $(\partial x^j / \partial \xi^i)$.

More general rules of transformation of tensors are proposed to avoid changes of units due to the change of coordinates. For tensorial quantities varying with the power W of the volume, the transformation rule becomes:

$$\tau^{\cdots n \cdots}_{\cdots m \cdots}(\xi) = \left| \frac{\partial x}{\partial \xi} \right|^W \left(\prod_i \frac{\partial x^i}{\partial \xi^m} \right) \left(\prod_j \frac{\partial \xi^n}{\partial x_j} \right) t^{\cdots j \cdots}_{\cdots i \cdots}(x) \qquad \text{(B.2-23)}$$

If $W = 1$, the quantity is a tensorial density; if $W = -1$, it is a tensorial capacitance. The metric tensor is dimensionless, hence $W = 0$. The rule (B.2-23) can also be written as:

$$\left(\prod_j \frac{\partial x^n}{\partial \xi^j} \right) \tau^{\cdots j \cdots}_{\cdots m \cdots}(\xi) = \left| \frac{\partial x}{\partial \xi} \right|^W \left(\prod_i \frac{\partial x^i}{\partial \xi^m} \right) t^{\cdots n \cdots}_{\cdots i \cdots}(x) \qquad \text{(B.2-24)}$$

B.2-5 Time derivative of a tensor

We are now in a position to derive the convected derivative in the formalism of
OLDROYD (1950).

Up to now, the curvilinear coordinates were unspecified and the transformation
was assumed to be time independent in a non moving continuum. Let us consider
now a medium that is deformed in time with positions occupied in space at t_0 and
at t defined respectively by x_0 and x_t. The kinematics of any point is:

$$x_t = d_t(x_0) = y(x_0, t) \qquad (\text{B.2-25})$$

The deformation is continuous by definition if y is a continuous differentiable
function with respect to x_0 and t. Then the correspondence between x_0 and x_t is
unique for any t, i.e.:

$$J(t) = \frac{\partial(x_t^1, x_t^2, x_t^3)}{\partial(x_0^1, x_0^2, x_0^3)} \neq 0 \ \forall t \qquad (\text{B.2-26})$$

As $J(t_0) = 1$, $J(t) > 0$ for all t. The physical interpretation of this inequality is
that a material element initially located inside a closed material surface cannot
cross the surface when deformed.

Let us consider a curvilinear coordinate system defined at time t_0 by a
continuous differentiable function:

$$\xi = \psi(x_0) \qquad (\text{B.2-27})$$

The system is also chosen such that it coincides with three families of surfaces:

$$\xi^1 = a^1$$
$$\xi^2 = a^2 \qquad (\text{B.2-28})$$
$$\xi^3 = a^3$$

where a^1, a^2, a^3 are constants. Let us consider now material surfaces that coincide
with the surfaces (B.2-28) at time t_0. These surfaces are defined at any time t in
a curvilinear frame of reference with respect to x_t by:

$$\xi = \psi[d_t^{-1}(x_t)] \qquad (\text{B.2-29})$$

This is the so-called convected frame of reference.

By definition a fixed point with respect to the material has constant convected
coordinates, independently of the deformation history, as shown by the point M
in two dimensions on Figure B.2-1. The convected coordinates of M are constants,
a^1, a^2.

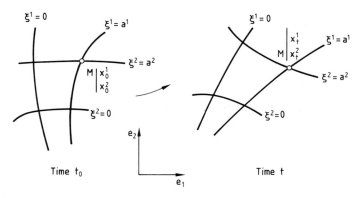

Figure B.2-1   Convected coordinate system

Example: The metric tensor

Let us consider the infinitesimal vector $d\boldsymbol{M}$ in the neighbourhood of M with convected components $d\xi^1$, $d\xi^2$, $d\xi^3$ and the convected covariant components of the metric tensor $\gamma_{ij}(t_0)$, $\gamma_{ij}(t)$ evaluated respectively at time t_0 and t. The square of the infinitesimal distance is given by:

$$dM^2(t_0) = \gamma_{ij}(t_0)\, d\xi^i\, d\xi^j$$

or:

$$dM^2(t) = \gamma_{ij}(t)\, d\xi^i\, d\xi^j$$

(B.2-30)

It is obvious that if $\gamma_{ij}(t) = \gamma_{ij}(t_0)$, $dM^2(t_0) = dM^2(t)$, i.e. the infinitesimal distance between two adjacent points is constant independently of the flow or deformation history. This implies that a volume element about point M is not deformed during the time interval $t_0 \rightarrow t$, it is subjected to pure rotation or translation only.

On the contrary, if $\gamma_{ij}(t) \neq \gamma_{ij}(t_0)$ a volume element is deformed and the convected components of the metric tensor are characteristics of the deformation. The relative deformation in the interval to $t_0 \rightarrow t$ is:

$$\gamma_{ij_r}(t, t_0) = \gamma_{ij}(t) - \gamma_{ij}(t_0)$$

(B.2-31)

More generally, a tensorial quantity is time independent with respect to the material, i.e. it remains a constant, if its convected components are constant. On the other hand, any variation of a physical tensorial quantity during the interval $t_0 \rightarrow t$ should be expressed by the variations of its convected components, i.e. following material elements. However, experimental quantities cannot be measured in a convected frame. We must thus establish rules of transfer from a convected to a fixed coordinate system.

Convected derivative

Let us consider a tensorial field $b(x,t)$ defined in fixed coordinates. To obtain a derivative of this quantity that has a physical sense we must proceed according to the following steps:

- The tensorial field $b(x,t)$ must first be written in a convected system with coordinates $\beta(\xi,t)$.
- The convected components are then differentiated with respect to time. The partial derivative of the convected components are by definition identical to the substantial derivative, i.e.:

$$\frac{\partial \beta(\xi,t)}{\partial t} = \frac{D\beta(\xi,t)}{Dt} \tag{B.2-32}$$

- The tensorial quantity $\partial\beta(\xi,t)/\partial t$ is finally transferred in a fixed coordinate system and written as $\delta b(x,t)/\delta t$. By definition, this expression is called the convected derivative of $b(x,t)$. The whole procedure is schematized in Table B.2-1.

Table B.2-1 The convected derivative

Fixed coordinates	Convected coordinates
$b(x,t)$ $\longrightarrow$	$\beta(\xi,t)$
convected derivative	partial derivative with respect to t
$\downarrow$	$\downarrow$
$\dfrac{\delta b(x,t)}{\delta t}$ $\longleftarrow$	$\dfrac{\partial\beta(\xi,t)}{\partial t}$

One must now obtain simple expressions for the convected derivative in terms of the velocity field.

Example: The convected derivative of the metric tensor

As a first example, we wish to derive the expression of the convected derivative for the metric tensor. The procedure summarized in Table B.2-2 is the following:

Table B.2-2 Convected derivative of the metric tensor

Fixed coordinates	Convected coordinates
$g_{kl}(x,t)$ $\longrightarrow$	$\gamma_{ij}(\xi,t) = \dfrac{\partial x^k}{\partial \xi^i}\dfrac{\partial x^l}{\partial \xi^j} g_{kl}(x,t)$
$\downarrow$	$\downarrow$
$\dfrac{\delta g_{kl}(x,t)}{\delta t} = \dfrac{\partial \xi^i}{\partial x^k}\dfrac{\partial \xi^j}{\partial x^l}\dfrac{\partial}{\partial t}\gamma_{ij}(\xi,t)$ $\longleftarrow$	$\dfrac{\partial}{\partial t}\gamma_{ij}(\xi,t)$

$$\frac{\partial}{\partial t}\,\gamma_{ij}(\xi,t) = \frac{D}{Dt}\,\gamma_{ij}(\xi,t)$$

$$= \left(\frac{D}{Dt}\,\frac{\partial x^k}{\partial \xi^i}\right)\frac{\partial x^l}{\partial \xi^j}\,g_{kl}(x,t) + \frac{\partial x^k}{\partial \xi^i}\left(\frac{D}{Dt}\,\frac{\partial x^l}{\partial \xi^j}\right)g_{kl}(x,t)$$

$$+ \frac{\partial x^k}{\partial \xi^i}\,\frac{\partial x^l}{\partial \xi^j}\left(\frac{D}{Dt}\,g_{kl}(x,t)\right) \tag{B.2-33}$$

But:

$$\frac{D}{Dt}\,\frac{\partial x^k}{\partial \xi^i} = \frac{\partial}{\partial \xi^i}\,\frac{dx^k}{dt} = \frac{\partial u^k}{\partial \xi^i} \tag{B.2-34}$$

where u^k is the fixed component of the velocity vector $\boldsymbol{u}$. Hence:

$$\frac{D}{Dt}\,\gamma_{ij}(\xi,t) = \frac{\partial u^k}{\partial \xi^i}\,\frac{\partial x^l}{\partial \xi^j}\,g_{kl}(x,t) + \frac{\partial x^k}{\partial \xi^i}\,\frac{\partial u^l}{\partial \xi^j}\,g_{kl}(x,t)$$

$$+ \frac{\partial x^k}{\partial \xi^i}\,\frac{\partial x^l}{\partial \xi^j}\left(\frac{D}{Dt}\,g_{kl}(x,t)\right) \tag{B.2-35}$$

Transferring the derivative into a fixed coordinate system, we have:

$$\frac{\delta}{\delta t}\,g_{mn}(x,t) = \frac{\partial \xi^i}{\partial x^m}\,\frac{\partial \xi^j}{\partial x^n}\,\frac{D}{Dt}\,\gamma_{ij}(\xi,t)$$

$$= \frac{\partial \xi^i}{\partial x^m}\,\frac{\partial \xi^j}{\partial x^n}\,\frac{\partial u^k}{\partial \xi^i}\,\frac{\partial x^l}{\partial \xi^j}\,g_{kl}(x,t) + \frac{\partial \xi^i}{\partial x^m}\,\frac{\partial \xi^j}{\partial x^n}\,\frac{\partial x^k}{\partial \xi^i}\,\frac{\partial u^l}{\partial \xi^j}\,g_{kl}(x,t)$$

$$+ \frac{\partial \xi^i}{\partial x^m}\,\frac{\partial \xi^j}{\partial x^n}\,\frac{\partial x^k}{\partial \xi^i}\,\frac{\partial x^l}{\partial \xi^j}\left(\frac{D}{Dt}\,g_{kl}(x,t)\right) \tag{B.2-36}$$

This result can also be written as:

$$\frac{\delta}{\delta t}\,g_{mn}(x,t) = \frac{\partial u^k}{\partial x^m}\,g_{kn}(x,t) + \frac{\partial u^l}{\partial x^n}\,g_{ml}(x,t) + \frac{D}{Dt}\,g_{mn}(x,t) \tag{B.2-37}$$

This is the expression of the convected derivative of the covariant components of a second-order tensor of zero weight. It is obviously independent of the choice of the convected coordinates introduced in the development.

If we now consider the particular case of the metric tensor written in a fixed Cartesian coordinate system, $g_{mn} = \delta_{mn}$ for all values of x and t, hence:

$$\frac{D}{Dt}\,g_{mn} = 0 \tag{B.2-38}$$

and the convected derivative is reduced to:

$$\frac{\delta}{\delta t}\,g_{mn} = \frac{\partial u^k}{\partial x^m}\,g_{kn} + \frac{\partial u^l}{\partial x^n}\,g_{ml} = \frac{\partial u_n}{\partial x^m} + \frac{\partial u_m}{\partial x^n} \tag{B.2-39}$$

We notice that the substantial derivative of g_{mn} is equal to zero, but not the convected derivative which is within a constant factor equal to the rate-of-deformation tensor. The convected derivative of the metric tensor is therefore a measure of the rate of deformation in the fluid.

General case

The previous development can be generalized to include tensors of any order or weight. From Equation (B.2-24) one can write the following general correspondence rule between convected and fixed coordinates:

$$\left(\prod_j \frac{\partial x^n}{\partial \xi^j} \right) \beta^{\cdots j \cdots}_{\cdots m \cdots}(\xi, t) = \left| \frac{\partial x}{\partial \xi} \right|^W \left(\prod_i \frac{\partial x^i}{\partial \xi^m} \right) b^{\cdots n \cdots}_{\cdots i \cdots}(x, t) \qquad (B.2\text{-}40)$$

Taking the substantial derivative:

$$\sum_j \frac{D}{Dt}\left(\frac{\partial x^n}{\partial \xi^j} \right) \left(\prod_{j' \neq j} \frac{\partial x^n}{\partial \xi^{j'}} \right) \beta^{\cdots j' \cdots}_{\cdots m \cdots}(\xi, t) + \left(\prod_j \frac{\partial x^n}{\partial \xi^j} \right) \frac{D}{Dt} \beta^{\cdots j \cdots}_{\cdots m \cdots}(x, t)$$

$$= \left(\frac{D}{Dt}\left| \frac{\partial x}{\partial \xi} \right|^W \right) \left(\prod_i \frac{\partial x^i}{\partial \xi^m} \right) b^{\cdots n \cdots}_{\cdots i \cdots}(x, t)$$

$$+ \left| \frac{\partial x}{\partial \xi} \right|^W \sum_i \frac{d}{dt}\left(\frac{\partial x^i}{\partial \xi^m} \right) \left(\prod_{i' \neq i} \frac{\partial x^{i'}}{\partial \xi^m} \right) b^{\cdots n \cdots}_{\cdots i' \cdots}(x, t)$$

$$+ \left| \frac{\partial x}{\partial \xi} \right|^W \left(\prod_i \frac{\partial x^i}{\partial \xi^m} \right) \frac{D}{Dt} b^{\cdots n \cdots}_{\cdots i \cdots}(x, t) \qquad (B.2\text{-}41)$$

As previously:

$$\frac{d}{dt} \frac{\partial x^i}{\partial \xi^j} = \frac{\partial u^i}{\partial \xi^j}$$

on the other hand:

$$\frac{D}{Dt}\left| \frac{\partial x}{\partial \xi} \right|^W = W \left| \frac{\partial x}{\partial \xi} \right|^{W-1} \frac{D}{Dt}\left| \frac{\partial x}{\partial \xi} \right| \qquad (B.2\text{-}42)$$

and in three dimensions:

$$\frac{D}{Dt}\left| \frac{\partial x}{\partial \xi} \right| = \left| \frac{\partial x}{\partial \xi} \right| \left(\frac{\partial u^1}{\partial x^1} + \frac{\partial u^2}{\partial x^2} + \frac{\partial u^3}{\partial x^3} \right) \qquad (B.2\text{-}43)$$

Hence:

$$\frac{D}{Dt}\left| \frac{\partial x}{\partial \xi} \right|^W = W \left| \frac{\partial x}{\partial \xi} \right|^{W-1} (\nabla \cdot \boldsymbol{u}) \qquad (B.2\text{-}44)$$

The convected derivative of tensor $\boldsymbol{b}$ may be written from Equation (3.2-23) and Table 3.2-1:

$$\frac{\delta b^{\cdots n \cdots}_{\cdots i \cdots}}{\delta t} = \left| \frac{\partial \xi}{\partial x} \right|^W \left(\prod_i \frac{\partial \xi^m}{\partial x^i} \right) \left(\prod_j \frac{\partial x^n}{\partial \xi^j} \right) \frac{D}{Dt} \beta^{\cdots j \cdots}_{\cdots m \cdots}(\xi, t) \qquad (B.2\text{-}45)$$

Replacing $\beta^{\cdots n\cdots}_{\cdots i\cdots}$ in (B.2-41) by the expression (B.2-40) we get the expression for the time derivative of β which can now be substituted in (B.2-45) to obtain:

$$\frac{\delta b^{\cdots n\cdots}_{\cdots i\cdots}}{\delta t} = \frac{D}{Dt} b^{\cdots n\cdots}_{\cdots i\cdots} + \sum_\alpha \frac{\partial u^{m_\alpha}}{\partial x^i} b^{\cdots n}_{\cdots m_\alpha\cdots}{}^{\cdots} - \sum_\beta \frac{\partial u^n}{\partial x^{m_\beta}} b^{\cdots m_\beta\cdots}_{\cdots i\cdots}$$
$$+ W(\nabla \cdot \boldsymbol{u}) b^{\cdots n\cdots}_{\cdots i\cdots} \tag{B.2-46}$$

Example: The convected derivative of the stress tensor ($W = 1$) is

$$\frac{\delta\sigma^{ij}}{\delta t} = \frac{D\sigma^{ij}}{Dt} - \frac{\partial u^i}{\partial x^m}\sigma^{mj} - \frac{\partial u^j}{\partial x^m}\sigma^{im} + \sigma^{ij}\nabla\cdot\boldsymbol{u} \tag{B.2-47}$$

Remarks

Incompressible fluids

For incompressible fluids $\nabla \cdot \boldsymbol{u} = 0$ and the weight of the tensor has no effect on the convective derivative. It is the case in the following points.

Second-order tensors

The commonly used tensors in fluid mechanics and in rheology of polymers are second-order tensors (stress and rate-of-deformation tensors). It is then possible to use tensorial or matricial notation in the expression of the convected derivative. The gradient of the velocity $\boldsymbol{u}$ is a tensor with components $(\partial u^i/\partial x^j)$ which can be written as $\nabla\boldsymbol{u}$. The transpose of the velocity gradient tensor is $(\nabla\boldsymbol{u})^\dagger$.

The convected derivative of a tensor written in covariant components $[b_{ij}]$ is:

$$\frac{\delta[bij]}{\delta t} = \frac{D[bij]}{Dt} + [b_{ij}][\nabla\boldsymbol{u}] + [\nabla\boldsymbol{u}]^\dagger[b_{ij}] \tag{B.2-48}$$

This is the lower convected derivative or the Rivlin form. For the contravariant components, we obtain:

$$\frac{\delta[b^{ij}]}{\delta t} = \frac{D[b^{ij}]}{Dt} - [\nabla\boldsymbol{u}][b^{ij}] - [b^{ij}][\nabla\boldsymbol{u}]^\dagger \tag{B.2-49}$$

This is the upper or Oldroyd convected derivative. The mixed form is written as:

$$\frac{\delta[b^i{}_j]}{\delta t} = \frac{D[b^i{}_j]}{Dt} - [\nabla\boldsymbol{u}][b^i{}_j] + [b^i{}_j][\nabla\boldsymbol{u}]^\dagger \tag{B.2-50}$$

It is interesting to note that the expression for the derivatives depend on the choices of components used to describe the tensorial quantity. But the components of any tensors in orthonormal coordinates are uniquely defined and these different expressions for the convected derivative appear to be a paradox. One must, however, keep in mind that the convected frame does not remain orthonormal and the deformation in the material is expressed differently if covariant, contravariant or mixed components are used. No mathematical reasoning can allow us to make a judicious choice and combinations of the different forms have been frequently used in the development of constitutive relations. One commonly used linear combination is the Jaumann derivative which is the half sum of the Rivlin and Oldroyd derivatives:

$$\frac{\mathcal{D}\boldsymbol{\sigma}}{\mathcal{D}t} = \frac{D\boldsymbol{\sigma}}{Dt} - \boldsymbol{\Omega} \cdot \boldsymbol{\sigma} + \boldsymbol{\sigma} \cdot \boldsymbol{\Omega} \qquad (B.2\text{-}51)$$

where the vorticity tensor, $\boldsymbol{\Omega}$, is defined by:

$$\boldsymbol{\Omega} = \frac{1}{2}\left[(\nabla\boldsymbol{u}) - (\nabla\boldsymbol{u})^\dagger\right] \qquad (B.2\text{-}52)$$

As the stress tensor is usually expressed in its contravariant form, it is then logical to use the Oldroyd convected derivative as given by Equation (B.2-50) or in tensorial notation as:

$$\frac{\delta\boldsymbol{\sigma}}{\delta t} = \frac{D\boldsymbol{\sigma}}{Dt} - \nabla\boldsymbol{u} \cdot \boldsymbol{\sigma} - \boldsymbol{\sigma} \cdot \nabla\boldsymbol{u}^\dagger \qquad (B.2\text{-}53)$$

The upper convected derivative appears naturally in the development of molecular theories of the rheological behaviour of macromolecules, as shown in Appendix C.3.

Appendix C
Viscosity of Polymeric Liquids

C.1   Purely viscous liquids

The behaviour of Newtonian liquids is a special case of purely viscous behaviour that can be defined by:

$$\sigma \text{ does depend only on } \dot{\varepsilon}.$$

From the general principles of continuum mechanics, the stress tensor of isotropic material should be an isotropic function of the rate-of-strain tensor, $\dot{\varepsilon}$. This implies that:

$$\sigma = A(J_1, J_2, J_3)\,\delta + B(J_1, J_2, J_3)\,\dot{\varepsilon} + C(J_1, J_2, J_3)\,\dot{\varepsilon}^2 \qquad \text{(C.1-1)}$$

where J_1, J_2 and J_3 are the three invariants of the rate-of-strain tensor; A, B and C are unspecified functions. For incompressible liquids $J_1 = 0$ and the constitutive relation (C.1-1) can be written in terms of σ' within an arbitrary pressure (isotropic) term as

$$\sigma' = B(J_2, J_3)\,\dot{\varepsilon} + C(J_2, J_3)\,\dot{\varepsilon}^2 \qquad \text{(C.1-2)}$$

This relation includes as special cases:

C.1-1 Newtonian liquid

$$B = 2\eta, \quad C = 0$$
$$\sigma' = 2\eta\dot{\varepsilon} \qquad \text{(C.1-3)}$$

C.1-2 The Rivlin liquid

$$B = \alpha, \quad C = \beta \quad (\alpha \text{ and } \beta \text{ are parameters})$$
$$\sigma' = \alpha\dot{\varepsilon} + \beta\dot{\varepsilon}^2 \qquad \text{(C.1-4)}$$

In *simple shear flow:*

$$\dot{\varepsilon} = \begin{bmatrix} 0 & \frac{\dot{\gamma}}{2} & 0 \\ \frac{\dot{\gamma}}{2} & 0 & 0 \\ 0 & 0 & 0 \end{bmatrix} \qquad \dot{\varepsilon}^2 = \begin{bmatrix} \frac{\dot{\gamma}^2}{4} & 0 & 0 \\ 0 & \frac{\dot{\gamma}^2}{4} & 0 \\ 0 & 0 & 0 \end{bmatrix}$$

and

$$\boldsymbol{\sigma}' = \begin{bmatrix} \beta \frac{\dot{\gamma}^2}{4} & \alpha \frac{\dot{\gamma}}{2} & 0 \\ \alpha \frac{\dot{\gamma}}{2} & \beta \frac{\dot{\gamma}^2}{4} & 0 \\ 0 & 0 & 0 \end{bmatrix} \qquad \text{(C.1-5)}$$

The parameter α is obviously the viscosity coefficient and β is related to the secondary normal stress difference. From (C.1-5) we get:

$$\sigma_{11} - \sigma_{22} = 0$$
$$\sigma_{22} - \sigma_{33} = \beta \frac{\dot{\gamma}^2}{4} \qquad \text{(C.1-6)}$$

One observes experimentally a totally different behavior, i.e.

$$\sigma_{22} - \sigma_{33} \simeq 0$$
$$\sigma_{11} - \sigma_{22} \gg \sigma_{22} - \sigma_{33}$$

The reader should refer to Chapter 6 to find simple constitutive equations or viscoelastic models that can predict at least qualitatively the normal stress differences. If the Rivlin equation is of interest from a theoretical point of view, it is of little use for applications.

For a *bi-directional incompressible flow*, with velocity components u an v,

$$\dot{\varepsilon}^2 = \left(\left(\frac{\partial u}{\partial x} \right)^2 + \frac{1}{4} \left(\frac{\partial u}{\partial y} + \frac{\partial v}{\partial x} \right)^2 \right) \boldsymbol{\delta} = J_2 \, \boldsymbol{\delta}$$

It follows that the term in $\dot{\varepsilon}^2$ is an isotropic tensor that can be included in the pressure term. The Rivlin equation reduces then to Newton's law of viscosity and both solutions for the velocity field in two dimension flow situations are identical.

C.1-3 Power-law model

The power-law model written for simple shear flow as:

$$\sigma = K|\dot{\gamma}|^{n-1}\dot{\gamma}$$

can be generalized by using the second invariant of the rate-of-strain tensor

$$\dot{\bar{\gamma}} = 2\sqrt{J_2(\dot{\varepsilon})}$$

Then

$$B = 2K\left(\dot{\bar{\gamma}}\right)^{n-1}$$
$$C = 0$$

and

$$\boldsymbol{\sigma}' = 2K\left(\dot{\bar{\gamma}}\right)^{n-1}\dot{\varepsilon} \qquad \text{(C.1-7)}$$

Equation of motion

Neglecting inertial and gravitational forces, the equation of motion reduces to:

$$-\nabla p + \nabla \cdot \boldsymbol{\sigma}' = 0$$

Replacing $\boldsymbol{\sigma}'$ by relation (C.1-7), we get

$$-\nabla p + K\left(\dot{\bar{\gamma}}\right)^{n-1} \nabla^2 \boldsymbol{u} + 2K(n-1)\left(\dot{\bar{\gamma}}\right)^{n-2} \dot{\boldsymbol{\varepsilon}} \cdot \nabla \dot{\bar{\gamma}} = 0 \qquad \text{(C.1-8)}$$

This equation reduces to the classical Navier-Stokes equation in the limit of Newtonian behavior: for $n = 1$ and $K = \eta$, we get

$$-\nabla p + \eta \nabla^2 \boldsymbol{u} = 0$$

The other limiting case of interest is the rigid plastic behavior of Von Misès obtained by setting $n = 0$.

C.1-4 Generalized shear-thinning liquids

Any constitutive relation defined for simple shear flow by any function between the shear stress and the shear rate can be easily generalized for more complex flow situations. For example, the following relation

$$\sigma = f(|\dot{\gamma}|)\dot{\gamma}$$

can be generalized to

$$\boldsymbol{\sigma}' = 2f\left(\dot{\bar{\gamma}}\right)\dot{\boldsymbol{\varepsilon}} \qquad \text{(C.1-9)}$$

This implies that $B = 2f\left(\dot{\bar{\gamma}}\right)$ and $C = 0$. The equation of motion is then:

$$-\nabla p + f\left(\dot{\bar{\gamma}}\right)\nabla^2 \boldsymbol{u} + 2f'\left(\dot{\bar{\gamma}}\right)\dot{\boldsymbol{\varepsilon}} \cdot \nabla \dot{\bar{\gamma}} = 0 \qquad \text{(C.1-10)}$$

C.2 Molecular aspects of viscosity

In this section we present some basic molecular aspects which are useful to correlate the viscosity of polymeric materials to molecular structure of the macromolecules.

C.2-1 Eyring theory

GLASSTONE, LAIDLER and EYRING (1941) are probably the first authors to propose a predictive theory for the viscosity based on statistical mechanics of liquids. This is their theory that is summarized here.

The molecules of simple liquids can be modeled as spheres with weak van der Waals interactions, randomly distributed and under Brownian motion. The frequency of the Brownian motion is approximately that of a harmonic oscillator of energy $k_b T$, i.e.:

$$\nu_b = \frac{k_b T}{h} \qquad \text{(C.2-1)}$$

where k_b and h are respectively the Boltzman and the Planck constants. At 300 K, $\nu_b \approx 10^{12}\,\mathrm{s}^{-1}$, i.e. each molecule collides or changes direction 10^{12} times per second. Thus molecules diffuse at random across the liquid changing position as illustrated in Figure C.2-1.

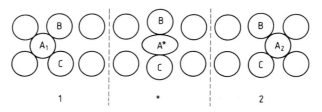

Figure C.2-1   Molecular diffusion in liquids

The molecule A changes position from A_1 to A_2. To do so, it must force its way through molecules B and C to occupy a position A^*, in an unstable state called activated complex. From A_1 to A_2, the enthalpy and entropy of the molecule vary:

– The enthalpy goes through a maximum corresponding to the activation energy H; this is illustrated in Figure C.2-2.

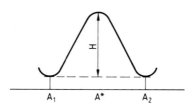

Figure C.2-2   Variation of enthalpy in a change of molecular configuration

– The entropy goes through a maximum or minimum due to the number of possible configurations which are not the same for the molecule in the state $A^*(W^*)$ as in the state $A_1(W)$ or $A_2(W)$:

$$\frac{W^*}{W} = \exp\left(\frac{S}{k_b}\right) \qquad \text{(C.2-2)}$$

where S is the entropy of activation.

The frequency of changes A_1 to A_2 or the inverse is then given by

$$\nu_{1\to2} = \nu_{2\to1} = \frac{k_b T}{h}\, e^{-\left(\frac{H}{k_b T}\right)} \frac{W^*}{W} = \frac{k_b T}{h}\, e^{-\left(\frac{G}{k_b T}\right)} \qquad \text{(C.2-3)}$$

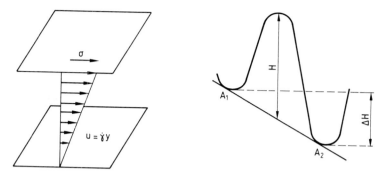

Figure C.2-3   Effect of shear on the energy barrier

where $G(= H - TS)$ is the free energy of activation.

The application of a force to the liquid such as a shear stress, σ, will unbalance the two levels of energy A_1 and A_2, as shown in Figure C.2-3. The enthalpy difference is:

$$\Delta H = \sigma V \qquad (\text{C.2-4})$$

where V is the volume of a molecule. This follows from the work done by shear, i.e.:

$$\Delta H = \underbrace{\sigma \times \text{surface area of molecule}}_{\text{applied force}} \times \underbrace{\text{distance between } A_1\, A_2}_{\simeq \text{molecule diameter}}$$

Hence the net frequency for the changes $A_1 \rightarrow A_2$ is given by:

$$r = \nu_{1\rightarrow 2} - \nu_{2\rightarrow 1} = \frac{k_b T}{h}\, e^{-\left(\frac{G}{k_b T}\right)} \left[e^{\left(\frac{\Delta H}{2k_b T}\right)} - e^{-\left(\frac{\Delta H}{2k_b T}\right)} \right]$$

$$= \frac{2k_b T}{h}\, e^{-\frac{G}{k_b T}}\, \sinh\left(\frac{\sigma V}{2k_b T}\right) \qquad (\text{C.2-5})$$

In most applications, the applied shear stress is of the order of 10^5 Pa, whereas V is of the order of 10^{-21} mm^3. Hence

$$\sigma V < 10^{-25}\ \text{J}$$

But since

$$k_b T \simeq 10^{-21}\ \text{J}$$

$$\sigma V \ll k_b T$$

Therefore:

$$r \simeq \sigma \frac{V}{h}\, e^{-\left(\frac{G}{k_b T}\right)} \qquad (\text{C.2-6})$$

The jump of a molecule from position A_1 to A_2 is equivalent to a local shear deformation of the liquid equal to 1. As the jump frequency is equal to r, the average shear rate is equal to that frequency, i.e.:

$$\dot{\gamma} = r \qquad (\text{C.2-7})$$

The application of a shear stress results in a macroscopic motion of the liquid in simple shear flow at the rate of deformation $\dot{\gamma}$. From the definition of the viscosity it follows that:

$$\eta = \frac{\sigma}{\dot{\gamma}} = \frac{\sigma}{r} = \frac{h}{V} e^{\left(\frac{G}{k_b T}\right)} \qquad (\text{C.2-8})$$

The following example shows that this expression predicts the correct order of magnitude for the viscosity of liquids.

Example: Magnitude of the viscosity of a liquid

Let us consider an organic liquid with molecules of volume equivalent to a cube of 5 Å(50 nm).

Hence

$$V \simeq 125 \cdot 10^{-30} \, \text{m}^3$$

$$h \simeq 6 \cdot 10^{-34} \, \text{J} \cdot \text{s}$$

and

$$\frac{h}{V} \simeq 5 \cdot 10^{-6} \, \text{Pa} \cdot \text{s}$$

Usually, G is of the order of $12 \cdot 10^4/\text{mole}$ and $G/(k_b T) = 5$ at $20°$ C. Therefore, $\eta \simeq 7 \cdot 10^{-4} \, \text{Pa} \cdot \text{s}$, that is close to 1 centipoise, order of magnitude of the viscosity of water and of low molecular weight organic liquids.

C.2-2 Molecular-weight dependence of polymer viscosity

The viscosity of polymer melts and polymer solutions is considerably different from that of low molecular weight liquids. All the experimental observations show that the viscosity of polymers is related to the molecular weight according to the following relations:

-   For polymers of molecular weight less than a critical value, M_c, the viscosity increases linearly with molecular weight. Moreover, these low molecular weight polymers (or oligomers) obey (or do not deviate much) Newton's law of viscosity.
-   Above M_c, the viscosity increases much more rapidly with molecular weight. Linear polymers, generally, follow the well known relation (see FERRY, 1970).

$$\eta \simeq K M^{3.4} \qquad (\text{C.2-9})$$

Moreover, high-molecular-weight polymers are non-Newtonian in simple shear flow. This will be discussed in Section C.2-3. The viscosity-molecular weight relation is sketched in Figure C.2-4. Most commercial polymers have large distribution of molecular weights (they are polydispersed) and the viscosity is a more complex function of the average molecular weight and of the distribution.

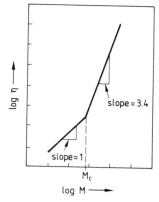

Figure C.2-4   Typical polymers viscosity dependence

Low molecular weight polymers

For polymers of molecular weight less than M_c or dilute polymer solutions, the macromolecules take a random configuration (in amorphous state) known as statistical coil for flexible enough chains. The radius, R, of the coil is approximately equal to $a\sqrt{x}$ where a is the length of a chain segment and x is the number of segments. Figure C.2-5 illustrates this molecular concept.

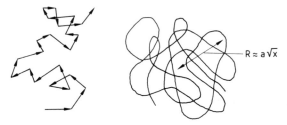

Figure C.2-5   Coil conformation of a macromolecule

In concentrated polymer solutions or in melts, the coils of adjacent molecules are interpenetrated. The following reasoning explains why polymeric liquids exhibit large viscosities.

Let us consider a liquid under a shear flow of magnitude $\dot{\gamma}$ (see Figure C.2-6). If the liquid has a low molecular weight, the molecules have a low but uniform relative velocity of the order of $\dot{\gamma}a/2$, where a is the molecules diameter. If we consider a polymeric liquid, each macromolecule is submitted to an average rotational speed of $\dot{\gamma}/2$ and the relative velocity of the periphery of the macromolecule, of the order of $\dot{\gamma}R/2$ ($\simeq \dot{\gamma}a\sqrt{x}/2$), increases with x (or molecular weight).

The molecular structure implies that the relative velocities between adjacent molecules (or chain segments) could be quite large and, if one assumes that the energy dissipated is proportional to the velocity square (hence R^2 or x), the viscosity then increases linearly with molecular weight. This can be explained as follows:

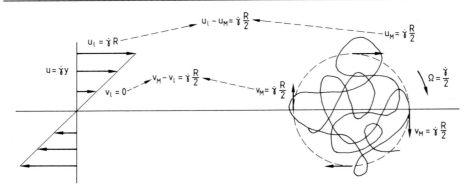

Figure C.2-6   Deformation of a macromolecule under shear flow

The rate of energy dissipated per unit volume is expressed by:

$$\dot{W} = N \int_{\text{molecule}} f_0 (v_{\text{molecule}} - v_{\text{average}})^2 \qquad (C.2\text{-}10)$$

where N is the concentration of molecules, f_0 is the drag coefficient of a chain segment defined such that the force necessary to bring a segment to the relative velocity v with respect to the surrounding is $F = f_0 v$.

Then, from the Debye calculation and by using Equation (C.2-10) and the fact that N is inversely proportional to xa^3, one obtains:

$$\dot{W} = \frac{1}{36} \frac{x}{a} f_0 \dot{\gamma}^2 \qquad (C.2\text{-}11)$$

But as shown in Chapter 2.1, the rate of energy dissipated by viscous forces is proportional to the product of the viscosity and the square of the shear rate (i.e. $W = \eta \dot{\gamma}^2$). From (C.2-11) we have

$$\eta = \frac{1}{36} \frac{x}{a} f_0 \qquad (C.2\text{-}12)$$

Using Stokes' law to express f_0:

$$F = f_0 v = 6\pi \eta_1 a v \qquad (C.2\text{-}13)$$

where η_1 is the viscosity of the liquid monomer, one obtains:

$$\eta \simeq x \eta_1 \qquad (C.2\text{-}14)$$

However, one would expect f_0 to increase slightly with x, thus explaining why η increases faster than linearly with x. This analysis is valid for $x < x_c$, i.e. the macromolecules are small enough so that there are no chain entanglements.

High-molecular-weight polymers

BUECHE (1952) proposed the following concept to explain the dependence of the viscosity on the molecular weight for values larger than M_c. His idea, based on chains entanglement, is now largely accepted.

The force necessary to bring a macromolecule containing x segments to the velocity v is:

$$F = x\,f_0 v = fv \qquad (C.2\text{-}15)$$

where $f = xf_0$ is the drag coefficient for the whole macromolecule. If the macromolecule is large enough, it is reasonable to assume that it drags along other macromolecules entangled at random positions due to the Brownian motion. The number of entangled molecules is approximately proportional to the number of chain segments x, i.e.

$$n_1 = Kx$$

The dragged molecules slip slightly with respect to the first one at the entanglement points. Their velocity is then in average smaller than v, i.e.

$$v_1 = sv \quad (\text{where } 0 < s < 1)$$

The net drag force is then given by the following expression:

$$F_1 = fv + n_1 fsv = f_0 vx(1 + Ksx) \qquad (C.2\text{-}16)$$

These molecules drag other molecules in similar conditions. Their number is:

$$n_2 = C_2 n_1 Kx \quad \text{where } C_2 < 1$$

since the entanglements with the first molecule should be included. Their velocity v_2 is given by:

$$v_2 = sv_1 = s^2 v$$

In this pattern illustrated in Figure C.2-7, each dragged molecule is surrounded by a cluster of other molecules dragged at a lower velocity with intensity decreasing more rapidly as they are less directly entangled with the first one. It follows that the total drag force is given by:

$$F = f_0 xv[1 + Ksx + C_2(Ksx)^2 + C_3(Ksx)^3 + \ldots] \qquad (C.2\text{-}17)$$

where $1, C_2, C_3 \ldots$ is a rapidly decreasing series.

If one assumes that the transition occurs at $x_c \simeq 500$, then $K \simeq 1/500$ and for $x < x_c$, Ksx is small and $F \simeq f_0 xv$. For $x > x_c$, BUECHE (1952) has computed the function F corresponding to various values of s (from 0.1 to 0.5). He has shown that, for $s \simeq 0.3$, F is related to x by:

$$F = Ax^{2.5}$$

This yields from Equations C.2-12 and C.2-13 a correlation for the viscosity of the type $\eta = Bx^{3.5}$.

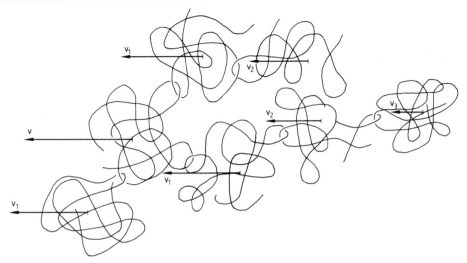

Figure C.2-7   Deformation of entangled macromolecules

C.2-3 Shear-thinning properties of polymers

The shear-thinning behaviour or the decrease of the shear viscosity with shear rate for polymers of molecular weight larger than M_c has been explained by GRAESSLEY (1965, 1967) and by BUECHE (1968) using the concept of chain entanglements. Their ideas are summarized here.

Two adjacent macromolecules can be entangled only if the distance between their centers of mass is less than twice their radius of gyration R. In shear flow, the spheres of influence of two molecules interact only during a limited time as shown by Figure C.2-8.

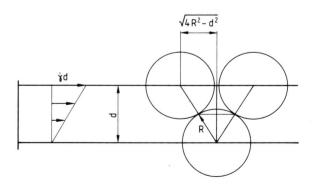

Figure C.2-8   Interaction between macromolecules in shear flow

For two molecules located in a plane normal to the shear plane, with their centers of mass at a distance d, the interaction or contact time is given by:

$$t = \frac{2\sqrt{4R^2 - d^2}}{\dot{\gamma}d}, \quad \text{for } d \leq 2R$$

$$t = 0, \qquad\qquad \text{for } d > 2R$$

(C.2-18)

Now one must consider the term associated with the molecular or Brownian motion. As discussed in Chapter 6.1, this molecular time seems to be of the same order as the relaxation time θ. But the creation or the loss of entanglements requires an overall motion of the macromolecules. Hence one can assume as a first approximation that

– if the interaction time of the spheres of influence of two molecules, t, is less than the relaxation time, θ, entanglements will not be created;
– if the interaction time is longer than θ, then entanglements are effectively created;
– also existing entanglements will require a time θ to be lost after interaction occurred.

From these phenomenological considerations, one can understand why the entanglement density depends on the shear rate, $\dot{\gamma}$:

– for $\dot{\gamma} \ll \dfrac{1}{\theta}$, $t \gg \theta$, $\forall d < 2R$

Under these conditions, the entanglement density is practically that existing in the polymeric liquid at rest (equilibrium value). The quantity M_e is the average molecular weight of the chain segments between two consecutive entanglements.

– for $\dot{\gamma} \gg \dfrac{1}{\theta}$, $t \ll \theta$, $\forall d > 0$

Under these conditions (very high shear rates), the entanglement density tends to zero.

As shown in Section C.2-2, the viscosity increases with the number of entanglements. As the entanglement density decreases with shear rate, the viscosity will decrease but this can be detected only when the shear rate, $\dot{\gamma}$, reaches and exceeds values of the order of $1/\theta$.

The main theories of GRAESSLEY and of BUECHE yield viscosity expressions of power-law type at high shear rates. Depending on the physics introduced different limiting power-law expressions are obtained. These are reported in Table C.2-1.

Table C.2-1   Asymptotic power-law expressions for the theories of GRAESSLEY (1965, 1967) and of BUECHE (1968)

		Power-law form $\sigma = K|\dot{\gamma}|^n$
BUECHE	$\eta \sim |\dot{\gamma}|^{-6/7}$	$n = 1/7 = 0.14$
GRAESSLEY	$\eta \sim |\dot{\gamma}|^{-3/4}$	$n = 1/4 = 0.25$
and	$\eta \sim |\dot{\gamma}|^{-9/11}$	$n = 2/11 = 0.18$

These results are in good agreement with the observation. Most polymer melts and concentrated polymer solutions show the typical shear-thinning behaviour illustrated in Figure C.2-9, obeying a power-law expression with a low value of the index n at high shear rate.

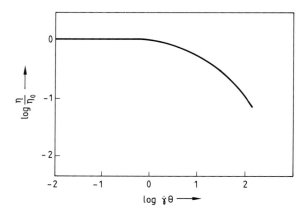

Figure C.2-9   Shear-thinning viscosity of polymers

C.3   Elastic dumbbell rheological models

C.3-1   Interest of dumbbell models

In the rheological models developed by ROUSE (1953) and by ZIMM (1956), the macromolecules are depicted as chains of simple elastic dumbbells. Although the physical description of the macromolecules is rather crude, the resulting constitutive equations lead to three main results of considerable interest:

– The convected derivative appears automatically from the use of physical concept and coherent mathematical treatment. This supports the mathematical development of the convected derivative presented in Appendix B.
– Both theories lead to the upper convected derivative (or Oldroyd form). This supports the choice of that derivative for the Maxwell convected model of Chapter 6.1, which predicts the following material functions in steady simple shear flow:

$$\sigma_{12} = \eta \dot{\gamma}$$

$$\sigma_{11} - \sigma_{22} > 0$$

$$\frac{\sigma_{11} - \sigma_{22}}{2\sigma_{12}} = \theta \dot{\gamma}$$

and

$$\sigma_{22} - \sigma_{33} = 0$$

– Finally, the dumbbell models predict markedly different rheological behaviour in elongational flow compared to the behaviour in simple shear flow.

C.3-2 Physics of dumbbell models

In the simplest case, the macromolecule is pictured as a dumbell consisting of two rigid spheres linked together by a spring, and immersed in a viscous fluid. The model is based on the following hypotheses:

– The drag exerted by the surrounding fluid (solvent) is concentrated on the extremities of the polymeric chain, taken as simple spheres as shown in Figure C.3-1.

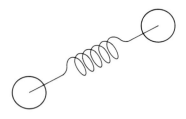

Figure C.3-1   Elastic dumbbell model

– The entropic forces that tend to keep the macromolecule in its original and most probable statistical coil shape have a unique resulting force acting in the direction of the end-to-end vector separating the two spheres.
  Considering simple shear flow, defined by

$$u = \dot{\gamma} y$$
$$v = w = 0 \tag{C.3-1}$$

we propose the following mechanism for the deformation of a macromolecule. A macromolecule with center of mass at y_g has an average velocity $\boldsymbol{u}_g = \dot{\gamma} y_g$. Let us take A_1 and A_2 as the coordinates of the macromolecule extremities (see Figure C.3-2). We can write then:

$$\boldsymbol{OA}_1 = \boldsymbol{OG} + \frac{\boldsymbol{r}}{2}$$
$$\boldsymbol{OA}_2 = \boldsymbol{OG} - \frac{\boldsymbol{r}}{2} \tag{C.3-2}$$

where $\boldsymbol{A}_1 \boldsymbol{A}_2 = \boldsymbol{r}\,(x, y, z)$. As the average velocity of the fluid is larger than that of the molecule at A_1 and smaller than that at point A_2, the viscous drag exerted on both spheres tend to pull the spheres apart. The spring force, however, is in the opposite direction. An equilibrium is then attained. In the next sections, we will establish expressions for the average position $\boldsymbol{r}$ in various flow situations. Although the dumbbell model is a crude representation of polymeric chains, the chain conformations predicted under different flow conditions are qualitatively correct.

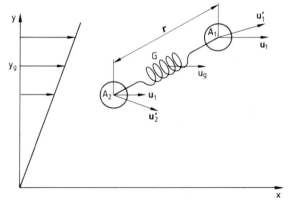

Figure C.3-2   Shear deformation of a dumbbell

C.3-3 Dumbbell deformation in shear flow

Let us define $\boldsymbol{u}'_1$ and $\boldsymbol{u}'_2$ as the velocity vector of A_1 and A_2 respectively. Then we can write:

$$
\boldsymbol{u}'_1 = \boldsymbol{u}_g + \frac{1}{2}\frac{d\boldsymbol{r}}{dt}\left(u_g + \frac{1}{2}\frac{dx}{dt}, \frac{1}{2}\frac{dy}{dt}, \frac{1}{2}\frac{dz}{dt}\right)
$$
$$
\boldsymbol{u}'_2 = \boldsymbol{u}_g - \frac{1}{2}\frac{d\boldsymbol{r}}{dt}\left(u_g - \frac{1}{2}\frac{dx}{dt}, -\frac{1}{2}\frac{dy}{dt}, -\frac{1}{2}\frac{dz}{dt}\right)
$$

(C.3-3)

The velocities of the solvent at A_1 and A_2 are:

$$
\boldsymbol{u}_1\left(u_g + \frac{1}{2}\dot{\gamma}y, 0, 0\right)
$$
$$
\boldsymbol{u}_2\left(u_g - \frac{1}{2}\dot{\gamma}y, 0, 0\right)
$$

(C.3-4)

In a frame attached at the center of gravity of the molecule, the relative velocities are:

$$
\boldsymbol{u}_1 - \boldsymbol{u}'_1\left(\frac{1}{2}\left(\dot{\gamma}y - \frac{dx}{dt}\right), -\frac{1}{2}\frac{dy}{dt}, -\frac{1}{2}\frac{dz}{dt}\right)
$$
$$
\boldsymbol{u}_2 - \boldsymbol{u}'_2\left(-\frac{1}{2}\left(\dot{\gamma}y - \frac{dx}{dt}\right), \frac{1}{2}\frac{dy}{dt}, \frac{1}{2}\frac{dz}{dt}\right)
$$

(C.3-5)

The *hydrodynamic force* resulting from the drag forces exerted on the two spheres is:

$$
\boldsymbol{F}_h = f(\boldsymbol{u}_1 - \boldsymbol{u}'_1) - f(\boldsymbol{u}_2 - \boldsymbol{u}'_2)
$$

with coordinates:

(C.3-6)

$$
\left(X_h = f\left(\dot{\gamma}y - \frac{dx}{dt}\right); \quad Y_h = -f\frac{dy}{dt}; \quad Z_h = -f\frac{dz}{dt}\right)
$$

where f is the drag coefficient. As mentioned above, this hydrodynamic force tends to pull the spheres apart.

In absence of hydrodynamic field, the end-to-end vector r of the macro-molecule follows a distribution function of the following type:

$$\psi_0(r) = k_0 \, \exp\left\{ -\frac{3r^2}{2r_0{}^2} \right\}$$

with:

$$\iiint_{R^3} \psi_0(r) \, d^3r = 1 \tag{C.3-7}$$

where r_0 is the quadratic average length of the molecule at rest.

In a flow field, the macromolecules are deformed and stretched, and the distribution function for r is modified. This function $\psi(r)$ is unknown, but remains normalized, i.e.:

$$\iiint_{R^3} \psi(r) \, d^3r = 1 \tag{C.3-8}$$

To maintain a non-equilibrium conformation, a force due to the Brownian motion is exerted on the macromolecule; this is expressed by:

$$F_b = -k_b T \, \nabla \, \ln\left(\frac{\psi}{\psi_0} \right) \tag{C.3-9}$$

Using (C.3-7), we can write

$$\frac{\partial}{\partial r} \ln \psi_0 = -\frac{3r}{r_0{}^2}$$

and express the *Brownian force* as:

$$F_b = -3k_b T \left(\frac{r}{r_0{}^2} \right) - k_b T \, \nabla \, \ln \psi \tag{C.3-10}$$

with coordinates given by:

$$X_b = -\frac{3k_b T}{r_0{}^2} x - k_b T \frac{1}{\psi} \frac{\partial \psi}{\partial x},$$

$$Y_b = -\frac{3k_b T}{r_0{}^2} y - k_b T \frac{1}{\psi} \frac{\partial \psi}{\partial y},$$

$$Z_b = -\frac{3k_b T}{r_0{}^2} z - k_b T \frac{1}{\psi} \frac{\partial \psi}{\partial z}$$

Assuming that the inertial forces are negligible, the hydrodynamic force has to be equal to the Brownian force. Hence defining

$$\sigma = \frac{3k_b T}{r_0{}^2} \quad \text{and} \quad D = \frac{k_b T}{f} \tag{C.3-11}$$

we obtain:

$$\frac{dx}{dt} = \dot{\gamma}y - \sigma x - D\frac{1}{\psi}\frac{\partial\psi}{\partial x}$$

$$\frac{dy}{dt} = -\sigma y - D\frac{1}{\psi}\frac{\partial\psi}{\partial y} \qquad \text{(C.3-12)}$$

$$\frac{dz}{dt} = -\sigma z - D\frac{1}{\psi}\frac{\partial\psi}{\partial z}$$

To obtain this result, we have assumed that the distribution function depends only on position (x, y, z) and not on velocities $(dx/dt, dy/dt, dz/dt)$. Hence for each vector $\boldsymbol{r}$, there is a unique vector $d\boldsymbol{r}/dt$ given by Equation (C.3-12).

We now apply the conservation principle for the molecules. At any instant, the number of molecules with configuration $\boldsymbol{r}$ is proportional to $\psi(\boldsymbol{r})$ and their velocity is $d\boldsymbol{r}/dt$. The divergence of the flux $(\psi d\boldsymbol{r}/dt)$ is hence equal to zero, i.e.:

$$\nabla \cdot \left[\psi\frac{d\boldsymbol{r}}{dt}\right] = \frac{\partial}{\partial x}\left[\psi\frac{dx}{dt}\right] + \frac{\partial}{\partial y}\left[\psi\frac{dy}{dt}\right] + \frac{\partial}{\partial z}\left[\psi\frac{dz}{dt}\right] = 0$$

$$= \psi\nabla\cdot\left(\frac{d\boldsymbol{r}}{dt}\right) + \frac{dx}{dt}\frac{\partial\psi}{\partial x} + \frac{dy}{dt}\frac{\partial\psi}{\partial y} + \frac{dz}{dt}\frac{\partial\psi}{\partial z} \qquad \text{(C.3-13)}$$

(particular case of the Liouville theorem).

Introducing result (C.3-12) in (C.3-13), we obtain the following partial differential equation for ψ:

$$3\sigma\psi - \dot{\gamma}y\frac{\partial\psi}{\partial x} + \sigma\left[x\frac{\partial\psi}{\partial x} + y\frac{\partial\psi}{\partial y} + z\frac{\partial\psi}{\partial z}\right] + D\left[\frac{\partial^2\psi}{\partial x^2} + \frac{\partial^2\psi}{\partial y^2} + \frac{\partial^2\psi}{\partial z^2}\right] = 0 \quad \text{(C.3-14)}$$

In general, no exact solution exists for ψ. However, solutions can be obtained for various moments such as:

$$\langle x^2 \rangle = \iiint_{R^3} x^2\,\psi(\boldsymbol{r})\,d^3\boldsymbol{r}, \quad \langle xy \rangle = \iiint_{R^3} xy\,\psi(\boldsymbol{r})\,d^3\boldsymbol{r}, \dots \qquad \text{(C.3-15)}$$

The results are obtained by multiplying Equation (C.3-14) by x^2, xy, ..., and integrating over R^3, for example in terms of x^2:

$$-3x^2\sigma\psi + \dot{\gamma}yx^2\frac{\partial\psi}{\partial x} - \sigma\left[x^3\frac{\partial\psi}{\partial x} + x^2y\frac{\partial\psi}{\partial y} + x^2z\frac{\partial\psi}{\partial z}\right]$$

$$- D\left[x^2\frac{\partial^2\psi}{\partial x^2} + x^2\frac{\partial^2\psi}{\partial y^2} + x^2\frac{\partial^2\psi}{\partial z^2}\right] = 0$$

The integration for terms containing ψ is direct; integration by parts has to be used for terms containing $\partial\psi/\partial x$ and $\partial^2\psi/\partial x^2$. We assume that the products $x^2\psi$,

$x^3\psi$, $x^2\partial\psi/\partial x$ go to zero as $\boldsymbol{r}$ goes to infinity. We obtain:

- $$\iiint\limits_{R^3} x^2\,\psi(\boldsymbol{r})\,d^3\boldsymbol{r} = \langle x^2\rangle$$

- $$\iiint\limits_{R^3} yx^2\frac{\partial\psi}{\partial x}(\boldsymbol{r})\,d^3\boldsymbol{r} = \iint\limits_{R^2} y\,dy\,dz\int\limits_{R} x^2\frac{\partial\psi}{\partial x}\,dx$$

$$\int\limits_{R} x^2\frac{\partial\psi}{\partial x}\,dx = [x^2\psi]^{+\infty}_{-\infty} - 2\int\limits_{R}\psi x\,dx = -2\int\limits_{R} x\psi(\boldsymbol{r})\,dx$$

Hence:

$$\iiint\limits_{R^3} yx^2\frac{\partial\psi}{\partial x}\,d^3\boldsymbol{r} = -2\langle xy\rangle$$

- $$\iiint\limits_{R^3} x^3\frac{\partial\psi}{\partial x}\,d^3\boldsymbol{r} = \iint\limits_{R^2} dy\,dz\int\limits_{R} x^3\frac{\partial\psi}{\partial x}\,dx$$

$$\int\limits_{R} x^3\frac{\partial\psi}{\partial x}\,dx = [x^3\psi]^{+\infty}_{-\infty} - 3\int\limits_{R} x^2\psi\,dx$$

Thus:

$$\iiint\limits_{R^3} x^3\frac{\partial\psi}{\partial x}\,d^3\boldsymbol{r} = -3\langle x^2\rangle$$

- $$\iiint\limits_{R^3} x^2\frac{\partial^2\psi}{\partial x^2}\,d^3\boldsymbol{r} = \iint\limits_{R^2} dy\,dz\int\limits_{R} x^2\frac{\partial^2\psi}{\partial x^2}\,dx$$

$$\int\limits_{R} x^2\frac{\partial^2\psi}{\partial x^2}\,dx = \left[x^2\frac{\partial\psi}{\partial x}\right]^{+\infty}_{-\infty} - 2\int\limits_{R} x\frac{\partial\psi}{\partial x}\,dx = -2[x\psi]^{+\infty}_{-\infty} + 2\int\limits_{R}\psi\,dx$$

Thus:

$$\iiint\limits_{R^3} x^2\frac{\partial^2\psi}{\partial x^2}\,d^3\boldsymbol{r} = 2$$

In a similar way, we can show that:

- $$\iiint\limits_{R^3} x^2\frac{\partial^2\psi}{\partial y^2}\,d^3\boldsymbol{r} = \iiint\limits_{R^3} x^2\frac{\partial^2\psi}{\partial z^2}\,d^3\boldsymbol{r} = 0$$

The relation in x^2 is:

$$-3\sigma\langle x^2\rangle - 2\dot\gamma\langle xy\rangle - \sigma[-3\langle x^2\rangle - \langle x^2\rangle - \langle x^2\rangle] - 2D = 0$$

Hence:

$$\sigma \langle x^2 \rangle - \dot{\gamma} \langle xy \rangle - D = 0 \qquad \text{(C.3-16)}$$

By definition, $D/\sigma = r_0^2/3$, and taking $\theta = 1/2\sigma$ (σ has the units of s^{-1}), we re-write (C.3-16) as:

$$\langle x^2 \rangle - 2\dot{\gamma}\,\theta \langle xy \rangle = \frac{r_0^2}{3} \qquad \text{(C.3-17)}$$

Five other relations are obtained in the similar way:

$$\langle y^2 \rangle = \frac{r_0^2}{3} \qquad \text{(C.3-18)}$$

$$\langle z^2 \rangle = \frac{r_0^2}{3} \qquad \text{(C.3-19)}$$

$$\langle xy \rangle - \dot{\gamma}\,\theta \langle y^2 \rangle = 0 \qquad \text{(C.3-20)}$$

$$\langle yz \rangle = 0 \qquad \text{(C.3-21)}$$

$$\langle xz \rangle - \dot{\gamma}\,\theta \langle yz \rangle = 0 \qquad \text{(C.3-22)}$$

and from these results, we obtain:

$$\langle xz \rangle = \langle yz \rangle = 0 \qquad \text{(C.3-23)}$$

$$\langle xy \rangle = \dot{\gamma}\,\theta\,\frac{r_0^2}{3} \qquad \text{(C.3-24)}$$

$$\langle x^2 \rangle = (1 + 2\dot{\gamma}^2\theta^2)\frac{r_0^2}{3} \qquad \text{(C.3-25)}$$

$$\langle y^2 \rangle = \langle z^2 \rangle = \frac{r_0^2}{3} \qquad \text{(C.3-26)}$$

Hence, the total deformation of the macromolecules is given by:

$$\langle r^2 \rangle = \langle x^2 \rangle + \langle y^2 \rangle + \langle z^2 \rangle = r_0^2\left[1 + \frac{2}{3}\dot{\gamma}^2\theta^2\right] \qquad \text{(C.3-27)}$$

From these results, we can draw the following conclusions:

- The macromolecules are stretched in the x-direction, but they are not deformed in the y and z-directions. This is consistent with measurements of normal stresses, i.e.

$$\sigma_{xx} > \sigma_{yy} \quad \text{and} \quad \sigma_{yy} \approx \sigma_{zz}$$

- The extension of the macromolecules increases with $\dot{\gamma}$; the relative extension depends only on $\dot{\gamma}\theta$:

$$\frac{\langle r^2 \rangle}{r_0^2} = 1 + \frac{2}{3}\dot{\gamma}^2\theta^2$$

- The term $\langle xy \rangle$ is different from zero. This shows that the principal directions for the macromolecule deformation are not x and y, but are at angles α and

α' with the x-axis, given by:

$$\tan \alpha = \sqrt{1 + \dot{\gamma}^2 \theta^2} - \dot{\gamma}\theta,$$

$$\tan \alpha' = -\sqrt{1 + \dot{\gamma}^2 \theta^2} - \dot{\gamma}\theta,$$

$$\text{If } \dot{\gamma}\theta \to 0, \quad \alpha \to \frac{\pi}{4} \text{ and } \alpha' \to -\frac{\pi}{4}$$

$$\text{If } \dot{\gamma}\theta \to \infty, \quad \alpha \to 0 \text{ and } \alpha' \to -\frac{\pi}{2}$$

(C.3-28)

C.3-4 Deformation in complex flow

All the relations and results of Section C.3-3 can be generalized to more complex flows of incompressible fluids. The velocity gradient tensor ∇u appears and the four basic expressions are:

Hydrodynamic force:

$$\boldsymbol{F}_h = f\left(\nabla u \cdot \boldsymbol{r} - \frac{d\boldsymbol{r}}{dt}\right)$$

(C.3-29)

Brownian force:

$$\boldsymbol{F}_b = -\frac{3k_b T}{r_0^2}\,\boldsymbol{r} - k_b T \nabla \ln \psi$$

(C.3-30)

The equality of these two forces yields:

$$\frac{d\boldsymbol{r}}{dt} = \nabla u \cdot \boldsymbol{r} - \sigma \boldsymbol{r} - D\frac{1}{\psi}\nabla\psi$$

(C.3-31)

which then, by using the conservation principle, implies:

$$3\sigma\psi - \nabla\psi \cdot (\nabla u \cdot \boldsymbol{r}) + \sigma \boldsymbol{r} \cdot \nabla\psi + D\nabla^2\psi = 0$$

(C.3-32)

Defining the conformation tensor $\boldsymbol{C}$ by:

$$[\boldsymbol{C}] = \begin{bmatrix} \langle x^2 \rangle & \langle xy \rangle & \langle xz \rangle \\ \langle xy \rangle & \langle y^2 \rangle & \langle yz \rangle \\ \langle xz \rangle & \langle yz \rangle & \langle z^2 \rangle \end{bmatrix}$$

(C.3-33)

Equation (C.3-32) implies the following equation for the conformation tensor:

$$\boldsymbol{C} - \theta(\nabla u \cdot \boldsymbol{C} + \boldsymbol{C} \cdot \nabla u^\dagger) - \frac{r_0^2}{3}\boldsymbol{\delta} = 0$$

(C.3-34)

where $\boldsymbol{\delta}$ is the unit tensor. This result, as that of section C.3-3, is valid only for steady-state flows. The diffusion equation (Liouville equation) for transient flows is:

$$\frac{d\psi}{dt} + \nabla \cdot \left(\psi\frac{d\boldsymbol{r}}{dt}\right) = 0$$

(C.3-35)

and the corresponding result for the conformation tensor is:

$$C + \theta\left(\frac{\partial C}{\partial t} - \nabla u \cdot C - C \cdot \nabla u^\dagger\right) = \frac{r_0{}^2}{3}\delta \qquad \text{(C.3-34)}$$

Here the derivative $\partial/\partial t$ is the partial derivative in a frame of reference based on the center of mass moving with the fluid. This is in fact the substantial derivative and result C.3-34 can be written in term of the convected derivative:

$$C + \theta\left(\frac{\delta C}{\delta t}\right) = \frac{r_0{}^2}{3}\delta \qquad \text{(C.3-35)}$$

where

$$\frac{\delta C}{\delta t} = \frac{DC}{Dt} - \nabla u \cdot C - C \cdot \nabla u^\dagger \qquad \text{(C.3-36)}$$

is the Oldroyd or upper convected derivative, introduced in Appendix B.2. The elastic properties of polymeric liquids may be explained by the elastic deformation of the polymer chains. Hencer, this result can serve as a foundation for the use of the convected derivative in the development of constitutive equations.

The parameter θ $(= 1/2\sigma)$ is a relaxation time of the polymeric system. For instance, if the liquid is suddenly stopped while the conformation tensor C is not in equilibrium state, C will reach the equilibrium value $(r_0{}^2/3)\delta$ according to the following equation:

$$C + \theta\left(\frac{\partial C}{\partial t}\right) = \frac{r_0{}^2}{3}\delta \qquad \text{(C.3-37)}$$

C.3-5 Deformation in extensional flow

Let us consider the following extensional flow field:

$$\left|\begin{array}{l} u(x, y, z) = \dot{\varepsilon}x \\ v(x, y, z) = -\dot{\varepsilon}x \\ w(x, y, z) = 0 \end{array}\right. \qquad \text{(C.3-38)}$$

The velocity gradient tensor is:

$$\nabla u = \begin{bmatrix} \dot{\varepsilon} & 0 & 0 \\ 0 & -\dot{\varepsilon} & 0 \\ 0 & 0 & 0 \end{bmatrix} \qquad \text{(C.3-39)}$$

For this steady, uniform flow, $DC/Dt = 0$ and:

$$\nabla u \cdot C = C \cdot \nabla u^\dagger = \begin{bmatrix} \dot{\varepsilon}\langle x^2\rangle & 0 & 0 \\ 0 & -\dot{\varepsilon}\langle y^2\rangle & 0 \\ 0 & 0 & 0 \end{bmatrix} \qquad \text{(C.3-40)}$$

Hence, Equation (C.3-35) reduces to:

$$
\begin{bmatrix}
(1 - 2\dot{\varepsilon}\theta)\langle x^2 \rangle & \langle xy \rangle & \langle xz \rangle \\
\langle xy \rangle & (1 + 2\dot{\varepsilon}\theta)\langle y^2 \rangle & \langle yz \rangle \\
\langle xz \rangle & \langle yz \rangle & \langle z^2 \rangle
\end{bmatrix}
=
\begin{bmatrix}
\frac{r_0^2}{3} & 0 & 0 \\
0 & \frac{r_0^2}{3} & 0 \\
0 & 0 & \frac{r_0^2}{3}
\end{bmatrix}
\quad \text{(C.3-41)}
$$

and the various components of the conformation tensor are:

$$\langle xy \rangle = \langle yz \rangle = \langle xz \rangle = 0 \tag{C.3-42}$$

$$\langle x^2 \rangle = \frac{r_0^2}{3} \frac{1}{1 - 2\dot{\varepsilon}\theta} \tag{C.3-43}$$

$$\langle y^2 \rangle = \frac{r_0^2}{3} \frac{1}{1 + 2\dot{\varepsilon}\theta} \tag{C.3-44}$$

$$\langle z^2 \rangle = \frac{r_0^2}{3} \tag{C.3-45}$$

and:

$$\langle r^2 \rangle = r_0^2 \frac{1 - \frac{4}{3}\dot{\varepsilon}^2\theta^2}{1 - 4\dot{\varepsilon}^2\theta^2} \tag{C.3-46}$$

This last result predicts that the trace of the tensor C goes to infinity as $\dot{\varepsilon}\theta \rightarrow 1/2$. A similar paradoxical result has been observed in Problem 6.1-D. Obviously, the model has no physical meaning for flows with extensional rates larger than $1/2\theta$. Physically, $\langle r^2 \rangle$ is the length of the macromolecules and cannot become infinity. This shortcoming can be overcome by using non-Gaussian distribution functions for ψ_0 and ψ, that would go to zero as $|r|$ reaches the length of the macromolecules. More realistic results are obtained by using non-linear springs (see BIRD, ARMSTRONG and HASSAGER, 1987; GRMELA and CARREAU, 1987).

C.3-6 Concluding remarks

The main results of Section C.3 are illustrated in Figure C.3-3 in terms of the reduced trace of the conformation tensor, C. It is interesting to note that under shear flow the macromolecules are continuously deformed but the deformation is considerably less than under extensional flow in a comparable condition. We also note for the extensional flow the sudden transition from coil to stretched state. This is the reason why extensional flows are considered as strong flows. Under shear flows, the macromolecules are rotating in the shear plane and stretching is not as efficient. Under extensional conditions, the polymeric chains are rapidly aligned in the flow direction and efficiently stretched.

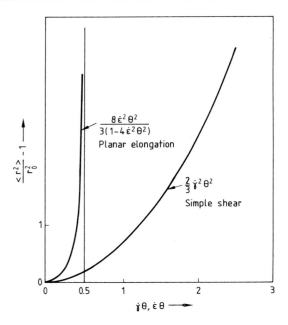

Figure C.3-3   Macromolecule deformation in flow

C.4   Variation of viscosity with temperature and pressure

C.4-1 The Andrade law

The expression C.2-8 can be written as

$$\eta = \frac{h}{V}\, e^{-S/k_b}\, e^{H/k_b T} = A e^{B/T} \tag{C.4-1}$$

or in terms of the activation energy, E, and of the ideal gas constant R:

$$\eta = \eta_0\, e^{\frac{E}{R}\left(\frac{1}{T}-\frac{1}{T_0}\right)} \tag{C.4-2}$$

where η_0 is the viscosity evaluated at the reference temperature T_0.

The activation energy for the viscosity of usual liquids is of the order of 10^4 J/mole. The activation energy for polymers is much larger and increases with both the chain rigidity and the size of lateral branches. Typical values are reported in Table C.4-1.

Table C.4-1   Activation energy values for polymer melts (from SCHOTT 1962)

Polymer melt	$E \cdot 10^4$ J/mole at $100°$ C above glass transition
Polyethylene	2.9
Polypropylene	3.7
Polyisobutylene	5.9
Polystyrene	9.6
Poly-α-Methylstyrene	13.3

The viscosity of polymer melts changes drastically with temperature. The variation is given by:

$$\frac{\Delta\eta}{\eta} = -\frac{E}{RT^2}\Delta T \qquad (C.4\text{-}2)$$

Considering E equal to 10^5 J and $T = 400$ K, $E/RT = 25$ and $\Delta\eta/\eta = 25\Delta T/T$. Therefore, a 1% error in the temperature measurement could lead to a 25% error in the viscosity.

C.4-2 The WLF expression

Glass transition can be experimentally observed by a net change in a narrow range of temperature of the thermal expansion coefficient, of the specific heat or of mechanical properties. This phenomenon is not specific to polymers, it has been observed for liquids that can exhibit important supercooling without crystallizing (liquids with strong Hydrogen bondings such as glycerine, silicates, ...). Glass transition temperature, T_g, is a key parameter in polymers since

- on the one hand, atactic polymers cannot crystallize and their only change of state occurs at the glass transition temperature, T_g;
- on the other hand, crystalline polymers always show an amorphous phase, either glassy or rubbery; in many cases, these polymers can be quenched in the glassy state.

The empirical relation proposed by WILLIAMS, LANDEL and FERRY (see FERRY, 1970) accounts for the rapid variation of the viscosity between the glass transition temperature, T_g and $T_g + 100°$ C. The WLF expression is:

$$\eta = \eta_g a_T \frac{T\varrho}{T_g\varrho_g} \qquad (C.4\text{-}4)$$

where η_g and ϱ_g are respectively the viscosity and the density at T_g and the shift factor, a_T, is given by:

$$\log a_T = \frac{-C_1{}^g(T - T_g)}{T - T_g + C_2{}^g} \qquad (C.4\text{-}5)$$

Values of the parameters $C_1{}^g$ and $C_2{}^g$ are presented in Table C.4-2 for selected polymers.

Table C.4-2 WLF parameters for selected polymers (from FERRY, 1970)

Polymer	T_g °C	$C_1{}^g$	$C_2{}^g$
Polystyrene	100	13	50
Polyvinyl acetate	132	13	47
Polymethyl methacrylate	115	32	80
Natural rubber	−73	16	53

For example, the viscosity of polystyrene varies between T_g and $T_g + 100°$ C according to:

$$\frac{\eta(T_g)}{\eta(T_g + 100°\ \text{C})} \simeq 4 \cdot 10^8$$

The viscosity changes over eight decades for a temperature variation of 100° C, for example η is of the order of 10^3 Pa $\cdot$ s at 200° C and increases to 10^{11} Pa $\cdot$ s at 100° C.

The WLF expression gives good predictions in the range of T_g and $T_g + 100°$ C. Above, it is preferable to use the Andrade expression (of the Arrhenius type). Both expression can be combined to get:

$$\log \eta = A + \frac{B}{RT} + \frac{C}{T - T_g + C_2{}^g} \tag{C.4-7}$$

C.4-3 Free volume theory

Evidently, the Andrade expression based on a unique activation energy term cannot describe the properties of polymers in the glass transition zone. This follows from the Eyring theory based on a unique molecular mechanism which is no longer valid. Indeed, when increasing the temperature above the glass transition temperature, the spectrum of the molecular motions, frozen in the glassy state, is progressively reactivated. Even for low molecular weight liquids, the Andrade expression is valid in a limited range of temperature. In fact the parameter B varies slightly with temperature. To correct that, a more recent viscosity theory based on the concept of free volume has been proposed. This is briefly described below.

DOOLITTLE (1951) has shown from precise data for decanes that the viscosity could be better correlated to the free volume than to the temperature (on a wider range of temperature). The relative free volume is defined by:

$$\frac{V_f(T)}{V_0} = \frac{V(T) - V_0}{V_0} \tag{C.4-8}$$

where V and V_0 are respectively the volume at temperature T and the volume extrapolated to 0 K without phase changes. Following COHEN and TURNBULL (1959, 1961), the expression for the viscosity can then be written as:

$$\eta = A' \exp \left\{ B' \frac{V_0}{V_f(T)} \right\} \tag{C.4-9}$$

This result is equivalent to the WLF expression. Figure C.4-1 shows a typical variation of the free volume as a function of the absolute temperature:

–   The variation of the free volume is slow below T_g with a slope given by α_g.
–   Above T_g, the variation is more rapid with a slope α.

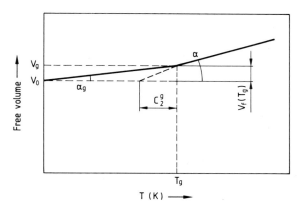

Figure C.4-1   Variation of the free volume with temperature

From the definition of the free volume and following the expression of Doolittle, we can write:

$$V_f(T) = V_f(T_g) + \alpha(T - T_g) \tag{C.4-10}$$

and

$$\log \frac{\eta}{\eta_g} = B' \left[\frac{V_0}{V_f(T)} - \frac{V_0}{V_f(T_g)} \right] \tag{C.4-11}$$

It follows that

$$\log \frac{\eta}{\eta_g} = B'' \left[\frac{1}{V_f(T_g) + \alpha(T - T_g)} - \frac{1}{V_f(T_g)} \right]$$

$$= -B''' \frac{T - T_g}{T - T_g + C_2{}^g} \tag{C.4-12}$$

This is clearly identical to the WLF expression with:

$$C_2{}^g = \frac{V_f(T_g)}{\alpha} \tag{C.4-13}$$

The magnitudes of V_f and α give a correct order of magnitude for $C_2{}^g$:

$$\left. \begin{array}{l} V_f(T_g) \simeq 0.025 \\ \alpha \simeq 10^{-3}\,^\circ\mathrm{C}^{-1} \end{array} \right| \quad \Rightarrow \quad C_2{}^g \simeq 25^\circ\,\mathrm{C}$$

C.4-4 Effect of pressure on the viscosity

The molecular mechanism for the viscosity as illustrated in Figure C.2-1 shows a hole formation of volume ΔV due to the separation of molecules B and C. The increase of volume results into an enthalpy of activation that depends on the pressure:

$$H(p) = H + \Delta V p \tag{C.4-14}$$

From Eq. C.2-8, one gets

$$\eta = \frac{h}{V} \, e^{-S/k_b} \, e^{H/k_b T} \, e^{(\Delta V/k_b T)p} \tag{C.4-15}$$

This shows that the viscosity increases exponentially with pressure. Using a pressure dependence coefficient defined by:

$$\gamma = \frac{\Delta V}{k_b T} = \frac{d \ln \eta}{dp} \tag{C.4-16}$$

the viscosity is then expressed by:

$$\eta(T, p) = A \, e^{B/T} e^{\gamma p} \tag{C.4-17}$$

POWELL et al. (1941) have reported that for all liquids,

$$\Delta V \approx \frac{V}{2.5} \simeq 5 \cdot 10^{-20} \, \text{mm}^3$$

and as $k_b T \simeq 4 \cdot 10^{-21}$ J, one gets:

$$\gamma \simeq 1.2 \cdot 10^{-8} \, \text{Pa}^{-1} \simeq 1.2 \cdot 10^{-3} \, \text{atm}^{-1}$$

This is in agreement with the experimental data, 2.2 to $3.5 \cdot 10^{-8}$ Pa^{-1} reported by GUPTA et al. (1969–70) for paraffinic and aromatic oils, and 2 to $6 \cdot 10^{-8}$ Pa^{-1} for commodity polymers as shown in Table C.4-3.

Table C.4-3   Values of the pressure dependence coefficient for selected polymers

Polymer	T °C	γ 10^{-8} Pa^{-1} or 10^{-3} atm^{-1}	Reference
PVC injection grade	180	6	WALES (1975)
	190	4	WALES (1975)
	200	3	WALES (1975)
Linear polyethylene	190	2.31	NAKAJIMA and COLLINS (1974)
Butadiene-acrylonitrile copolymer (33% acrylonitrile)	125	2.45 to 3.17	

As an example, it is useful to examine the variation of the viscosity with pressure for an average value of the coefficient γ of $3.3 \cdot 10^{-8}$ Pa^{-1}. The viscosity is then expressed by:

$$\eta = \eta_0 e^{P/30} \quad \text{(with } p[=]\,\text{MPa)}$$
$$= \eta_0 e^{P/300} \quad \text{(with } p[=]\,\text{atm)}$$

Table C.4-4 shows how drastically the viscosity increases with pressure for pressure larger than 100 atm.

Table C.4-4 Variation of the viscosity of polymers with pressure for an average value of the pressure dependence coefficient equal to 0.0033 atm^{-1}

P atm	η/η_0
30	1.11
100	1.39
300	2.70
500	5.29
1000	27.9
3000	22026.

Appendix D
Complements on Heat Transfer

D.1   Convection

D.1-1 Free and forced convection

Let us consider a solid body immersed in a fluid in motion. The fluid elements stay in contact with the solid during a very short time interval. During this interval, heat is exchanged between the solid and the fluid elements by conduction and the fluid motion drags these elements far from the solid body. Other fluid elements are then brought in contact with the solid surface. This is the mechanism of convection which associates heat conduction with the bulk motion of a fluid.

The convection is called natural or free convection when the bulk motion of the fluid is due to no external forces other than gravity (buoyancy effects). If the flow is imposed, for example as in cooling of an object with an air jet, this is forced convection.

Both free and forced convection mechanisms may be encountered in most processes. However, forced convection is usually preferred in order to increase production rates. Then free convection plays a negligible role althrough it is never absent.

D.1-2 The Bénard problem

BÉNARD (1900) has contributed greatly to the understanding of free convection first by his experiments and secondly, by introducing appropriate dimensionless numbers.

Let us consider a cavity containing a liquid between two regulated horizontal walls. The lower wall is at a higher temperature, ΔT, than the upper wall as illustrated in Figure D.1-1.

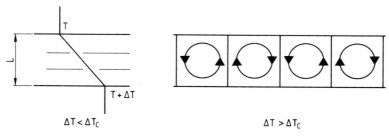

Figure D.1-1   Heat exchange by conduction and free convection

In so far as ΔT is not too large, the lower layers of liquid which are at higher temperature (and less dense) can stay stagnant under the more dense upper layers. If the temperature difference exceeds, however, a critical value ΔT_c the warmer layers will rise creating chimney effects. The ascending warm liquid is progressively cooled down and feeds the downstream liquid. These convective currents create stable cells or vortices, periodically spaced. This is laminar convection. If ΔT is further increased the flow will become turbulent characterized by irregular and unstable vortices.

If one measures the heat fluxes across the liquid, for example from the power needed to control the lower and upper wall temperatures, one observes (SILVERSTON, 1958) that:

—   provided that $\Delta T < \Delta T_c$, the heat flux is that given by simple conduction:

$$q = k\frac{\Delta T}{L} \qquad\qquad (\text{D.1-1})$$

where k is the thermal conductivity of the liquid;
—   if $\Delta T > \Delta T_c$, the heat transfer is larger than that of pure conduction. This is due to the fact that a larger part of heat is transferred by convection. The convective heat transfer coefficient, h, is defined by:

$$q = h\,\Delta T \qquad\qquad (\text{D.1-2})$$

The ratio of the total flux to that of pure conduction is:

$$\frac{qL}{k\Delta T} = \frac{hL}{k} = \text{Nu} \qquad\qquad (\text{D.1-3})$$

This dimensionless number is the Nusselt number which will be generalized to any flow geometries.

Determination of ΔT_c

Following RAYLEIGH (1916), the problem is to determine the value of ΔT_c for which the equilibrium shown on the left side of Figure D.1-1 is no longer stable.

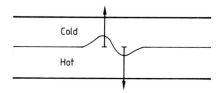

Figure D.1-2   Generation of instabilities

The instabilities occur from perturbations induced in the system: a small quantitity of warm liquid tends to rise while a small quantitity of cool liquid tends to go down, as illustrated in Figure D.1-2. Then two series of factors tend to enhance the perturbations or to dampen the effects.

- *Instability factors*

These are essentially bouyancy forces that exert an ascending force on warm liquid elements surrounded by cold liquid and a downstream force on cold elements surrounded by warm liquid. The buoyancy force is proportional to:

- $\Delta\varrho$, given by:

$$\Delta\varrho = -\frac{d\varrho}{dT}\,\Delta T = \varrho\beta\Delta T \qquad \text{(D.1-4)}$$

where the coefficient of volume expansion for the liquid is:

$$\beta = -\frac{1}{\varrho}\frac{d\varrho}{dT} \qquad \text{(D.1-5)}$$

- g, the acceleration due to gravity;
- the volume affected by the perturbations, that can be taken to be proportional to L^3.

- *Stabilizing factors*

- the thermal diffusivity, α, of the liquid; indeed, heat conduction tends to dampen the fluctuations;
- the viscosity of the liquid which is a characteristic of the resistance to any motion in the liquid.

All these factors can be considered in a single dimensionless number, called the Rayleigh number, of which the numerator expresses the instability factors and the denominator the stabilizing ones:

$$\text{Ra} = \frac{\varrho g\beta\Delta T L^3}{\eta\alpha} = \frac{g\beta\Delta T L^3}{\nu\alpha} \qquad \text{(D.1-6)}$$

RAYLEIGH (1916) has shown from experiments that free convection appears when:

$$\text{Ra} > \text{Ra}_c \quad \text{with} \quad \text{Ra}_c \approx 1700$$

Hence:

$$\Delta T_c \approx 1700\frac{\eta\alpha}{\varrho g\beta L^3} \qquad \text{(D.1-7)}$$

Figure D.1-3 shows the variation of the Nusselt number as a function of the Rayleigh number (from SILVERSTON, 1958).

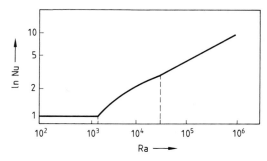

Figure D.1-3 Variation of the Nusselt number as a function of the Rayleigh number in the Bénard experiment

D.1-3 Other usual dimensionless numbers

a) Grashof number

The Grashof number is associated to the magnitude of the convective streams induced by horizontal temperature gradients. For a temperature gradient ΔT on a width L, the number is written as:

$$\text{Gr} = \frac{g\beta\Delta T L^3}{\nu^2} \tag{D.1-8}$$

Example D.1-1: Free convection in a vertical cylinder

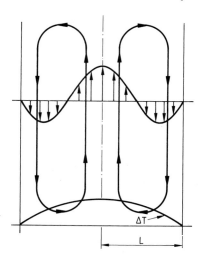

Figure D.1-4 Free convection due to horizontal temperature gradients

Let us consider a vertical cylinder of radius L filled with a liquid such that the central temperature is ΔT higher than the wall temperature (see Figure D.1-4). Due to the buoyancy forces the liquid in the center will rise and will come down at the wall vicinity. This is observed for any positive values of ΔT and no stable zone is detected. One can show that the characteristic velocity of the liquid is given by:

$$V = \frac{\varrho g \beta \Delta T L^2}{\eta} \qquad (D.1\text{-}9)$$

The Grashof number is then equivalent to the Reynolds number for a flow induced by a thermal gradient.

b) Prandtl number

The Prandtl number is the ratio of the momentum diffusivity or kinematic viscosity, ν, and the thermal diffusivity:

$$\mathrm{Pr} = \frac{\nu}{\alpha} = \frac{\eta c_p}{k} \qquad (D.1\text{-}10)$$

We notice that, if the characteristic length L is taken to be the same, the Rayleigh number is the product of the Prandtl and the Grashof number, i.e.:

$$\mathrm{Ra} = \mathrm{Gr} \cdot \mathrm{Pr} \qquad (D.1\text{-}11)$$

c) Nusselt number

The Nusselt number was first introduced for the particular case of the Bénard experiment. For more general flow situations, the heat exchange from a warm solid surface S at temperature T placed in a cold liquid at T_f is given by:

$$Q = hS(T - T_f) \qquad (D.1\text{-}12)$$

Where h is an average heat transfer coefficient. In absence of convection, heat would be lost only by conduction, proportionally to $T - T_f$:

$$Q_0 = k\frac{S}{l}(T - T_f) \qquad (D.1\text{-}13)$$

where l is a characteristic distance away from the surface at which the liquid temperature is T_f (see Figure D.1-5); it can be related to a characteristic length L of the solid and is often of the same order of magnitude. Then the ratio of Equations (D.1-12) and (D.1-13) is the Nusselt number as defined previously:

$$\mathrm{Nu} = \frac{hL}{k} \qquad (D.1\text{-}3)$$

Hence the Nusselt number represents approximately the ratio of the heat exchanged by convection and the heat that would be transferred by pure conduction.

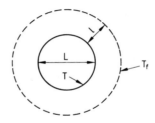

Figure D.1-5   Definition of the Nusselt number

D.1-4 Physical properties of air and water

In most applications, free or forced convection cooling is achieved in air or in water. We present in this section the main properties of air and water as a function of temperature.

For the computation of the dimensionless numbers, the physical properties of the fluids are evaluated at the so-called film temperature, $T_f{'}$, which is the average temperature between the solid wall surface and the fluid temperature far from the interface:

$$T_f{'} = \frac{T_f + T_s}{2} \tag{D.1-14}$$

Figure D.1-6 reports the physical properties of air and Figure D.1-7 those of water.

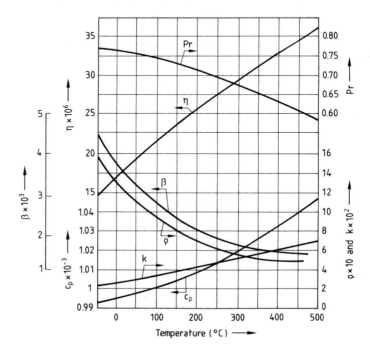

Figure D.1-6   Physical properties of air as a function of temperature

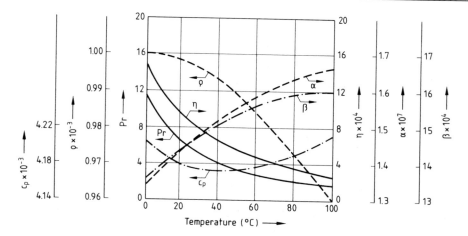

Figure D.1-7 Physical properties of water as a function of temperature

D.1-5 Free convection correlations

The dimensionless analysis of free convection leads to the general relationship (see
MC ADAMS, 1954, BIRD, STEWART and LIGHTFOOT, 1960):

$$Nu = Nu(Gr, Pr) \qquad (D.1\text{-}15)$$

In many situations, and the analysis of the Bénard problem shows why, the
correlation reduces to:

$$Nu = Nu(Gr \cdot Pr) = Nu(Ra) \qquad (D.1\text{-}16)$$

Three zones of interest have been reported:

- $Ra < 10^3 - 10^5$: Nu varies slightly with Ra. This corresponds to the stable
  zone of the Bénard problem with:

$$Ra < 1700, \quad Nu \simeq 1 \qquad (D.1\text{-}17)$$

  The heat is mostly transferred by conduction.
- $10^3 - 10^5 < Ra < 10^7 - 10^9$ with:

$$Nu = const \, Ra^{1/4} \qquad (D.1\text{-}18)$$

  In this zone the convective flow is laminar with stable vortices as observed in
  the Bénard experiment.
- $Ra > 10^7 - 10^9$ with:

$$Nu = const \, Ra^{1/3} \qquad (D.1\text{-}19)$$

  The flow is now turbulent.

a) Horizontal cylinders

The dependence of Nu with the Rayleigh number for horizontal cylinders is reported in Figure D.1-8. The characteristic length L is the diameter. Two useful correlations are:

$$10^3 < \text{Ra} < 10^9 \ , \quad \text{Nu} = 0.53 \ \text{Ra}^{1/4} \tag{D.1-20}$$

$$10^9 < \text{Ra} < 10^{12}, \quad \text{Nu} = 0.125 \, \text{Ra}^{1/3} \tag{D.1-21}$$

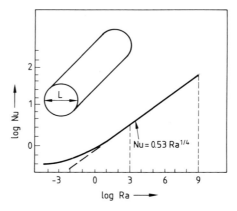

Figure D.1-8 Nusselt number for free convection about a horizontal cylinder

b) Vertical plates or cylinders

The characteristic length L is now the height of the plates or cylinders and the dependence of the Nusselt number on the Rayleigh number is shown in Figure D.1-9. The corresponding correlations are:

$$10^4 < \text{Ra} < 10^9 \ , \quad \text{Nu} = 0.59 \, \text{Ra}^{1/4} \tag{D.1-22}$$

$$10^9 < \text{Ra} < 10^{12}, \quad \text{Nu} = 0.13 \, \text{Ra}^{1/3} \tag{D.1-23}$$

c) Horizontal plates

In the case of horizontal plates one must distinguish two different patterns depending on the direction of the heat flux. Typical streamlines are illustrated in Figure D.1-10.

Taking the characteristic length L as the smaller dimension of the plate, the correlations are the followings:

–   For rising heat flux

$$10^5 < \text{Ra} < 2 \times 10^7, \quad \text{Nu} = 0.54 \, \text{Ra}^{1/4} \tag{D.1-24}$$

$$2 \times 10^7 < \text{Ra} < 10^{11}, \quad \text{Nu} = 0.14 \, \text{Ra}^{1/3} \tag{D.1-25}$$

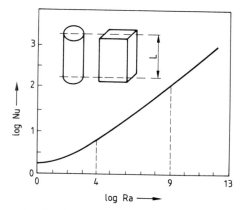

Figure D.1-9   Nusselt number for free convection about a vertical cylinder or plate

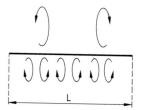

Figure D.1-10   Streamlines for free convection about a horizontal plate

— For downward flux

$$\mathrm{Nu} = 0.27\,\mathrm{Ra}^{1/4} \tag{D.1-26}$$

Remarks

— The rising heat flux is enhanced by buoyancy forces and is much larger than the downward flux (twice as large in laminar convection).
— For rising heat flux, the streamlines become turbulent at Rayleigh numbers larger than 2×10^7, whereas the downward flux streamlines do not become turbulent up to Rayleigh number equal to 3×10^{11}. Clearly heating a fluid from above is more stabilizing than heating from below. It is the reverse for cooling.

d) Sphere

According to RANZ and MARSHALL (1952), the case of the sphere is more complex than the other ones due to the different dependence of the Nusselt number on the Grashof and Prandtl numbers:

$$\mathrm{Nu} = 2 + 0.6\,\mathrm{Gr}^{1/4}\,\mathrm{Pr}^{1/3}, \quad \text{for} \quad \mathrm{Gr}^{1/4}\,\mathrm{Pr}^{1/3} < 200 \tag{D.1-27}$$

Example D.1-2 Free convection about a vertical cylinder

Let us consider a vertical cylinder (or plate) of height equal to 50 mm. The wall temperature is at 220° C and is cooled in ambient air at 20° C by free convection. Determine the value of the heat transfer coefficient.

Solution

The physical properties of the air are evaluated at 120° C (the film temperature), the characteristic length is the height equal to 50 mm, and ΔT is equal to 200° C. The Rayleigh number is:

$$\mathrm{Ra} = 0.93 \times 10^6$$

and from Equation (D.1-22):

$$\mathrm{Nu} = 0.59 \, \mathrm{Ra}^{1/4} = 18$$

Hence:

$$h = 11 \cdot 3 \, \mathrm{W}/(\mathrm{m}^2 \cdot {}^\circ\mathrm{C})$$

D.2 Forced convection

D.2-1 General expressions

In forced convection the fluid motion is caused by external forces other than gravitational or buoyancy forces. One can show from dimensional analysis that the Nusselt number is a function of the Reynolds and Prandtl numbers, i.e.:

$$\mathrm{Nu} = \mathrm{Nu}(\mathrm{Re}, \mathrm{Pr}) \tag{D.2-1}$$

For a given geometry, the function $\mathrm{Nu}(\mathrm{Re}, \mathrm{Pr})$ is not the same if the flow is laminar or turbulent. The laminar flow is characterized by a small value of the Reynolds number and at large Reynolds numbers, the flow becomes turbulent. For intermediate values, the flow is in the transition regime. The ranges for the three regimes depend on the geometry and on the roughness of the surface. The flow around a cylindrical tube is laminar for Reynolds numbers smaller than 20 and turbulent for values above 10^4. It is therefore essential to verify in what range of Reynolds number a given expression is valid before using it. The main cases are presented below and can be found in most textbooks (see BIRD, STEWART and LIGHTFOOT, 1960).

D.2-2 Forced convection correlations

a) Sphere

The correlation of the Nusselt number for forced convection about a sphere is shown in Figure D.2-1. The characteristic length is the sphere diameter. The curve shown in the figure is described by the following expression:

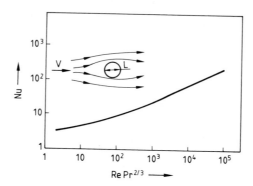

Figure D.2-1   Forced convection about a sphere

$$Nu = 2 + 0.60\,Re^{1/2}\,Pr^{1/3} \qquad (D.2-2)$$

which is valid for Reynolds numbers ranging from 0 to 10^5.

b) Cylinder (perpendicular to the flow stream)

The characteristic length L is the diameter of the cylinder and the correlation is shown in Figure D.2-2.

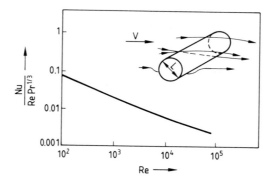

Figure D.2-2   Forced convection about a cylinder (perpendicular to the flow)

c) Plate or cylinder (parallel to the flow stream)

The Reynolds number varies with the distance x according to the definition:

$$\mathrm{Re}_x = \frac{V\varrho}{\eta}x \qquad (\text{D.2-3})$$

where x is the distance from the incipient point as shown in Figure D.2-3. The transition from the laminar to the turbulent flow regime in this flow geometry occurs at a very high value of the Reynolds number (of the order of 10^5). For higher Reynolds numbers, the determination of the Nusselt number may be highly inaccurate due to the long unstable transition regime (Re ranging from 10^5 to $5 \cdot 10^5$).

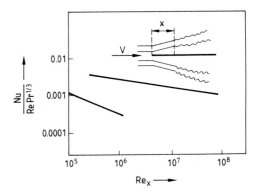

Figure D.2-3   Forced convection about a thin plate or a cylinder parallel to the flow

The Reynolds number increases with x, and if the plate is sufficiently long, it is possible for a flow regime initially laminar to become turbulent at a critical distance. The Nusselt number depends also on x. In the laminar regime, the following theoretical result holds (BIRD, STEWART and LIGHTFOOT, 1960):

$$\mathrm{Nu} = 0.66\,\mathrm{Re}_x{}^{1/2}\,\mathrm{Pr}^{1/3} \qquad (\text{D.2-4})$$

Example D.2-1 Forced convection about a cylinder

Let us consider a cylinder placed perpendicularly to a flow stream of air at velocity equal to 1 m/s. The wall temperature of the cylinder is at 150° C and the air is initially at 50° C. The diameter of the cylinder is $D = 10$ mm. Determine the heat transfer coefficient.

Solution

The physical properties of the air are evaluated at the so-called film temperature T_f', equal to 100° C and the characteristic length is taken as the diameter $D = 10$ mm. The Reynolds number is then equal to:

$$\mathrm{Re} = \frac{DV}{\eta_f}\varrho_f = 437$$

From Figure D.2-2, one obtains:

$$\frac{\text{Nu}}{(\text{Re Pr}^{1/3})} = 0.025$$

The Prandtl number for air at $100°$ C is $\text{Pr} = 0.74$, hence:

$$\text{Nu} = h\frac{D}{k_f} = 9.9$$

The air conductivity at $100°$ C is $k_f = 0.045$ W/m·°C. The heat transfer coefficient is therefore:

$$h = 45 \text{ W/m}^2 \cdot °\text{C}$$

D.3 Radiative heat transfer

Heat transfer by radiation plays an important role in many polymer processes. For example, in the extrusion of thin films the radiative heat transfer is comparable to the heat losses by convection to air (see Appendix D.4).

D.3-1 Black body

The basic laws for heat radiation have been developed by STEFAN and BOLTZMANN for a black body which, by definition, absorbs all radiations received at its surface. The theory also stipulates that the emission of electromagnetic radiations from a black body is proportional to the fourth power of the absolute temperature of the body, i.e.:

$$E^b = \sigma T^4 \tag{D.3-1}$$

where E^b is the total energy flux integrated on all directions and for all wave lengths of the radiations. It is the radiation intensity (W/m^2). T is the absolute temperature of the black body expressed in degrees Kelvin (K) and σ is the Stefan-Boltzmann constant:

$$\sigma = 5.664 \cdot 10^{-8} \text{ W/m}^2 \cdot K^4$$

The radiative flux E^b in the θ-direction with respect to the normal to the surface is given by Lambert's law:

$$E_\theta^b = \frac{E^b}{\pi} \cos \theta \tag{D.3-2}$$

On the other hand, the radiative flux for a given wave length λ is given by Plank's law:

$$E_\lambda^b = \frac{2\pi c^2 h}{\lambda^5} \frac{1}{e^{ch/\lambda k_b T} - 1} \tag{D.3-3}$$

where h is Planck's constant, c is the light velocity in vacuum and k_b is Boltzmann's constant. The radiative intensity in the θ-direction and for a wave length λ is then given by:

$$E_{\lambda\theta}^b = E_\lambda^b \frac{\cos\theta}{\pi} \qquad (D.3\text{-}4)$$

This function can be integrated over all wave lengths and over the whole space to obtain a relation between E^b and T. The value of the Stefan–Boltzmann constant is then obtained in terms of the other constants:

$$\sigma = \frac{2}{15} \frac{\pi^5 k_b^4}{c^2 h^3} \qquad (D.3\text{-}5)$$

The value computed from this expression is found to be 1.5% smaller than the experimentally determined one.

D.3-2 Nonblack bodies

In reality, most bodies do not behave as black bodies and the nonblack body is defined as one which does not absorb all incident radiations. Its absorbance capacity or absorptivity for a given wave length λ and a given direction θ is defined by:

$$\alpha_{\lambda\theta} = \frac{\text{Radiative flux absorbed } (\lambda, \theta)}{\text{Incident radiative flux } (\lambda, \theta)} \qquad (D.3\text{-}6)$$

From the definition of a black body, $\alpha_{\lambda\theta} = 1$. If, on the contrary, there is no absorption the body is called white. The grey body is defined as one for which the absorption is independent of λ and θ. The total absorbance capacity or absorptivity is defined by:

$$\alpha = \frac{\text{Total radiative flux absorbed}}{\text{Total incident radiative flux}} \qquad (D.3\text{-}7)$$

The radiation emitted by a nonblack body, $E_{\lambda\theta}$, in the θ direction and for a wave length λ is proportional to the fourth power of the body temperature, i.e.:

$$E_{\lambda\theta} = \varepsilon_{\lambda\theta}\, \sigma\, T^4 \qquad (D.3\text{-}8)$$

where $\varepsilon_{\lambda\theta}$ is the emissivity or emission factor for the wave length λ and for the θ-direction. Kirchhoff's law states that:

$$\varepsilon_{\lambda\theta} = \alpha_{\lambda\theta} \qquad (D.3\text{-}9)$$

Since the absorptivity $\alpha_{\lambda\theta}$ is by definition smaller than 1, it follows from Kirchhoff's law that the emissivity $\varepsilon_{\lambda\theta}$ must also be smaller than 1. Hence, at the same temperature the black body will emit a larger radiation flux than a nonblack body. The total emissivity ε is the ratio between radiative fluxes of the nonblack and black body, i.e.:

$$\varepsilon = \frac{E}{E^b} = \frac{\text{Total radiations emitted by the nonblack body at } T}{\text{Total radiations emitted by a black body at } T} \qquad (D.3\text{-}10)$$

The emissivity is a dimensionless parameter less than 1 for a nonblack body and equal to 1 for a black body.

Remarks

The emissivity ε of a nonblack body is independent of the surroundings. This is not the case for the absorptivity α which depends on the incident radiations and hence on the surroundings.

One can describe the emissivity of a body without making reference to the surroundings. On the contrary, the absorptivity cannot in general be described without knowing the characteristics of the incident radiations.

D.3-3 Radiant heat exchange between grey bodies

The complexity of the radiant heat exchange between two bodies comes from the dependence of $\varepsilon_{\lambda\theta}$ on T, λ and θ (Planck's and Lamber's laws). The problem is simplified in assuming that $\varepsilon_{\lambda\theta}$ is independent of λ and θ ($\varepsilon_{\lambda\theta}$ is taken to be equal to ε). This approximation is acceptable is the case of nonmetallic materials. This is the same as considering the materials as grey bodies. The radiative flux of a grey body is proportional to that emitted by a black body at the same temperature:

$$E_{\lambda\theta} = \varepsilon\, E_{\lambda\theta}^b \qquad (D.3\text{-}11)$$

This can be integrated over λ and for all directions θ to obtain:

$$E_\lambda = \varepsilon E_\lambda^b$$
$$E_\theta = \varepsilon E_\theta^b \qquad (D.3\text{-}12)$$
$$E = \varepsilon E^b$$

From Kirchhoff's law $\varepsilon_{\lambda\theta} = \alpha_{\lambda\theta}$ and it follows that the absorptivity of a grey body, $\alpha_{\lambda\theta} = \alpha$ and is equal to its emissivity ε.

The solution of problems of radiant heat transfer between two grey bodies could be quite complex. Here, we will present only results applicable to polymer processing examples.

Radiant heat transfer between two infinite surfaces

Let us define the radiant heat transfer rate Q_{12} as the heat exchange per unit time between two bodies and q_{12}, the radiant heat flux, i.e. the heat transferred per unit time and per unit surface area. The results for grey bodies are the following;

a) *Infinite parallel plates:*

$$q_{12} = \sigma \frac{1}{1/\varepsilon_1 + 1/\varepsilon_2 - 1}(T_1^4 - T_2^4) \qquad (D.3\text{-}13)$$

$$q_{12} = -q_{21}$$

b) Concentric spheres or concentric infinite cylinders:

$$S_1 = \text{Surface area of inside body}$$

$$S_2 = \text{Surface area of outside body}$$

$$Q_{12} = \sigma \frac{S_1}{\dfrac{1}{\varepsilon_1} + \dfrac{S_1}{S_2}\left(\dfrac{1}{\varepsilon_2} - 1\right)} (T_1^4 - T_2^4) \qquad \text{(D.3-14)}$$

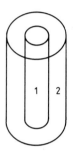

If the outside surface is much larger than the inner one $(S_1/S_2 \to 0)$, then:

$$Q_{12} = \sigma S_1 \varepsilon_1 (T_1^4 - T_2^4) \qquad \text{(D.3-15)}$$

This last expression will be used to compute the heat exchange from any convex body inside a room or cavity of any geometry but of large dimensions with respect to the body.

This will be the case for heat transfer by radiation between a polymer filament or film and the walls of a surrounding room, provided that the reflexion, absorption and emission from other objects (die, etc. ...) are negligible.

Example D.3-1 Radiant heat transfer from a polymeric filament

Let us consider a liquid polymeric filament of emissivity equal to $\varepsilon = 0.9$. The filament is at temperature $T_1 = 220°$ C and the walls of the surrounding room are at $T_2 = 20°$ C. Determine the heat transfer coefficient by radiation.

Solution

The heat transfer coefficient for radiation, h_r, can be defined by:

$$q_{12} = h_r(T_1 - T_2) \qquad \text{(D.3-16)}$$

By comparing this to Equation (D.3-15) expressed per unit surface area:

$$q_{12} = \sigma\,\varepsilon(T_1^4 - T_2^4)$$

one obtains:

$$h_r = \sigma\,\varepsilon\frac{T_1^4 - T_2^4}{T_1 - T_2} \tag{D.3-17}$$

Substituting the numerical values, one gets:

$$h_r = 13\,\mathrm{W/m^2 \cdot {}^\circ C}$$

This value found for h_r is equivalent to the value that one would get for the free convection heat transfer coefficient (under identical temperature conditions) between a vertical cylindrical polyethylene filament of 50 mm long (see Example D.1-2).

In polymer processing situations where the molten polymer is in contact with air, it will be necessary to account for heat losses due to radiation. Even when forced convection in air is used, radiant heat transfer is responsible for a non negligible part of the overall heat losses (the convective heat transfer coefficient computed in Example D.2-1 is only three times larger than the value of h_r obtained above).

D.4   Cooling of polymeric films in air or water

D.4-1 Problem statement

This heat transfer problem appears in many polymer processing situations where the hot polymer melt is cooled in air or water. The case of the cooling of a film at the exit of a die is illustrated in Figure D.4-1.

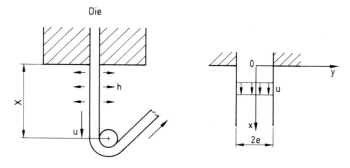

Figure D.4-1   Cooling of a vertical polymeric film

The initial following approximations are made:

- The film thickness remains constant (equal to $2e$) and its width is large enough that the film can be taken as infinite is the z-direction.
- The only velocity component of the polymeric film is u (in direction x) and taken as a constant (in reality, the film is stretched, the velocity increases and the thickness decreases).
- At the exit of the die, the polymer temperature is uniform and equal to T_0.

We wish to determine the temperature profile in the film as a function of x and y. The cooling of the film is essentially achieved by convection in air or water and by radiation with the surrounding walls of the room. We will use h for the heat transfer coefficient between the film and the fluid. As done in Chapter 2.3, the axial conduction in the film will be considered negligible with respect to the convective term. These approximations allow us to reduce the mathematical problem to the following simple partial differential equation:

$$u\frac{\partial T}{\partial x} = \alpha\frac{\partial^2 T}{\partial y^2} \tag{D.4-1}$$

to be solved with the following boundary conditions:

$$\text{at } x = 0, \qquad T = T_0$$
$$\text{at } y = \pm e, \ k\frac{\partial T}{\partial y} = \pm h[T(x) - T_f] \tag{D.4-2}$$

As u is a constant in Equation (D.4-1), the change of variable $x = ut$ allows to reduce the problem to the one described by Equation (2.2-1) in Chapter 2.2:

$$\frac{\partial T}{\partial t} = \alpha\frac{\partial^2 T}{\partial y^2}$$

D.4-2 General solution

The exact solution is hence the following:

$$\frac{T - T_f}{T_0 - T_f} = \sum_{n=1}^{\infty}\left[\frac{4\sin M_n}{2M_n + \sin(2M_n)}\right]\exp\left\{-M_n^2\frac{\alpha x}{e^2 u}\right\}\cos\left(\frac{M_n y}{e}\right) \tag{D.4-3}$$

where M_n is the nth positive root of the following equation:

$$M_n\tan M_n = h\frac{e}{k} = \text{Bi} \tag{D.4-4}$$

and k the thermal conductivity of the polymer. We have shown in Chapter 2 that for small values of the Biot number the first term of the series in Equation (D.4-3) is sufficient (this approximation will be verified below for a thin polymeric film) and the result reduces to:

$$\frac{T(x) - T_f}{T_0 - T_f} = \exp\left\{-\frac{hx}{\varrho c_p e u}\right\} \tag{D.4-5}$$

As discussed in Chapter 2.2, the assumption of a small Biot number is equivalent to assume that the temperature is a unique function of x (no or very small temperature gradient in the y-direction). Under this assumption, the exact solution is given by Equation (D.4-5).

D.4-3 Cooling in air

The problem is now restricted to the determination of the heat transfer coefficient as a function of the physical properties of air on one hand, and of the emissivity of the polymer on the other hand.

Heat losses by radiation and by convection are independent mechanisms and the overall heat transfer coefficient can be taken as the sum of the radiative heat transfer coefficient, h_r, and the convective heat transfer coefficient, h_c:

$$h = h_c + h_r \qquad\qquad (\text{D.4-6})$$

Convective heat transfer coefficient

This situation corresponds to forced convection since the polymeric liquid is moving in the ambient air. The flow of air along the film may be laminar or turbulent. For the case discussed here, the Reynolds number is of the order of 2000, that is considerably less than the critical value for turbulent flow in air, $\text{Re} = 3 \cdot 10^5$. The Nusselt number is given by Equation (D.2-4):

$$\text{Nu} = 0.66\,\text{Re}_x^{1/2}\,\text{Pr}^{1/3}$$

with, from the definitions:

$$\text{Nu} = h_c x/k_a, \quad \text{Re}_x = x u \varrho_a / \eta_a \text{ and } \text{Pr} = \eta_a c_{pa}/k_a$$

where Pr, η_a, c_{pa}, ϱ_a and k_a are respectively the Prandtl number, the viscosity, the specific heat, the density and the thermal conductivity of air evaluated at the temperature, $T_f{}'$, defined as the average between the temperature of the surface and that of the ambient fluid, i.e. $T_f{}' = (T_f + T_s)/2$. It follows for the problem considered that:

$$h_c = 0.66 k_a\,\text{Pr}^{1/3}\left(\frac{\varrho_a}{\eta_a}\right)^{1/2}\left(\frac{u}{x}\right)^{1/2}$$

Radiative heat transfer coefficient

We assume that the walls of the surrounding room are at T_f, the temperature of ambient air. The radiant heat flux is then obtained from:

$$q = \sigma\varepsilon[T_s^4 - T_f^4]$$

where ε and σ are respectively the emissivity of the polymeric filament and the Stefan-Boltzmann constant. It follows from the definition of h_r that:

$$h_r = \sigma\varepsilon\frac{T_s^4 - T_f^4}{T_s - T_f}$$

Expression for cooling

Equation (D.4-5) becomes after substitution on h by $h_c + h_r$:

$$\frac{T(x) - T_f}{T_0 - T_f} = \exp\left\{ -\frac{1}{\varrho c_p e} \frac{x}{u} \left(K\sqrt{\frac{u}{x}} + h_r \right) \right\}$$ (D.4-7)

where K is a constant depending only on the physical properties of the ambient fluid (air in this case) at the temperature considered:

$$K = 0.66 k_a \, \mathrm{Pr}^{1/3} \left(\frac{\varrho_a}{\eta_a} \right)^{1/2}$$

The physical properties of the air are evaluated at the temperature T_f'. Strictly speaking, T_f' is a function of x since the polymeric film or sheet is cooled down and T_f' decreases. This complicated considerably the solution and in general one assumes that the air properties are constant. This is justified when the polymer temperature varies only by a few tens of degrees.

Numerical applications

Let us consider the following conditions:

- Cooling of a variable thickness polymeric film extruded at the velocity equal to 0.5 m/s; the initial polymer temperature is 220° C and the ambient air is at 20° C.
- The material is a polyethylene with the following properties:
  $c_p = 2.5 \ \mathrm{kJ/(°C \cdot kg)}, \quad k = 0.33 \ \mathrm{W/m \cdot °C}$
  $\varrho = 900 \ \mathrm{kg/m^3}$
  $\varepsilon = 0.9$
- The physical properties of air at 100° C are:
  $k_a = 0.035 \ \mathrm{W/m \cdot °C}$
  $\varrho_a = 0.94 \ \mathrm{kg/m^3}$
  $\eta_a = 2.15 \cdot 10^{-5} \ \mathrm{Pa \cdot s}$
  $c_{pa} = 1 \ \mathrm{kJ/(°C \cdot kg)}$
  $\mathrm{Pr} = 0.74$

This leads to $h_r = 13 \ \mathrm{W/(°C \cdot m^2)}$.

Then reporting the different values in Equation (D.4-7), we get:

$$\frac{T(x) - T_f}{T_0 - T_f} = \exp\left\{ -10^{-6} \frac{x}{e u} \left(1.9 \sqrt{\frac{u}{x}} + 5.8 \right) \right\}$$

We can now verify that this approximate solution is acceptable. The values of h ($h = h_c + h_r$) and of k are such that he/k ($= \mathrm{Bi}$) remains smaller than 0.1 for film thickness less than 2 mm.

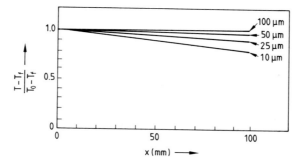

Figure D.4-2 Axial temperature profile of a polymeric film extruded in air

Temperature profiles for different film thickness are reported in Figure D.4-2. These are almost straight lines and the temperature at the end of the fall ($x = X$) is T_1 given by the following expression:

$$\frac{T_1 - T_f}{T_0 - T_f} = \exp\left\{ -\frac{1}{\varrho c_p e}\left(K\sqrt{\frac{X}{u}} + h_r\frac{X}{u}\right)\right\} \tag{D.4-8}$$

The temperature of the film at the end of the fall, T_1, is a unique function of the residence time X/u of a polymer element for a given film thickness.

It is interesting to distinguish the percentage of reduction of the reduced temperature, $(T_1 - T_f)/(T_0 - T_f)$, due to convection compared to radiant losses. From (D.4-8), one can write:

$$\frac{T_1 - T_f}{T_0 - T_f} = \underbrace{\exp\left\{ -\frac{1}{\varrho c_p e}\left(K\sqrt{\frac{X}{u}}\right)\right\}}_{Y_c} \underbrace{\exp\left\{ -\frac{1}{\varrho c_p e}\,h_r\frac{X}{u}\right\}}_{Y_r}$$

convective term   radiant term

For small values of the residence time, X/u, cooling is mostly achieved by radiation. For values larger than a critical residence time, t_l, convective losses become more important than the radiant losses. The critical residence time is defined by:

$$t_l = \left(\frac{K}{h_r}\right)^2 \tag{D.4-9}$$

It is independent of the film thickness and of the polymer properties (h_r is a function of the polymer emissivity which, however, is known to be around 0.9 for most polymers). The critical residence time, t_l, is then a unique function of the thermal conditions (average temperature of the film, ambient air temperature) and the physical properties of air at those conditions. Figure D.4-3 reports the dimensionless temperature as a function of the residence time. The operating conditions and the properties are those reported in the text above.

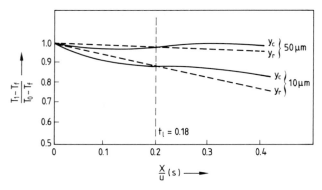

Figure D.4-3 Convective and radiant contributions to the dimensionless temperature of a polymer film

D.4-4 Cooling in water

The cooling of a polymeric film in a water bath proceeds by forced convection (since the film is displaced at an imposed velocity u). In practice, the velocity u is small enough that the water flow regime about the film is laminar (small Reynolds number). We have shown that in the case of cooling in air the convective and the radiant heat losses could be comparable. In water, however, the convective heat transfer coefficient is much larger and the radiant heat losses can be neglected. In summary, we will assume that the heat transfer in a water bath proceeds only by forced convection in laminar regime.

From the results from the previous section, one can readily write:

$$\frac{T(x) - T_f}{T_0 - T_f} = \exp\left\{ -\frac{K'}{\varrho c_p e}\sqrt{\frac{x}{u}} \right\} \tag{D.4-10}$$

with:

$$K' = 0.66 k_w\, \mathrm{Pr}^{1/3} \left(\frac{\varrho_w}{\eta_w}\right)^{1/2}$$

where Pr, ϱ_w, c_{pw}, η_w and k_w are respectively the Prandtl number, density, specific heat, viscosity and thermal conductivity of water evaluated at the temperature $T_f{}'$, the average between the polymeric film and the water temperatures.

The polymer properties are those of the example of the previous section. The physical properties of water are evaluated at 50° C since cooling is much more rapid in water than in air.

$k_w = 0.634 \ \mathrm{W/m \cdot {}^\circ C}$
$\varrho_w = 990 \ \mathrm{kg/m}^3$
$\eta_w = 0.55 \ \mathrm{m\,Pa \cdot s}$
$c_{pw} = 4.18 \ \mathrm{kJ/kg \cdot {}^\circ C}$
$\mathrm{Pr} = 6.0$

Then reporting these values in Equation (D.4-10), we get:

$$\frac{T(x) - T_f}{T_0 - T_f} = \exp\left\{ - \frac{4.5 \cdot 10^{-4}}{e} \sqrt{\frac{x}{u}} \right\}$$

Figure D.4-4 reports the results for the dimensionless axial temperature as a function of the axial distance. As in the previous case, the total distance, X, is equal to 0.1 m and the film velocity is 0.5 m/s.

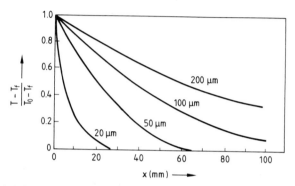

Figure D.4-4   Axial dimensionless temperature profile for a polymeric film extruded in water

Remark

This mathematical solution is valid only if the Biot number, he/k, is smaller than 0.1. For this particular case:

$$h = h_c = K'\sqrt{\left(\frac{u}{x}\right)} = 1000\sqrt{\frac{u}{x}}$$

The velocity u is of the order of 0.5 m/s, x of the order of 0.1 m; hence h is approximately equal to 2200 W/(m²·°C). The thermal conductivity of polyethylene is $k = 0.33$ W/m · °C and the restriction $he/k < 0.1$ implies that the solution is valid only for thickness, $2e$, smaller than 30 μm.

D.5   Analogy between heat and momentum transfer

D.5-1 Momentum transfer

The mechanism of momentum transfer by viscous effect is explained in the following example. Let us consider a semi-infinite medium (above a horizontal plate, $y > 0$) containing a fluid of viscosity η which is initially at rest. At time $t = 0$, the plate at $y = 0$ is set in motion at constant velocity U in the x-direction.

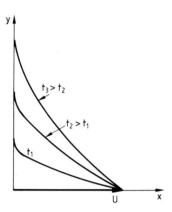

Figure D.5-1   Momentum transfer by viscous effects

The velocity is then progressively transmitted to the fluid as illustrated in Figure D.5-1.

Since the pressure gradient in the x-direction is equal to zero, the x-component of the Navier-Stokes equation reduces to:

$$\varrho\frac{\partial u}{\partial t} = \eta\frac{\partial^2 u}{\partial y^2}$$

or   (D.5-1)

$$\frac{\partial u}{\partial t} = \nu\frac{\partial^2 u}{\partial y^2}$$

This equation is identical to the diffusion equation for thermal energy with ν (kinematic viscosity or momentum diffusivity) playing the equivalent role of α (thermal diffusivity). The solution for the velocity profile is then:

$$u = U\left\{1 - \mathrm{erf}\left(\frac{y}{2\sqrt{\nu t}}\right)\right\}$$ (D.5-2)

where the time for which the momentum has "penetrated" the fluid at a distance y is:

$$t = \frac{y^2}{\nu}$$ (D.5.3)

Example

at $y = 10$ mm, for a polymer of $\nu = 1$ m^2/s, $t = 10^{-4}$ s

and for water $(\nu = 1$ mm^2/s$)$, $t = 100$ s

Momentum is transferred much more rapidly in the viscous polymer than in water. This explains why inertial terms are usually neglected in solving problems applied to polymeric materials. For low viscosity fluids such as water, the transient phenomena play a much more important role.

D.5-2 Flow of two immiscible liquids

When two immiscible liquids, one with velocity u_1 and the second with velocity u_2 are brought into contact, the interfacial velocity is u which can be solved again by analogy to the heat transfer problem. The boundary conditions are:

$$u_1(0,t) = u_2(0,t) = u \tag{D.5-4}$$

$$\eta_1 \frac{\partial u_1}{\partial y}(0,t) = \eta_2 \frac{\partial u_2}{\partial y}(0,t) \tag{D.5-5}$$

These conditions follow from the continuity of the velocity and of the momentum transfer at the interface.

By analogy to the heat transfer problem, we define the following velocity coefficients

$$B_1 = \sqrt{\varrho_1 \eta_1}, \quad B_2 = \sqrt{\varrho_2 \eta_2} \tag{D.5-6}$$

and the interfacial velocity is given by:

$$u = \frac{u_1 B_1 + u_2 B_2}{B_1 + B_2} \tag{D.5-7}$$

Hence, the more viscous liquid at comparable density or the more dense liquid at comparable viscosity tends to "impose" its velocity.

Appendix E
Complements on Extrusion

E.1 Hydrodynamical lubrication approximations

The flow field frequently encountered in lubrication is illustrated in Figure E.1-1.

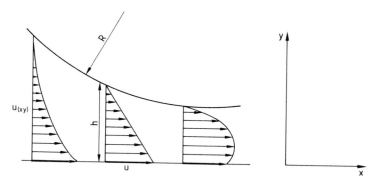

Figure E.1-1   Velocity profile in lubrication

1st *approximation*:

We assume that the gap h in the contact area varies slowly, i.e. $dh/dx \ll 1$. The flow rate per unit width may be written as:

$$Q = \bar{u}h \qquad \text{(E.1-1)}$$

where $\bar{u}$ is the average velocity at a given position (given h). Taking the derivative with respect to x, we get

$$\frac{dQ}{dx} = 0 = \frac{d\bar{u}}{dx}h + \bar{u}\frac{dh}{dx}$$

or

$$\frac{1}{\bar{u}}\frac{d\bar{u}}{dx} + \frac{1}{h}\frac{dh}{dx} = 0 \qquad \text{(E.1-2)}$$

The first approximation implies then:

$$\frac{d\bar{u}}{dx} \ll \frac{\bar{u}}{h} \qquad \text{(E.1-3)}$$

But away from the bearing surfaces,

$$\frac{\partial u}{\partial x} \simeq \frac{d\bar{u}}{dx} \qquad \text{(E.1-4)}$$

and also:

$$\frac{\bar{u}}{h} \simeq \frac{\partial u}{\partial y} \qquad \text{(E.1-5)}$$

Inequality (E.1-3) can then be written as:

$$\frac{\partial u}{\partial x} \ll \frac{\partial u}{\partial y} \qquad \text{(E.1-6)}$$

2nd *approximation*:

The solid surface curvatures are mild, i.e. $h/R \ll 1$ where the curvature radius is defined by

$$\frac{1}{R} = \frac{d^2 h}{dx^2} \qquad \text{(E.1-7)}$$

Taking the derivative of Equation (E.1-2) with respect to x, we get:

$$\frac{d^2 \bar{u}}{dx^2} + \frac{1}{h}\frac{d\bar{u}}{dx}\frac{dh}{dx} + \frac{\bar{u}}{h}\frac{d^2 h}{dx^2} - \frac{\bar{u}}{h^2}\left(\frac{dh}{dx}\right)^2 = 0 \qquad \text{(E.1-8)}$$

and eliminating $d\bar{u}/dx$ with the help of Equation (E.1-2) we obtain:

$$\frac{d^2 \bar{u}}{dx^2} + \frac{\bar{u}}{h}\frac{d^2 h}{dx^2} - 2\frac{\bar{u}}{h^2}\left(\frac{dh}{dx}\right)^2 = 0 \qquad \text{(E.1-9)}$$

or:

$$\frac{d^2 \bar{u}}{dx^2} = \frac{\bar{u}}{h^2}\left[2\left(\frac{dh}{dx}\right)^2 - h\frac{d^2 h}{dx^2}\right]$$

Since $d^2 h/dx^2 = 1/R$ and

$$\frac{\bar{u}}{h^2} \simeq \frac{\partial^2 u}{\partial y^2}, \qquad \frac{d^2 \bar{u}}{dx^2} \simeq \frac{\partial^2 u}{\partial x^2}$$

Equation (E.1-9) can be written as

$$\frac{\partial^2 u}{\partial x^2} \simeq \frac{\partial^2 u}{\partial y^2}\left[2\left(\frac{dh}{dx}\right)^2 - \frac{h}{R}\right] \qquad \text{(E.1-10)}$$

The first approximation implies $dh/dx \ll 1$; the second implies $h/R \ll 1$, and from (E.1-10)

$$\frac{\partial^2 u}{\partial x^2} \ll \frac{\partial^2 u}{\partial y^2} \qquad \text{(E.1-11)}$$

The continuity equation is

$$\frac{\partial u}{\partial x} + \frac{\partial v}{\partial y} = 0 \qquad \text{(E.1-12)}$$

One can write

$$v \simeq \bar{u}\,\frac{h}{L} \tag{E.1-13}$$

where L is the characteristic length of the flow. The first approximation implies $dh/dx \simeq h/L \ll 1$. Equation (E.1-12) may be written as

$$v \simeq \bar{u}\,\frac{h}{L} \leq \bar{u} \tag{E.1-14}$$

Hence the v component is small in comparison with the u component.

E.2 Complements on extrusion

E.2-1 Simultaneous plastication and melt flow in the compression zone of an extruder

Let us consider the unwrapped screw channel as illustrated in Figure E.2-1.

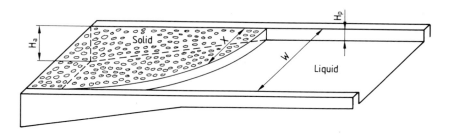

Figure E.2-1   Simultaneous plastication and melt flow in the compression zone of an extruder

–   At any point in the melt, the Reynolds equation can be written as:

$$Q_m = \left(W'\,V_z\,\frac{h}{2} - \frac{W'\,h^3}{12\eta}\,\frac{dp}{dz} \right)\varrho_l \tag{E.2-1}$$

where $W' = W - X$ and X is the solid bed width obtained from (see Chapter 4.3-3):

$$\frac{d}{dz}\left[\varrho_s\,uhX\right] = -\phi\sqrt{X} \tag{E.2-2}$$

If we assume that the density of the solid bed is constant (i.e. that the bed is compacted right at the beginning of the compression zone) and that the velocity of the bed, u is constant, Equation (E.2-2) can be integrated to obtain:

$$\sqrt{X} = \frac{\alpha}{\sqrt{h}} + \frac{\phi}{\varrho_s\,uA} \tag{E.2-3}$$

where A is the slope of the channel in the compression zone. The constant α is evaluated from $X = W$ at $h = H_a$ and (E.2-3) becomes:

$$X = \left[\sqrt{\frac{H_a}{h}} \left(\sqrt{W} - \frac{\phi}{\varrho_s uA} \right) + \frac{\phi}{\varrho_s uA} \right]^2 = \left[\frac{\sqrt{W} - (\phi/\varrho_s uA)}{\sqrt{1 - (zA/H_a)}} + \frac{\phi}{\varrho_s uA} \right]^2 \quad (\text{E.2-4})$$

The melt flow rate is obtained from the solution of X and from $dQ_m/dz = \phi \sqrt{X}$:

$$Q_m = \frac{2\phi H_a}{A} \left(\sqrt{W} - \frac{\phi}{\varrho_s uA} \right) \left(1 - \sqrt{1 - \frac{z}{H_a} A} \right) + \frac{\phi^2}{\varrho_s uA} z \quad (\text{E.2-5})$$

Eliminating Q_m, the melt flow rate, from Equation (E.2-1) and (E.2-5) and integrating, the pressure is expressed by:

$$p = \left[\frac{6\eta V_z}{AH_a} + \frac{Cu(2\,\text{Bu} - 1)}{2(\text{Bu} - 1)^2} \right] \frac{1}{\lambda^2} - \frac{2B^2\,Cu^3}{(\text{Bu} - 1)^3} \frac{1}{\lambda}$$
$$+ \frac{2B^2\,Cu^3(\text{Bu} + 1)}{(\text{Bu} - 1)^4} \ln \frac{(\text{Bu} + 1)\lambda + (\text{Bu} - 1)}{\lambda} + E \quad (\text{E.2-5})$$

where,

$$B = \frac{\varrho_s A \sqrt{W}}{\phi} \qquad C = \frac{24\eta \varrho_s}{AH_a \varrho_l} \qquad \lambda = \sqrt{\frac{h}{H_a}} \quad (\text{E.2-6})$$

The integration constant E is evaluated from the pressure at the beginning of the compression zone. The velocity u is known from the throughput of the extruder. The final expression for the pressure profile in the compression zone is:

$$p = p_c + \left[\frac{6\eta V_z}{AH_a} + \frac{Cu(2\,\text{Bu} - 1)}{2(\text{Bu} - 1)^2} \right] \left[\frac{1}{\lambda^2} - 1 \right] - \frac{2B^2\,Cu^3}{(\text{Bu} - 1)^3} \left[\frac{1}{\lambda} - 1 \right]$$
$$+ \frac{2B^2\,Cu^3(\text{Bu} + 1)}{(\text{Bu} - 1)^4} \log \frac{(\text{Bu} + 1)\lambda + (\text{Bu} - 1)}{2\,\text{Bu}\,\lambda} \quad (\text{E.2-7})$$

E.2-2 Leakage flow

The following analysis includes leakage flow across the flights for Newtonian isothermal flow in rectangular channels. The geometry and the pressure profile across the channel width and flights are shown in Figure E.2-2.

a) The throughput of the extruder is obtained from the flow rate across section AA_1. In absence of leakage

$$Q_{AA_1} = Q_{AC_1} = Q_{AB_1} \quad (\text{E.2-8})$$

Accounting for leakage (flow across the flights)

$$Q_{AA_1} = Q_{AB_2} - Q_{B_2 A_1} \quad (\text{E.2-9})$$

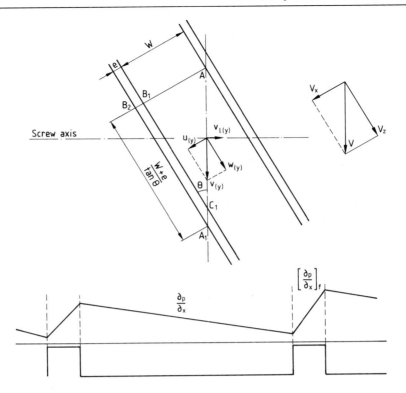

Figure E.2-2   Effect of flight clearance

and we can write

$$Q_{AB_2} = Q_{AB_1} + Q_{B_1 B_2} \tag{E.2-10}$$

where Q_{AB_1} is the expression obtained in Chapter 4.4-1, i.e.

$$Q_{AB_1} = \frac{W H_p}{2} V_z - \frac{W H_p^3}{12\eta} \left(\frac{\partial p}{\partial z} \right) \tag{E.2-11}$$

and $Q_{B_1 B_2}$ is a similar expression, but for a polymer of viscosity η_f flowing between two parallel plates at a distance δ_f apart and of width e:

$$Q_{B_1 B_2} = \frac{e\delta_f}{2} V_z - \frac{e\,\delta_f^3}{12\eta_f} \left(\frac{\partial p}{\partial z} \right) \tag{E.2-12}$$

$Q_{B_2 A_1}$ is the leakage flow in the x-direction. The previous expression can still be used considering now a width equal to $(W + e)/\tan\theta$ and a gap δ_f, the velocity of the plate is V_x and the pressure gradient $-(\partial p/\partial x)_f$. The expression is then:

$$Q_{B_2 A_1} = \left(\frac{W + e}{\tan\theta} \right) \frac{\delta_f}{2} V_x + \left(\frac{W + e}{\tan\theta} \right) \frac{\delta_f^3}{12\eta_f} \left(\frac{\partial p}{\partial x} \right)_f \tag{E.2-13}$$

Noting that $V_x = V_z \tan \theta$, the total flow rate is

$$Q_{AA_1} = \frac{1}{2} V_z W (H_p - \delta_f) - \frac{W H_p^3}{12\eta} \left(\frac{\partial p}{\partial z} \right) \left[1 + \frac{\eta}{\eta_f} \frac{e}{W} \left(\frac{\delta_f}{H_p} \right)^3 \right.$$

$$\left. + \frac{1 + e/W}{\tan \theta} \left(\frac{\delta_f}{H_p} \right)^3 \left(\frac{\eta}{\eta_f} \right) \frac{(\partial p/\partial x)_f}{(\partial p/\partial z)} \right] \qquad (E.2\text{-}14)$$

The pressure gradient in the z-direction, $\partial p/\partial z$, is related to flow conditions and die characteristics. The pressure gradient across the flight $(\partial p/\partial x)_f$ is, however, unknown.

b) To express $(\partial p/\partial x)_f$ as a function of $(\partial p/\partial z)$, we will assume that the gradient across AB_2 is the same as across $B_2 A_1$ (A and A_1 are the same points of the unwrapped screw), i.e.

$$\left[x \frac{\partial p}{\partial x} \right]_A^{B_2} = \left[z \frac{\partial p}{\partial z} \right]_{B_2}^{A_1} \qquad (E.2\text{-}15)$$

hence:

$$W \left(\frac{\partial p}{\partial x} \right) + e \left(\frac{\partial p}{\partial x} \right)_f - \frac{W + e}{\tan \theta} \left(\frac{\partial p}{\partial z} \right) = 0 \qquad (E.2\text{-}16)$$

c) Equation (E.2-16) shows a pressure gradient in the x-direction, unknown in the case of leakage flow. We need another relation:

$$Q_{AC_1} = Q_{AB_1} - Q_{B_1 C_1} \qquad (E.2\text{-}17)$$

where Q_{AB_1} is given by (E.2-11). The expression for $Q_{B_1 C_1}$ is similar to that of $Q_{B_2 A_1}$ with the width given by $W/\tan \theta$, i.e.:

$$Q_{B_1 C_1} = \left(\frac{W}{\tan \theta} \right) \frac{\delta_f}{2} V_x + \frac{W}{\tan \theta} \frac{\delta_f^3}{12\eta_f} \left(\frac{\partial p}{\partial x} \right)_f \qquad (E.2\text{-}18)$$

The derivation of the expression for Q_{AC_1} is more difficult. First, we have:

$$Q_{AC_1} = \frac{W}{\sin \theta} \int_0^{H_p} v_l \, dy \qquad (E.2\text{-}19)$$

where v_l is the polymer velocity in the direction of the screw axis (see Figure E.2-2) given by:

$$v_l = w \sin \theta - u \cos \theta$$

$$= \left[V_z \frac{y}{H_p} + \frac{1}{2\eta} \left(\frac{\partial p}{\partial z} \right) y(y - H_p) \right] \sin \theta$$

$$- \left[V_x \frac{y}{H_p} + \frac{1}{2\eta} \left(\frac{\partial p}{\partial x} \right) y(y - H_p) \right] \cos \theta \qquad (E.2\text{-}20)$$

Hence:

$$Q_{AC_1} = \left[\frac{V_z W H_p}{2} + \frac{W H_p{}^3}{12\eta} \left(-\frac{\partial p}{\partial z} \right) \right]$$
$$- \left[\frac{V_x W H_p}{2 \tan \theta} + \frac{W H_p{}^3}{12\eta \tan \theta} \left(\frac{\partial p}{\partial x} \right) \right] \qquad (E.2\text{-}21)$$

Eliminating Q_{AC_1} from Equations (E.2-17) and (E.2-21), we obtain the required relation:

$$\frac{V_x W H_p}{2 \tan \theta} + \frac{W H_p^3}{12\eta \tan \theta} \left(\frac{\partial p}{\partial x} \right) = \frac{V_x W \delta_f}{2 \tan \theta} + \frac{W \delta_f^3}{12\eta_f \tan \theta} \left(\frac{\partial p}{\partial x} \right)_f \qquad (E.2\text{-}22)$$

that can be written as:

$$\frac{\partial p}{\partial x} = \frac{\eta}{\eta_f} \left(\frac{\delta_f}{H_p} \right)^3 \left(\frac{\partial p}{\partial x} \right)_f - \frac{6\eta_f V_x (H_p - \delta_f)}{H_p^3} \qquad (E.2\text{-}23)$$

d) Substituting (E.2-23) in (E.2-16), we get:

$$\frac{(\partial p/\partial x)_f}{\partial p/\partial z} = \frac{\left(\dfrac{\eta_f}{\eta} \right) \left(\dfrac{H_p}{\delta} \right)^3 \dfrac{1 + e/W}{\tan \theta} + \dfrac{6\eta_f V_x(H_p - \delta_f)}{\delta_f^3 (\partial p/\partial z)}}{1 + \dfrac{e}{W} \dfrac{\eta_f}{\eta} \left(\dfrac{H_p}{\delta_f} \right)^3} \qquad (E.2\text{-}24)$$

and expression (E.2-14) becomes:

$$Q_{AA_1} = \frac{1}{2} V_z W(H_p - \delta_f) + \frac{W H_p^3}{12\eta} \left(-\frac{\partial p}{\partial z} \right)$$
$$\times \left\{ 1 + \frac{\eta}{\eta_f} \frac{e}{W} \left(\frac{\delta_f}{H_p} \right)^3 + \left(1 + \frac{e}{W} \right) \frac{\left[\dfrac{1 + e/W}{\tan^2 \theta} + \dfrac{6\eta V_z(H_p - \delta_f)}{H_p^3 (\partial p/\partial z)} \right]}{1 + \dfrac{e}{W} \dfrac{\eta_f}{\eta} \left(\dfrac{H_p}{\delta_f} \right)^3} \right\} \qquad (E.2\text{-}25)$$

This expression can be written as:

$$Q = Q_{1p} \left(1 - \frac{\delta_f}{H_p} \right) - Q_{2p}[1 + f_L] \qquad (E.2\text{-}26)$$

with:

$$f_L = \frac{\eta}{\eta_f} \frac{e}{W} \left(\frac{\delta_f}{H_p}\right)^3 + \left(1 + \frac{e}{W}\right) \frac{\dfrac{1 + e/W}{\tan^2 \theta} + \dfrac{6\eta \, V_z (H_p - \delta_f)}{H_p^3 \, (\partial p / \partial z)}}{1 + \dfrac{e}{W} \dfrac{\eta_f}{\eta} \left(\dfrac{H_p}{\delta_f}\right)^3} \qquad (E.2\text{-}27)$$

Q_{1p} and Q_{2p} are respectively the drag and pressure flows in absence of leakage defined below Equation (4.4-13).

Appendix F
Solutions to Problems

F.1   Solutions to problems of Chapter 1

Problem 1.2-B

a) *sphere:*

$$\sigma_{\theta\theta} = \sigma_{\phi\phi} = p\frac{R}{2e}, \quad \sigma_{rr} \approx 0$$

b) *cylinder:*

$$\sigma_{\theta\theta} = p\frac{R}{e}, \quad \sigma_{rr} \approx 0$$

Problem 1.2-C

$$S(z) = S(0)\exp\left(-\frac{\varrho g z}{\sigma}\right)$$

Problem 1.3-A

Let us consider the change of coordinates (x, y, z) to $(x, y, -z)$, as illustrated in Figure F.1-1. In the coordinates (x, y, z) the rate-of-strain tensor is:

$$[\dot{\varepsilon}] = \begin{bmatrix} 0 & \frac{1}{2}\dot{\gamma} & 0 \\ \frac{1}{2}\dot{\gamma} & 0 & 0 \\ 0 & 0 & 0 \end{bmatrix} \tag{F.1-1}$$

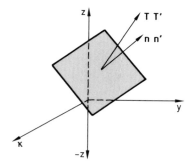

Figure F.1-1   Coordinate systems for Problem 1.3-A

And the corresponding stress tensor is:

$$[\boldsymbol{\sigma}] = \begin{bmatrix} \sigma_{xx} & \sigma_{xy} & \sigma_{xz} \\ \sigma_{xy} & \sigma_{yy} & \sigma_{yz} \\ \sigma_{xz} & \sigma_{yz} & \sigma_{zz} \end{bmatrix} \tag{F.1-2}$$

The force acting on the surface defined by the unit normal vector $\boldsymbol{n}$ (i.e. surface normal to $\boldsymbol{n}$), is:

$$\boldsymbol{T} = \boldsymbol{\sigma} \cdot \boldsymbol{n} \tag{F.1-3}$$

or:

$$\begin{aligned} T_x &= \sigma_{xx} n_x + \sigma_{xy} n_y + \sigma_{xz} n_z \\ T_y &= \sigma_{xy} n_x + \sigma_{yy} n_y + \sigma_{yz} n_z \\ T_z &= \sigma_{xz} n_x + \sigma_{yz} n_y + \sigma_{zz} n_z \end{aligned}$$

If the tensor $\dot{\boldsymbol{\varepsilon}}$ is not modified when written in coordinates $(x, y, -z)$, then tensor $\boldsymbol{\sigma}$ is also unmodified and the force acting on plane normal to $\boldsymbol{n}'$ $(n_x, n_y, -n_z)$ is:

$$\boldsymbol{T}' = \boldsymbol{\sigma} \cdot \boldsymbol{n}' \tag{F.1-4}$$

with components given by:

$$\begin{aligned} T_x &= \sigma_{xx} n_x + \sigma_{xy} n_y - \sigma_{xz} n_z \\ T_y &= \sigma_{xy} n_x + \sigma_{yy} n_y - \sigma_{yz} n_z \\ -T_z &= \sigma_{xz} n_x + \sigma_{yz} n_y - \sigma_{zz} n_z \end{aligned}$$

The equality of T_x, T_y, T_z in these two sets of results for any values of n_x, n_y and n_z implies that $\sigma_{xz} = \sigma_{yz} = 0$.

Problem 1.4-H: Converging flow in a dihedron or in a cone

I) Dihedron:

$$u = \frac{A}{r}[\cos 2\alpha - \cos 2\theta] \tag{F.1-5}$$

$$p = p_0 - \frac{2\eta A}{r^2}\cos 2\theta \tag{F.1-6}$$

II) Cone:

$$u = \frac{A}{r^2}[\cos 2\alpha - \cos 2\theta] \tag{F.1-7}$$

$$p = p_0 - \frac{2\eta A}{3r^3}[1 + 3\cos 2\theta] \tag{F.1-8}$$

Problem 1.4-I: Expansion of a cylindrical element

For the specified flow field, we write:

$$[\dot{\varepsilon}] = \begin{bmatrix} -\frac{A}{r^2} & 0 & 0 \\ 0 & \frac{A}{r^2} & 0 \\ 0 & 0 & 0 \end{bmatrix} \tag{F.1-9}$$

and

$$[\sigma] = \begin{bmatrix} -p - 2\eta\frac{A}{r^2} & 0 & 0 \\ 0 & -p + 2\eta\frac{A}{r^2} & 0 \\ 0 & 0 & -p \end{bmatrix} \tag{F.1-10}$$

The r-component of the equations of motion reduces to:

$$\frac{\partial p}{\partial r} = 0 \tag{F.1-11}$$

This result appears to be in contradiction with the boundary conditions, since $p_1 \neq p_2$.

In fact, the correct boundary conditions are:

$$\sigma_{rr}(R_1) = -p - 2\eta\frac{A}{R_1{}^2} = -p_1$$
$$\sigma_{rr}(R_2) = -p - 2\eta\frac{A}{R_2{}^2} = -p_2 \tag{F.1-12}$$

The constant A is evaluated from (F.1-12), i.e.:

$$A = \frac{1}{2\eta}\frac{p_1 - p_2}{\dfrac{1}{R_1{}^2} - \dfrac{1}{R_2{}^2}} \tag{F.1-13}$$

The velocity profile is then:

$$u = \frac{p_1 - p_2}{2\eta}\frac{R_1{}^2 R_2{}^2}{R_2{}^2 - R_1{}^2}\frac{1}{r} \tag{F.1-14}$$

and p is given by:

$$p = \frac{p_2 R_2{}^2 - p_1 R_1{}^2}{R_2{}^2 - R_1{}^2} \tag{F.1-15}$$

Hence the stresses at any points are given by:

$$\sigma_{rr}(r) = -p - 2\eta\frac{A}{r^2} \tag{F.1-16}$$

$$\sigma_{\theta\theta}(r) = -p + 2\eta\frac{A}{r^2} \tag{F.1-17}$$

$$\sigma_{zz}(r) = -p \tag{F.1-18}$$

In the case of a thin-wall shell, $R_1 \approx R_2 \approx R$ and $R_2 = R_1 + e$ with $e \ll R$. Then, the following results are obtained:

$$A = \frac{p_1 - p_2}{4\eta} \frac{R^3}{e} \tag{F.1-19}$$

$$p = (p_2 - p_1)\frac{R}{2e} \tag{F.1-20}$$

$$\sigma_{\theta\theta} = (p_1 - p_2)\frac{R}{e} \tag{F.1-21}$$

$$\sigma_{rr} = 0 \tag{F.1-22}$$

Problem 1.4-J: Expansion of a spherical element

For an incompressible fluid, the flow field is:

$$u = \frac{A}{r^2}, \quad v = 0, \quad w = 0 \tag{F.1-23}$$

and the stress tensor can be written as:

$$[\sigma] = \begin{bmatrix} -p - 4\eta\frac{A}{r^3} & 0 & 0 \\ 0 & -p + 2\eta\frac{A}{r^3} & 0 \\ 0 & 0 & -p + 2\eta\frac{A}{r^3} \end{bmatrix} \tag{F.1-24}$$

The r-component of the equation of motion shows that the pressure is constant across the shell thickness, and the boundary conditions are:

$$\sigma_{rr}(R_1) = -p - 4\eta\frac{A}{R_1{}^3} = -p_1$$

$$\sigma_{rr}(R_2) = -p - 4\eta\frac{A}{R_2{}^3} = -p_2 \tag{F.1-25}$$

One then obtains:

$$A = \frac{1}{4\eta} \frac{p_1 - p_2}{\dfrac{1}{R_1{}^3} - \dfrac{1}{R_2{}^3}} \tag{F.1-26}$$

and:

$$p = \frac{p_2 R_2{}^3 - p_1 R_1{}^3}{R_2{}^3 - R_1{}^3} \tag{F.1-27}$$

For the particular case of a thin-wall spherical shell, the results are:

$$\sigma_{\theta\theta} = \sigma_{\phi\phi} = (p_1 - p_2)\frac{R}{2e}, \quad \sigma_{rr} = 0 \tag{F.1-28}$$

F.2   Solutions to problems of Chapter 3

Problem 3.1-A

a) In the gap:

$$h(z) = h_1 + (h_2 - h_1)\frac{z}{L} \tag{F.2-1}$$

$$\frac{dp}{dz} = \frac{dp}{dh}\frac{dh}{dz} = \frac{dp}{dh}\frac{h_2 - h_1}{L} \tag{F.2-2}$$

Hence:

$$\frac{dp}{dh} = 6\eta U \frac{L}{h_2 - h_1}\frac{h - h^*}{h^3} \tag{F.2-3}$$

b) integrating the Reynolds equation, we obtain:

$$p(h) = 6\eta U \frac{L}{h_2 - h_1}\frac{h^* - 2h}{2h^2} + \text{const} \tag{F.2-4}$$

The constant and h^* can be eliminated with the help of the following boundary conditions:

B.C. 1) $p(0) = 0 \Leftrightarrow p(h_1) = 0$
B.C. 2) $p(L) = 0 \Leftrightarrow p(h_2) = 0$

Then from B.C. 1), we obtain:

$$\text{const} = -6\eta U \frac{L}{h_2 - h_1}\frac{h^* - 2h_1}{2h_1{}^2}$$

hence:

$$p(h) = 3\eta U \frac{L}{h_2 - h_1}\left[\frac{h^* - 2h}{h^2} - \frac{h^* - 2h_1}{h_1{}^2}\right] \tag{F.2-5}$$

and h^* is determined from B.C. 2):

$$p(h_2) = 0 \ \Rightarrow \ \frac{h^* - 2h_2}{h_2{}^2} = \frac{h^* - 2h_1}{h_1{}^2}$$

Hence:

$$h^* = \frac{2h_1 h_2}{h_1 + h_2} \tag{F.2-6}$$

This is the harmonic mean of h_1 and h_2. From the definition of h^* (see Section 3.1)

$$q = \frac{Q}{W} = \frac{Uh^*}{2} \tag{F.2-7}$$

and

$$q = U \frac{h_1 h_2}{h_1 + h_2}$$

The pressure is maximum for $h = h^*$; hence:

$$p_{\max} = p(h^*) = 3\eta U \frac{L}{h_2 - h_1} \left[-\frac{1}{h^*} - \frac{h^* - 2h_1}{h_1{}^2} \right]$$

Eliminating h^*, we obtain:

$$p_{\max} = 3\eta U L \frac{h_1 - h_2}{2 h_1 h_2 (h_1 + h_2)} \qquad (\text{F.2-8})$$

The expression for the load is:

$$F = W \int_0^L p(z)\, dz = \frac{W L}{h_1 - h_2} \int_{h_1}^{h_2} p(h)\, dh \qquad (\text{F.2-9})$$

and if we assume that the pressure profile is linear between A and B and between B and C, result (F.2-9) reduces to:

$$F = p_{\max} \frac{W L}{2} \qquad (\text{F.2-10})$$

c)   *Numerical applications*

I)   *polymer:*

$$h^* = 0.66 \text{ mm}$$
$$q = 330 \text{ mL/(m} \cdot \text{s)}$$
$$p_{\max} = 10^2 \text{ MPa}$$
$$F = 5 \cdot 10^5 \text{ N (50 tons)}$$

II)   *oil:*

$$h^* = 0.286 \text{ mm}$$
$$q = 429 \text{ mL/(m} \cdot \text{s)}$$
$$p_{\max} = 1.93 \cdot 10^4 \text{ Pa}$$
$$F = 48.2 \text{ N}$$

Problem 3.3-A: Flow in a conical channel

a) The hypotheses are:

$$
u \begin{cases} u(r,z) \\ v = 0 \\ w(r,z) \end{cases} \Rightarrow [\dot{\varepsilon}] = \begin{bmatrix} \frac{\partial u}{\partial r} & 0 & \frac{1}{2}\left(\frac{\partial u}{\partial z} + \frac{\partial w}{\partial r}\right) \\ 0 & \frac{u}{r} & 0 \\ \frac{1}{2}\left(\frac{\partial u}{\partial z} + \frac{\partial w}{\partial r}\right) & 0 & \frac{\partial w}{\partial z} \end{bmatrix}
\tag{F.2-11}
$$

and the Navier-Stokes equations reduce to:

$$
\begin{cases} -\dfrac{\partial p}{\partial r} + \eta\left[\dfrac{\partial}{\partial r}\left(\dfrac{1}{r}\dfrac{\partial}{\partial r}(ru)\right) + \dfrac{\partial^2 u}{\partial z^2}\right] = 0 \\[2mm] -\dfrac{\partial p}{\partial z} + \eta\left[\dfrac{1}{r}\dfrac{\partial}{\partial r}\left(r\dfrac{\partial w}{\partial r}\right) + \dfrac{\partial^2 w}{\partial z^2}\right] = 0 \end{cases}
\tag{F.2-12}
$$

By assumption:

$-\quad \dfrac{R_0 - R_1}{L} \ll 1 \quad \Leftrightarrow \quad \dfrac{dR}{dz} \ll 1$

$\qquad\qquad\qquad\qquad\qquad\qquad\qquad\qquad\qquad\qquad$ (F.2-13)

$-\quad$ The tangential curvatures of the surface are equal to zero. The assumptions of hydrodynamic lubrication are hence valid, i.e.

$-\quad u \ll w$ and their derivatives

$-\quad \dfrac{\partial^2 w}{\partial z^2} \ll \dfrac{1}{r}\dfrac{\partial}{\partial r}\left(r\dfrac{\partial w}{\partial r}\right)$

$\qquad\qquad\qquad\qquad\qquad\qquad\qquad\qquad\qquad\qquad$ (F.2-14)

Therefore, it is locally a simple shear flow with the rate-of-deformation tensor given by:

$$
[\dot{\varepsilon}] = \begin{bmatrix} 0 & 0 & \frac{1}{2}\frac{\partial w}{\partial r} \\ 0 & 0 & 0 \\ \frac{1}{2}\frac{\partial w}{\partial r} & 0 & 0 \end{bmatrix}
\tag{F.2-15}
$$

and the z-component of the Navier-Stokes becomes:

$$
\frac{dp}{dz} = \eta\frac{1}{r}\frac{\partial}{\partial r}\left(r\frac{\partial w}{\partial r}\right)
\tag{F.2-16}
$$

This is locally the expression for the Poiseuille flow in a tube:

$$
\boxed{Q = -\frac{\pi}{8\eta}\frac{dp}{dz}R^4(z)}
\tag{F.2-17}
$$

This represents a particular case of the Reynolds equation.

b) To integrate Equation (F.2-16), we first make the following change of variable:

$$R = \frac{R_1 - R_0}{L} z + R_0, \quad dz = -\frac{L}{R_0 - R_1} dR \qquad \text{(F.2-18)}$$

The Reynolds equation becomes:

$$\frac{dp}{dR} = \frac{8\eta\, Q\, L}{\pi(R_0 - R_1)} \frac{1}{R^4} \qquad \text{(F.2-19)}$$

and after integration:

$$p = -\frac{8\eta\, Q\, L}{\pi(R_0 - R_1)} \frac{1}{3R^3} + \text{const} \qquad \text{(F.2-20)}$$

Two boundary conditions are used to eliminate the constant and to obtain the expression for the flow rate. The pressure is P_0 at R_0 and it is taken as zero at the exit, $R = R_1$. Hence:

$$Q = \frac{3\pi P_0}{8\eta L} \frac{R_0{}^3 R_1{}^3 (R_0 - R_1)}{R_0{}^3 - R_1{}^3} \qquad \text{(F.2-21)}$$

One can define a mean or equivalent radius by:

$$Q = \frac{\pi}{8\eta} \frac{P_0}{L} R_m{}^4 = \frac{3\pi P_0}{8\eta L} \frac{R_0{}^3 R_1{}^3 (R_0 - R_1)}{R_0{}^3 - R_1{}^3} \qquad \text{(F.2-22)}$$

$$R_m{}^4 = \frac{3R_0{}^3 R_1{}^3 (R_0 - R_1)}{R_0{}^3 - R_1{}^3} \qquad \text{(F.2-23)}$$

If $R_1 \ll R_0$,

$$Q = \frac{3\pi P_0}{8\eta L} R_0 R_1{}^3 \quad \text{and} \quad R_m{}^4 = 3R_0 R_1{}^3 \qquad \text{(F.2-24)}$$

Problem 3.3-B: Flow in a dihedron

If $(h_0 - h_1)/L \ll 1$, the simplifications of hydrodynamic lubrication can be used and the Navier-Stokes equations reduce to:

$$\frac{dp}{dx} = \eta\frac{\partial^2 u}{\partial y^2} \qquad \text{(F.2-25)}$$

This can be easily integrated to obtain when using non slip conditions at the wall:

$$u = \frac{1}{2\eta} \frac{dp}{dx} (y^2 - h^2) \qquad \text{(F.2-26)}$$

and the Reynolds equation is obtained in terms of the flow rate by integrating over the cross section:

$$\frac{dp}{dx} = -\frac{3\eta Q}{2h^3 W} \qquad \text{(F.2-27)}$$

This result shows that in contrast to the case of the Reynolds bearing the pressure is a continuously decreasing function. The expression for the pressure is obtained by integration and taken $p = P_0$ at h_0:

$$p = P_0 + \frac{3\eta QL}{4(h_0 - h_1)W} \left[\frac{1}{h_0{}^2} - \frac{1}{h^2} \right] \tag{F.2-28}$$

The expression for the flow rate is then obtained from Equation (F.2-28) knowing the pressure at the exit. Taken for example $p = 0$ at $h = h_1$, one gets:

$$Q = \frac{4W P_0 h_0 h_1}{3\eta L(h_0 + h_1)} \tag{F.2-29}$$

A typical pressure profile is illustrated in Figure F.2-1. It shows that the pressure drop is concentrated near the exit.

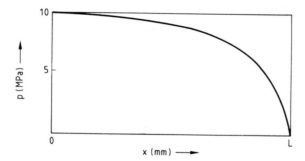

Figure F.2-1   Pressure profile for the flow in a dihedron

Problem 3.3-C: Flow in a sheeting die – First Part

We assume that the flow along the lips is not affected by a pressure gradient in the x-direction.

a) For the first increment, this is planar Poiseuille flow (between parallel plates) with a pressure p_0 at the entrance and a pressure p_s (that can be taken as equal to zero) at the exit. Hence,

$$\boxed{Q = \frac{h_0^3}{12\eta} (p_0 - p_s)\frac{L}{l}} \tag{F.2-30}$$

b) At any position x, the flow rate is given by:

$$Q(x) = Q - xq = Q - x\frac{Q}{L} = Q\left(1 - \frac{x}{L}\right) \tag{F.2-31}$$

and if the Poiseuille expression for a tube is locally valid, we can write:

$$Q(x) = -\frac{\pi}{8\eta} \frac{dp}{dx} R^4 \qquad \text{(F.2-32)}$$

$$\frac{dp}{dx} = -\frac{8\eta Q}{\pi R^4} \left(1 - \frac{x}{L}\right) \qquad \text{(F.2-33)}$$

$$p(x) = -\frac{8\eta Q}{\pi R^4} \left(x - \frac{x^2}{2L}\right) + \text{const} \qquad \text{(F.2-34)}$$

The integration constant is evaluated using $p = p_0$ at $x = 0$ and Equation (F.2-34) becomes:

$$\boxed{p(x) = p_0 - \frac{8\eta Q}{\pi R^4} x \left(1 - \frac{x}{2L}\right)} \qquad \text{(F.2-35)}$$

The extrusion is possible provided that the pressure remains positive (or larger than p_s) at the end of the channel. Hence, $p(L) > 0$ implies that:

$$p_0 - \frac{8\eta Q}{\pi R^4} \frac{L}{2} > 0 \qquad \text{(F.2-36)}$$

Combining with result (F.2-30), one obtains the following criterion for acceptable extrusion conditions:

$$\boxed{\frac{3\pi R^4 l}{h_0^3 L^2} > 1}$$

or: $\qquad\qquad\qquad\qquad\qquad\qquad\qquad\qquad\qquad\qquad$ (F.2-37)

$$h_0 < \sqrt[3]{\frac{3\pi R^4 l}{L^2}}$$

c) At any position, one can write from part a):

$$q = \frac{Q}{L} = \frac{h(x)^3}{12\eta} \frac{p(x)}{l} \qquad \text{(F.2-38)}$$

$$h(x)^3 = 12\eta Q \frac{l}{L} \frac{1}{p(x)} = \frac{12\eta Q \frac{l}{L}}{p_0 - \frac{8\eta Q}{\pi R^4} x \left(1 - \frac{x}{2L}\right)} \qquad \text{(F.2-39)}$$

And replacing Q by its expression (F.2-30), taking p_s equal to zero, one obtains:

$$h(x) = \frac{h_0}{\sqrt[3]{1 - \frac{2}{3\pi} \frac{h_0^3}{R^4} \frac{L}{l} x \left(1 - \frac{x}{2L}\right)}} \qquad \text{(F.2-40)}$$

and:

$$h_L = h(L) = \frac{h_0}{\sqrt[3]{1 - \dfrac{h_0{}^3 L^2}{3\pi R^4 l}}} \tag{F.2-41}$$

This yields the criterion for possible flow along the die width (question b).

d) *Numerical results*

- $h_{0\ max} = 3.35$ mm;
- if $h_0 = 1$ mm, $h_L = 1.009$ mm and the lips can be considered as parallel;
- if $h_0 = 2$ mm, $h_L = 2.17$ mm and the lips divergence remains finite;
- if $h_0 = 3.35$ mm, h_L becomes infinite.

Problem 3.3-D: Flow in a sheeting die – Second part

a) Let us consider a central point in the die. The flow rate per unit width is expressed by:

$$q = \frac{1}{12\eta} \frac{h_0{}^3}{l_0 + l_1} p_0 \tag{F.2-42}$$

Assuming that the flow is uniform with $q = Q/L$, the pressure p_0 is given by:

$$P_0 = \frac{12\eta Q(L\tan\alpha + l_1)}{L\, h_0^3} \tag{F.2-43}$$

b) The same type of relation can be written for any position z. Hence:

$$q = \frac{1}{12\eta} \frac{h_0^3}{l_1 + L\tan\alpha - z\sin\alpha} p(z) = \frac{Q}{L} \tag{F.2-44}$$

Hence:

$$p(z) = \frac{12\eta Q(l_1 + L\tan\alpha - z\sin\alpha)}{L\, h_0^3} = p_0 \frac{l_1 + L\tan\alpha - z\sin\alpha}{l_1 + L\tan\alpha} \tag{F.2-45}$$

c) For the equalizing channel, the flow rate can be written as:

$$Q(z) = Q - z\cos\alpha \, q = -\frac{\pi}{8\eta} \frac{dp}{dz} R(z)^4 \tag{F.2-46}$$

and the pressure gradient is evaluated using Equation (F.2-45):

$$\frac{dp}{dz} = -\frac{p_0 \sin\alpha}{l_1 + L\tan\alpha} \tag{F.2-47}$$

Combining with (F.2-46), the radius $R(z)$ is:

$$R(z) = \sqrt[4]{\frac{8\eta Q(l_1 + L\tan\alpha)}{\pi\, p_0 \sin\alpha}\left(1 - \frac{z\cos\alpha}{L}\right)} \tag{F.2-48}$$

It can be expressed in terms of geometrical factors by eliminating p_0 using (F.2-43):

$$R(z) = \sqrt[4]{\frac{2L\,h_0{}^3}{3\pi \sin \alpha}\left(1 - \frac{z \cos \alpha}{L}\right)}$$
(F.2-49)

The radius is approximately constant on most of the die width and equal to:

$$R_0 = \sqrt[4]{\frac{2L\,h_0{}^3}{3\pi \sin \alpha}}$$
(F.2-50)

Numerical results

$$\tan \alpha = \frac{l_0}{L} = \frac{1}{5}$$

$$h_0 = 1 \text{ mm}, \quad R_0 = 4.8 \text{ mm}$$

$$h_0 = 2 \text{ mm}, \quad R_0 = 8.1 \text{ mm}$$

Problem 3.4-A: Injection molding in a twin rectangular cavity mold

a) For the runners:

$$Q = \frac{\pi}{8\eta}\frac{\Delta p}{l}R^4$$
(F.2-51)

For the rectangular cavities:

$$Q = \frac{W}{12\eta}\frac{\Delta p}{l}h^3$$
(F.2-52)

b-I) If only the runners are partly filled with polymer, the relation between the flow rate and the injection pressure, p_I, is:

$$Q(t) = \frac{\pi}{8\eta}\frac{p_I}{l(t)}R^4$$
(F.2-53)

where $l(t)$ is the runner length filled with polymer at time t. In the interval between t and $t + dt$, the differential length filled due to the flow rate is given by $Q(t)\,dt = \pi R^2\,dl$, from which we obtain:

$$\boxed{t(l) = \frac{8\eta\,l^2}{p_I\,R^2}}$$
(F.2-54)

In particular, the time required to fill the distance to reach point A and B ($l = L$) is:

$$t(L) = \frac{8\eta\,L^2}{p_I\,R^2}$$
(F.2-55)

The numerical solution is $t(L) = 0.133$ s.

b-II) As the rectangular cavities are starting to the filled, the flow rate is related to the pressure drop in the runners by:

$$Q(t) = \frac{\pi}{8\eta} \frac{p_I - p_e(t)}{L} R^4 \tag{F.2-56}$$

where $p_e(t)$ is a variable pressure at point A or B. For the cavities:

$$Q(t) = \frac{W}{12\eta} \frac{p_e(t)}{l'(t)} h^3 \tag{F.2-57}$$

where $l'(t)$ is the length of the cavities filled by the polymer at time t. Combining Equations (F.2-56) and (F.2-57) we get:

$$p_e(t) = \frac{\pi p_I R^4}{2L} \left[\frac{1}{\dfrac{Wh^3}{3l'(t)} + \dfrac{\pi R^4}{2L}} \right] \tag{F.2-58}$$

The flow rate is related to the filled length of the cavities by $Q(t)\,dt = hW\,dl'$. Replacing in (F.2-58), we have:

$$t(l') = \frac{24\eta L}{\pi p_I h^2 R^4} \left[\frac{Wh^3 l'}{3} + \frac{\pi R^4}{2} \frac{l'^2}{L} \right] \tag{F.2-59}$$

and the total filling time of the cavities is given by:

$$t(L') = t(L) + \frac{24\eta L}{\pi p_I h^2 R^4} \left[\frac{Wh^3 L'}{3} + \frac{\pi R^4}{2} \frac{L'^2}{L} \right] \tag{F.2-60}$$

Substituting the numerical values, we obtain $t(L') = 0.82$ s. This is quite acceptable for injection molding.

c-I) The Cameron number is a function of time:
- For the runners, it is expressed by:

$$Ca(t) = \frac{\alpha l(t)}{V(t)R^2} = \frac{\alpha l(t)}{Q(t)} \pi \tag{F.2-61}$$

For the filling of the runners, the Cameron number can be written as:

$$Ca(t) = \frac{8\alpha\eta l(t)^2}{p_I R^4}, \quad 0 < Ca(t) < 3 \cdot 10^{-3} \tag{F.2-62}$$

and after the runners are filled:

$$Ca(t) = \frac{8\alpha\eta L^2}{p_I R^4} + \frac{12\pi\alpha\eta L l'(t)}{W p_I h^3}, \quad 3 \cdot 10^{-3} < Ca(t) < 5 \cdot 10^{-3} \tag{F.2-63}$$

The thermal regime is in all the filling stage under linear conditions. However, as the polymer temperature at the inlet is different from the mold temperature, it is not possible to make use of the expressions obtained for shear heating under adiabatic conditions.

- For the filling of the cavities:

$$\mathrm{Ca}(t) = \frac{\alpha l'(t)}{V(t)h^2} = \frac{\alpha W l'(t)}{Q(t)h} = \frac{12\alpha\eta l'(t)^2}{p_e(t)h^4} \tag{F.2-64}$$

or using (F.2-58) to eliminate $p_e(t)$:

$$\mathrm{Ca}(t) = \frac{8\alpha\eta W L l'(t)}{\pi p_I h R^4} + \frac{12\alpha\eta l'(t)^2}{p_I h^4}, \quad 0 < \mathrm{Ca}(t) < 2 \cdot 10^{-2} \tag{F.2-65}$$

Here again the conditions are very close to the linear regime.

c-II) The temperature profile in the runners is expressed by:

$$\bar{T} - T_w = (T_i - T_w)e^{-12\,\mathrm{Ca}(t)\frac{z}{i}} + \frac{2}{3}\eta\frac{V^2}{k}\left[1 - e^{-12\,\mathrm{Ca}(t)\frac{z}{i}}\right] \tag{F.2-66}$$

where T_i is the inlet temperature of the polymer and T_w is the wall temperature of the mold. Results are presented in Figure F.2-2 at various times of the filling stage.

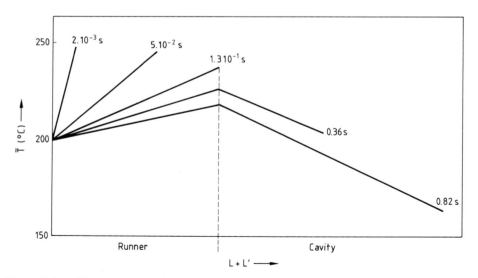

Figure F.2-2   Temperature profile during the filling stage of a rectangular mold

In particular, the temperature at the end of the runners is obtained from:

$$T_e - T_w = (T_i - T_w)e^{-12\,\mathrm{Ca}(t)} + \frac{2}{3}\eta\frac{V^2}{k}\left[1 - e^{-12\,\mathrm{Ca}(t)}\right] \tag{F.2-67}$$

c-III) The temperature profile in the cavities is expressed by:

$$\bar{T} - T_w = (T_e - T_w)e^{-20\,Ca(t)} + \frac{3}{5}\eta\frac{V^2}{k}\left[1 - e^{-20\,Ca(t)}\right]$$

(F.2-68)

The results are presented in the right part of Figure F.2-2.

Comments

This solution does not take into account heat accumulation with time. The more "correct" differential equation for the mean temperature of the polymer in the runners is:

$$\varrho\,c_p\pi R^2\,dx\left(\frac{\partial\bar{T}}{\partial t} + V\frac{\partial\bar{T}}{\partial x}\right) = -12k\frac{\bar{T} - T_w}{R}\,\pi\,R\,dx + 8\pi\,\eta V^2\,dx$$

(F.2-69)

This equation has to be solved by a numerical method.

F.3   Solutions to problems of Chapter 4

Problem 4.3-A: Initiation of back flight plastication

Following the analysis of the Rayleigh bearing in Chapter 3.1, we will consider on the one hand the flow between the screw flight and the barrel, and on the other hand the flow between the solid bed and the barrel. Since δ and δ_f are small compared to $e/\sin\theta$ and $W/\sin\theta$, we will assume parallel flow situation. The flow rates per unit length of the screw are:

- for zone 1:

$$q_1 = \underbrace{V\frac{\delta_f}{2}}_{\text{drag flow}} + \underbrace{\frac{\delta_f^3\sin\theta}{12\eta}\frac{\Delta p}{e}}_{\text{pressure flow}}$$

(F.3-1)

- for zone 2:

$$q_2 = \underbrace{V\frac{\delta}{2}}_{\text{drag flow}} - \underbrace{\frac{\delta^3\sin\theta}{12\eta}\frac{\Delta p}{W}}_{\text{pressure flow}}$$

(F.3-2)

From a mass balance, we have: $q_1 = q_2$, hence from (F.3-1) and (F.3-2):

$$V\frac{\delta_f}{2} + \frac{\delta_f^3\sin\theta}{12\eta}\frac{\Delta p}{e} = V\frac{\delta}{2} - \frac{\delta^3\sin\theta}{12\eta}\frac{\Delta p}{W}$$

Therefore, the pressure drop is:

$$\Delta p = \frac{6\eta V}{\sin\theta}\frac{\delta - \delta_f}{\dfrac{\delta_f{}^3}{e} + \dfrac{\delta^3}{W}} \qquad (\text{F.3-3})$$

Defining $\delta = \lambda\delta_f$, the previous expression can be written as:

$$\Delta p = \frac{6\eta V(\lambda - 1)}{\delta_f{}^2 \sin\theta \left(\dfrac{1}{e} + \dfrac{\lambda^3}{W}\right)} \qquad (\text{F.3-4})$$

Numerical results

The pitch is equal to the screw diameter which implies $\theta = 17°40'$.
The channel width is $W = B\cos\theta - e = 0.109\,\text{m}$.
The relative linear velocity of the barrel is $V = \Omega\pi D_2 = 0.314\,\text{m/s}$.

The pressure as a function of λ is shown in Figure F.3-1 for three different values of δ_f.

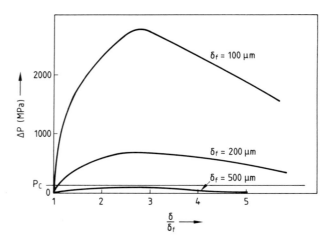

Figure F.3-1   Pressure generation along the back flight

The pressure necessary for deforming the solid bed along the back flight has to be larger than a critical value P_c (equal to 6 to 8 times the yield stress of the solid polymer σ_0, depending on the ratio H/W). Taking a typical yield value for a solid polymer, $\sigma_0 = 20\,\text{N/mm}^2$, the bed can be deformed only if the pressure exceeds 160 MPa.

In the present case, the solid bed will be deformed as soon as the film thickness δ is slightly larger than the flight clearance δ_f, and in all cases, much before the build-up of maximum pressure is attained when:

$$\frac{1}{e} + 3\frac{\lambda^2}{W} - 2\frac{\lambda^3}{W} = 0 \qquad (\text{F.3-5})$$

The solution of Equation (F.3-5) is:

$$\lambda = \frac{\delta}{\delta_f} = 2.85$$

If the flight clearance is increased, the pressure will be less, but the maximum will still occur for the same value of λ. The bed will be deformed for larger values of λ. For very large values of the clearance (for example $\delta_f = 500~\mu m$), the maximum pressure that could be attained would be smaller than 160 MPa. Then, one can consider that back flight plastication will not be initiated:

Problem 4.3-B: Back flight plastication

a) The polymer flow rate is equal to $Q/\varrho_s = 92.6$ mL/s, hence the average velocity is:

$$u = \text{flow rate/flow section} = 85~\text{mm/s}$$
$$V_z = 0.30~\text{m/s, hence larger than } u\,.$$

The angle φ is obviously small and we may assume than the plastication occurs in the direction of vector V, i.e. in the plane perpendicular to the screw axis.

b) The unknowns are V_s and δ. The two equations represent an energy balance at the solid-liquid interface and a mass balance.

c) *In the liquid film*, the equation of energy reduces to:

$$0 = k_l \frac{d^2 T}{dy^2} + \eta \left(\frac{V}{\delta}\right)^2 \tag{F.3.6}$$

for which the solution is:

$$T(y) = T_m + \frac{T_1 - T_m}{\delta} y + \frac{\eta V^2}{2 k_l \delta^2} y(\delta - y) \tag{F.3-7}$$

and the heat flux at the interface is:

$$q_{l \to i} = k_l \left.\frac{dT}{dy}\right|_i = \underbrace{k_l \frac{T_1 - T_m}{\delta}}_{\text{due to the conduction}} + \underbrace{\frac{\eta V^2}{2\delta}}_{\text{due to the viscous } \textit{dissipation}} \tag{F.3-8}$$

This situation is analyzed in BIRD, STEWART and LIGHTFOOT (1960, Section 9.4) in terms of the Brinkman number defined by:

$$\text{Br} = \frac{\eta V^2}{k_l(T_1 - T_m)} \tag{F.3-9}$$

Noting that the derivative of the temperature evaluated at the barrel wall is:

$$\frac{dT}{dy}\bigg|_{\delta} = \frac{T_1 - T_m}{\delta} - \frac{\eta V^2}{2k_l\,\delta} \tag{F.3-10}$$

we get the following limiting cases

$$Br < 2 \quad \Rightarrow \quad \frac{dT}{dy}\bigg|_{\delta} > 0 \qquad \begin{array}{l}\text{viscous dissipation}\\\text{is negligible}\end{array}$$

$$Br = 2 \quad \Rightarrow \quad \frac{dT}{dy}\bigg|_{\delta} = 0 \qquad \begin{array}{l}\text{adiabatic condition} -\\\text{no heat conduction}\\\text{at the wall}\end{array}$$

$$Br > 2 \quad \Rightarrow \quad \frac{dT}{dy}\bigg|_{\delta} < 0 \qquad \begin{array}{l}\text{viscous dissipation}\\\text{is of importance}\end{array}$$

d) *For the solid bed* the equation of energy reduces to:

$$\varrho_s c_{p_s} V_s \frac{dT}{dy} = k_s \frac{d^2 T}{dy^2} \tag{F.3-11}$$

for which the solution is:

$$T = A\exp\left\{\varrho_s c_{p_s} V_s \frac{y}{k_s}\right\} + B = A\exp\left\{V_s \frac{y}{\alpha_s}\right\} + B \tag{F.3-12}$$

For a typical polymer, $\alpha_s \approx 0.1~\text{mm}^2/\text{s}$, $V_s \approx 1~\text{mm/s}$. Hence $\alpha_s/V_s \approx 0.1$ mm and the temperature drops in a very short distance from the surface. The temperature of the solid bed can be taken as equal to T_s at a short distance from the surface and equal to the melting temperature at the liquid-solid interface, i.e. $B = T_s$ and $A = T_m - T_s$. Equation (F.3-12) becomes:

$$T = (T_m - T_s)\exp\left\{V_s \frac{y}{\alpha_s}\right\} + T_s \tag{F.3-13}$$

and the heat flux at the interface is expressed by:

$$q_{i \to s} = k_s \frac{dT}{dy}\bigg|_i = \varrho_s c_{p_s} V_s (T_m - T_s) \tag{F.3-14}$$

e) *Heat flux balance at the interface*

The heat flux at the liquid-solid interface is larger than the heat transferred by conduction in the solid bed, due to the heat required for the fusion of the polymer. This is expressed by:

$$q_{l \to i} - q_{i \to s} = k_l \frac{T_1 - T_m}{\delta} + \frac{\eta V^2}{2\delta} - \varrho_s c_{p_s} V_s (T_m - T_s) = \varrho_s V_s \lambda \tag{F.3-15}$$

This result can be written as:

$$\left[k_l(T_1 - T_m) + \eta \frac{V^2}{2}\right]\frac{1}{\delta} = \varrho_s V_s[c_{p_s}(T_m - T_s) + \lambda] \tag{F.3-16}$$

f) The mass balance in the film under steady-state conditions is:

$$\varrho_s V_s X' = \varrho_l \frac{V}{2}\delta \tag{F.3-17}$$

The problem can be solved by noting that we have two unknowns, δ and V_s with two equations:
- one of the type: $\delta V_s = $ const (from the energy balance)
- the other of the type: $\delta = V_s \cdot$ const (from a mass balance).

The solution is:

$$\delta = \sqrt{\frac{2k_l(T_1 - T_m) + \eta V^2}{\varrho_l V[c_{p_s}(T_m - T_s) + \lambda]}}\, X' = \sqrt{\frac{2k_l(T_1 - T_m) + \eta V^2}{\varrho_l V_x[c_{p_s}(T_m - T_s) + \lambda]}}\, X \tag{F.3-18}$$

Numerical results

$$k_l(T_1 - T_m) = 8.36 \text{ W/m}$$
$$V = 0.32 \text{ m/s}; \quad \eta V^2/2 = 51 \text{ W/m}$$
$$c_{ps}(T_m - T_s) = 0.226 \text{ MJ/kg}$$
$$\lambda = 0.125 \text{ MJ/kg}$$

At the beginning of plastication:

$$\frac{X}{V_x} = \frac{W}{V_x} = \frac{D_1}{V \tan \theta_1} = 1.14 \text{ s}$$

Hence:

$$\delta = 0.69 \text{ mm} = 690 \ \mu\text{m}$$

Problem 4.4-A: Design criteria of an extruder

We present here the solution for the compression and the metering zones. The system is unwrapped as shown in Figure F.3-2. We assume that the flow is unidirectional along the channel axis.

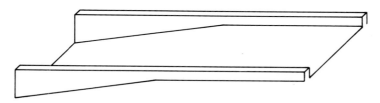

Figure F.3-2 Unwrapping of the screw channel compression and metering zones

a) *Basic equations*

Flow in the die:

$$Q = k_f(p_C - p_D) \tag{F.3-19}$$

where the die characteristics, k_f, is a function of the die geometry:

Circular die:

$$k_f = \frac{\pi}{8\eta} \frac{R^4}{L} \tag{F.3-20}$$

Rectangular die:

$$k_f = \frac{W}{12\eta} \frac{h^3}{L} \tag{F.3-21}$$

Flow in the metering or pumping zone:

$$Q = Q_{1p} - k_p(p_C - p_B) \tag{F.3-22}$$

where

$$Q_{1p} = \frac{WV_z H_p}{2}, \quad k_p = \frac{WH_p^3}{12\eta Z_p} \tag{F.3-23}$$

In the metering zone the pressure is a linear function.

Flow in the compression zone:

$$Q = Q_{1c} - k_c(p_B - p_A) \tag{F.3-24}$$

where

$$Q_{1c} = WV_z \frac{H_a H_p}{H_a + H_p} \quad k_c = \frac{W}{6\eta Z_c} \frac{H_a^2 H_p^2}{H_a + H_p} \tag{F.3-25}$$

The pressure in the compression zone is expressed by:

$$p = p_A + 6\eta V_z Z_c \frac{(H_a - h)(h - H_p)}{h^2(H_a^2 - H_p^2)} + (p_B - p_A)\left(\frac{H_p}{h}\right)^2 \frac{H_a^2 - h^2}{H_a^2 - H_p^2} \tag{F.3-26}$$

b) *Analytical solution*

We have to solve the following three equations:

$$Q = k_f(p_C - p_D) \tag{F.3-19}$$
$$Q = Q_{1p} - k_p(p_C - p_B) \tag{F.3-22}$$
$$Q = Q_{1c} - k_c(p_B - p_A) \tag{F.3-24}$$

for five unknowns, Q, p_A, p_B, p_C, and p_D. The problem is closed provided that two pressures are known or fixed, for example taking p_A, the pressure at the inlet of the compression zone, and p_D, the pressure at the die exit, equal to zero. The solution is then:

$$Q = \frac{\dfrac{Q_{1p}}{k_p} + \dfrac{Q_{1c}}{k_c}}{\dfrac{1}{k_p} + \dfrac{1}{k_c} + \dfrac{1}{k_f}} \tag{F.3-27}$$

$$p_C = \frac{1}{k_f} \frac{\dfrac{Q_{1p}}{k_p} + \dfrac{Q_{1c}}{k_c}}{\dfrac{1}{k_p} + \dfrac{1}{k_c} + \dfrac{1}{k_f}} \tag{F.3-28}$$

$$p_B = \frac{1}{k_c} \frac{Q_{1c}\left[\dfrac{1}{k_p} + \dfrac{1}{k_f}\right] - \dfrac{Q_{1p}}{k_p}}{\dfrac{1}{k_p} + \dfrac{1}{k_c} + \dfrac{1}{k_f}} \tag{F.3-29}$$

In general, if we consider n extruder zones placed in a series, each with a drag flow equal to Q_i and of characteristics k_i, the overall flow rate of the extruder will be given by the following expression:

$$Q = \frac{\displaystyle\sum_{i=1}^{n} \frac{Q_{1i}}{k_i}}{\displaystyle\sum_{i=1}^{n} \frac{1}{k_i}} \tag{F.3-30}$$

For the reference extruder:

$$Q_{1p} = \ 97.8 \ \mathrm{mL/s}, \quad k_p = 3.59 \cdot 10^{-13} \ \mathrm{mL/(Pa \cdot s)}$$
$$Q_{1c} = 130.4 \ \mathrm{mL/s}, \quad k_c = 8.50 \cdot 10^{-13} \ \mathrm{mL/(Pa \cdot s)}$$

The results for the flow rate and the pressures p_C and p_B are reported in Figure F.3-3 as a function of the die characteristics.

c) Figure F.3-3 shows that some of the operating conditions are not practical nor possible:

–   For small values of k_f, the pressure built up in the extruder becomes excessive and the flow rate goes down to zero.
–   For large values of k_f, the throughput of the extruder is maximum, but the pressure decreases in the metering zone to approach zero at the die inlet.

•   The design criterion for the mechanical parts restricts the pressure, $p_{\max} <$ 100 MPa. The criterion can be written as:

$$D_i p_{\max} < \sigma_0 (D_e - D_i) \tag{F.3-31}$$

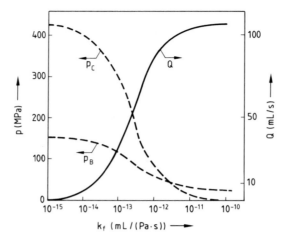

Figure F.3-3 Flow rate and pressures at the end of the compression and metering zones for the reference extruder as a function of the die characteristics

where σ_0 is the yield stress for steel. Taking:

$$\sigma_0 = 400 \text{ MPa}$$
$$D_i = 120 \text{ mm}$$
$$D_e = 180 \text{ mm}$$

we obtain $P_{\max} < 200 \,\text{MPa}$. Hence, a safety factor of 2 is included in the design criterion.

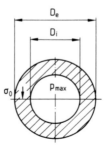

Figure F.3-4 Stress in the steel barrel of an extruder

• In practice, the pressure is not always increasing in the metering zone. In the case of a decreasing pressure, this zone no longer acts as a metering section, but rather as a mixing zone. However, if plastication was not completed in the compression zone, gas entrapped with the solid granules would not flow backwards and generate bubbles in the extrudate.

We notice in Figure F.3-5 that the behaviour of increasing or decreasing pressure in the metering zone is related to the position h^* in the compression zone where the pressure is maximum.

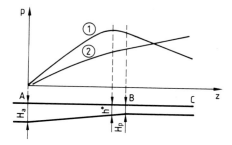

Figure F.3-5   Two different pressure profiles along the screw channel

– In the first case, $H_p < h^* < H_a$, and the pressure decreases in the metering zone (curve 1).
– In the second case, $h^* < H_p$, and the pressure increases in the compression zone (curve 2).

•   The criterion that $p > 1$ MPa in the feeding zone of the extruder is related to the operation of the extruder. If the pressure is too low, any small variation of the screw rotational speed (due for example to a variation in the power line voltage) could lead to instabilities and large variations in the flow rate.

•   A narrow residence time distribution corresponds to the criterion:

$$\frac{Q_{2p}}{Q_{1p}} < \frac{1}{3} \qquad (F.3-32)$$

or:

$$\frac{Q}{Q_{1p}} > \frac{2}{3} \qquad (F.3-33)$$

For this problem, we find:

$$Q > 64.5 \text{ mL/s}$$

•   The selection for possible dies is restricted. Using the results of Figure F.3-3, we obtain:

– $p_C < 100$ MPa $\Rightarrow k_f > 7 \cdot 10^{-13}$ mL/(s $\cdot$ Pa)
– $p_B < 100$ MPa $\Rightarrow k_f > 1.9 \cdot 10^{-13}$ mL/(s $\cdot$ Pa)
– Increasing pressure in the metering zone:

$$\Rightarrow k_f < 3.2 \cdot 10^{-12} \text{ mL/(s} \cdot \text{Pa)}$$

– inlet pressure larger than 1 MPa:

$$\Rightarrow k_f < 10^{-10} \text{ mL/(s} \cdot \text{Pa)}$$

– narrow residence time distribution:

$$\Rightarrow k_f > 4.3 \cdot 10^{-13} \text{ mL/(s} \cdot \text{Pa)}$$

To respect simultaneously all these conditions we have:

$$7 \cdot 10^{-13} < k_f < 3.2 \cdot 10^{-12} \; \text{mL}(/(\text{s} \cdot \text{Pa})$$

If we consider that the die consists of a circular tube for which the Poiseuille flow is valid, given by Eq. (F.3-20), then:

$$3.6 \; \text{mm} < R < 5.3 \; \text{mm}$$

This appears to be quite restrictive. However, it should be noted that the calculations were made for one single rotational speed, equal to 60 rev/s. For other values of the rotational speed, the results will cover other ranges.

This problem is oversimplified and the results are only qualitative. They, nevertheless, suggest that it may not be appropriate to install any type of die on a given extruder.

F.4   Solutions to problems of Chapter 5

Problem 5.2-A: Blowing of a spherical shell

a) *Expressions for the stresses*

We assume that the shear stresses are negligible compared to the extensional stresses $\sigma_{\theta\theta}$, $\sigma_{\phi\phi}$ and σ_{rr}. The sphere illustrated in Figure F.4-1 is first cut in two by a surface P and the resulting forces are examined.

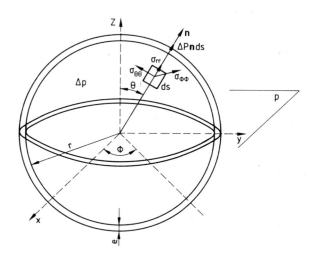

Figure F.4-1   Blowing of a sphere

The force acting on the surface is $2\pi r e \sigma_{\theta\theta}$. The force acting on the upper part of the sphere due to the applied pressure is:

$$F = \int_{\text{surface}} \Delta p\, ds \cos\theta = \int_0^{2\pi} d\phi \int_0^{\pi/2} r^2 \Delta p \sin\theta \cos\theta = \pi r^2 \Delta p \qquad \text{(F.4-1)}$$

A force balance yields:

$$\sigma_{\theta\theta} = \Delta p \frac{r}{2e} \qquad \text{(F.4-2)}$$

We now cut the sphere by a surface P' perpendicular to P. The same approach leads to:

$$\sigma_{\phi\phi} = \Delta p \frac{r}{2e} \qquad \text{(F.4-3)}$$

The normal stress σ_{rr} is undetermined, but we have $0 < \sigma_{rr} < \Delta p$. If $r \gg e$, then $\sigma_{rr} \ll \sigma_{\theta\theta} = \sigma_{\phi\phi}$.

Extensional rates

Neglecting shear effects, the deformation is axisymmetrical with respect to the radial direction. As shown in Figure F.4-2, we define a coordinate system by:

– M axis, the tangent to the meridian at point M, θ-direction
– P axis, the tangent to the parallel at M, ϕ-direction
– e axis, the normal to the shell surface, r-direction.

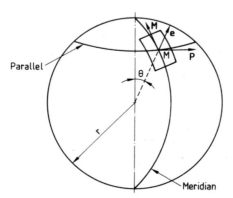

Figure F.4-2   Coordinates on the sphere surface

The extensional rates are then:

along the e-axis:
$$\dot{\varepsilon}_{rr} = \frac{1}{e}\frac{de}{dt} \qquad \text{(F.4-4)}$$

along the M-axis:
$$\dot{\varepsilon}_{\theta\theta} = \frac{1}{2\pi r}\frac{d(2\pi r)}{dt} = \frac{1}{r}\frac{dr}{dt} \qquad \text{(F.4-5)}$$

along the P-axis:
$$\dot{\varepsilon}_{\phi\phi} = \frac{1}{2\pi r \sin\theta}\frac{d(2\pi r \sin\theta)}{dt} = \frac{1}{r}\frac{dr}{dt} \qquad \text{(F.4-6)}$$

(θ is not a function of time since the flow field is symmetrical).

Remark

For an incompressible fluid, $\text{tr}\,\dot{\varepsilon} = 0$, i.e.

$$\frac{1}{e}\frac{de}{dt} + \frac{2}{r}\frac{dr}{dt} = 0 \qquad (\text{F.4-7})$$

This implies that $r^2 e = \text{const}$, result that can be obtained from a mass balance on the shell:

$$4\pi r^2 e = 4\pi r_0^2 e_0 \qquad (\text{F.4-8})$$

Stress – extensional rate relation

From the generalized Newton law for the viscosity:

$$\boldsymbol{\sigma} = -p\boldsymbol{\delta} + 2\eta\dot{\boldsymbol{\varepsilon}} \qquad (\text{F.4-9})$$

the normal stress components are:

$$\sigma_{rr} = -p + \frac{2\eta}{e}\frac{de}{dt} \qquad (\text{F.4-10})$$

$$\sigma_{\theta\theta} = -p + \frac{2\eta}{r}\frac{dr}{dt} \qquad (\text{F.4-11})$$

$$\sigma_{\phi\phi} = -p + \frac{2\eta}{r}\frac{dr}{dt} \qquad (\text{F.4-12})$$

But as discussed above, $\sigma_{rr} \ll \sigma_{\theta\theta} = \sigma_{\phi\phi}$. Taking σ_{rr} equal to zero and using the continuity equation, we obtain:

$$p = \frac{2\eta}{e}\frac{de}{dt} = -\frac{4\eta}{r}\frac{dr}{dt} \qquad (\text{F.4-13})$$

and from (F.4-11) and (F.4-12) the normal stresses reduce to:

$$\sigma_{\theta\theta} = \sigma_{\phi\phi} = \frac{6\eta}{r}\frac{dr}{dt} = 6\eta\dot{\varepsilon}_{\theta\theta} \qquad (\text{F.4-14})$$

This is the situation of biaxial stretching analyzed in Section 5.2.

b) Combining Equations (F.4-2 or 3) and (F.4-14), the normal stress $\sigma_{\theta\theta}$ is given by:

$$\sigma_{\theta\theta} = \frac{\Delta p\,r}{2e} = \frac{6\eta}{r}\frac{dr}{dt} \qquad (\text{F.4-15})$$

which can be integrated to obtain, using the initial condition $r = r_0$ at $t = 0$:

$$\frac{1}{r_0^3} - \frac{1}{r^3} = \frac{\Delta p}{4\eta e_0 r_0^2}\,t \qquad (\text{F.4-16})$$

The time t required for the sphere to attain the radius R is:

$$t = \frac{4\eta e_0 r_0^2}{\Delta p}\left(\frac{1}{r_0^3} - \frac{1}{R^3}\right) \approx \frac{4\eta e_0}{r_0 \Delta p} \qquad (\text{F.4-17})$$

The numerical solution is $t = 4$ s.

Problem 5.2-B: Cast-film process

a) The flow field is assumed to be the following:

$$\begin{cases} u(x) \\ v = 0 \\ w(x,z) \end{cases} \tag{F.4-18}$$

Hence the rate-of-strain tensor is:

$$[\dot{\varepsilon}] = \begin{bmatrix} \frac{du}{dx} & 0 & \frac{1}{2}\frac{\partial w}{\partial x} \\ 0 & 0 & 0 \\ \frac{1}{2}\frac{\partial w}{\partial x} & 0 & \frac{\partial w}{\partial z} \end{bmatrix} \tag{F.4-19}$$

Neglecting shear effects and using the continuity equation, we obtain:

$$[\dot{\varepsilon}] = \begin{bmatrix} \frac{du}{dx} & 0 & 0 \\ 0 & 0 & 0 \\ 0 & 0 & -\frac{du}{dx} \end{bmatrix} \tag{F.4-20}$$

For a Newtonian polymer, the stress tensor is:

$$[\sigma] = \begin{bmatrix} -p + 2\eta\frac{du}{dx} & 0 & 0 \\ 0 & -p & 0 \\ 0 & 0 & -p - 2\eta\frac{du}{dx} \end{bmatrix} \tag{F.4-21}$$

If one neglects the interfacial and drag forces at the film-air interface, the stress σ_{zz} is equal to zero at the surface. Since the film is very thin, it may be assumed to be zero everywhere. Hence:

$$\sigma_{zz} = -p - 2\eta\frac{du}{dx} = 0 \;\Rightarrow\; p = -2\eta\frac{du}{dx} \tag{F.4-22}$$

and the stress tensor is reduced to:

$$[\sigma] = \begin{bmatrix} 4\eta\frac{du}{dx} & 0 & 0 \\ 0 & 2\eta\frac{du}{dx} & 0 \\ 0 & 0 & 0 \end{bmatrix} \tag{F.4-23}$$

A lateral stress needs to be exerted:

$$\sigma_{yy} = \frac{1}{2}\sigma_{xx} \tag{F.4-24}$$

This is achieved with the help of tongs.

b) *Review of the equations:*

– Rheology:

$$\sigma_{xx} = 4\eta \frac{du}{dx} \qquad \text{(F.4-25)}$$

– Continuity: (mass balance) $ueL = u_0 e_0 L$ (F.4-26)
– Force balance: $F = \sigma_{xx} eL = \text{const}$ (F.4-27)

(for negligible gravitational and inertial forces)

Combining these equations, we obtain:

$$u = u_0 \exp\left(\frac{F}{4\eta u_0 e_0 L} x \right) \qquad \text{(F.4-28)}$$

where the force is expressed by:

$$F = \frac{4\eta u_0 e_0 L}{X} \ln \frac{u_1}{u_0} \qquad \text{(F.4-29)}$$

The numerical solution is $F = 23$ N.

Problem 5.2-C: Tube extrusion

The tube is first unrolled and then the problem reduced to the previous case. This is sketched in Figure F.4-3.

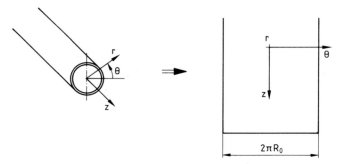

Figure F.4-3 Unrolling of a tubular extrudate

Forcing the tube to be of the same diameter as that of the die is similar to imposing a given width in cast-film processing. Hence we can directly write:

$$\sigma_{\theta\theta} = \frac{1}{2} \sigma_{zz} \qquad \text{(F.4-30)}$$

The normal stress σ_{zz} is obtained from the drawing force, i.e.:

$$\sigma_{zz} = \frac{F}{2\pi R_0 e} \qquad \text{(F.4-31)}$$

whereas the lateral normal stress, $\sigma_{\theta\theta}$, is related to the internal pressure. A force balance per unit length (cutting the tube in two longitudinal halves, as illustrated in Figure F.4-4) yields:

$$2e\sigma_{\theta\theta} = \int_{-\frac{\pi}{2}}^{+\frac{\pi}{2}} \Delta p R_0 \, d\theta \cos\theta = 2\Delta p R_0 \qquad \text{(F.4-32)}$$

hence:

$$\sigma_{\theta\theta} = \Delta p \frac{R_0}{e} \qquad \text{(F.4-33)}$$

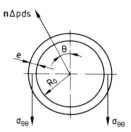

Figure F.4-4   Force balance on the tube shell

Combining Equations (F.4-30), (F.4-31) and (F.4-33), the following relation between F and Δp is obtained:

$$\Delta p \frac{R_0}{e} = \frac{F}{4\pi R_0 e} \qquad \text{(F.4-34)}$$

or

$$F = 4\pi R_0^2 \Delta p \qquad \text{(F.4-35)}$$

It is interesting to note that the force is not a function of the tube wall thickness. The numerical result is $F = 5.024\,\text{kN}$.

F.5   Solutions to problems of Chapter 6

Problem 6.1-A: Simple shear of a Maxwell fluid

a) The velocity vector is $u = (\dot{\gamma}y, 0, 0)$ and the velocity gradient and transpose of the velocity gradient tensor are:

$$[\nabla u] = \begin{bmatrix} 0 & \dot{\gamma} & 0 \\ 0 & 0 & 0 \\ 0 & 0 & 0 \end{bmatrix}, \quad [\nabla u]^{\dagger} = \begin{bmatrix} 0 & 0 & 0 \\ \dot{\gamma} & 0 & 0 \\ 0 & 0 & 0 \end{bmatrix} \qquad \text{(F.5-1)}$$

Moreover, since this is a steady and fully-developed flow,

$$\frac{D\boldsymbol{\sigma}'}{Dt} = \frac{\partial \boldsymbol{\sigma}'}{\partial t} + u\frac{\partial \boldsymbol{\sigma}'}{\partial x} \equiv 0 \tag{F.5-2}$$

and:

$$[\nabla \boldsymbol{u}] \cdot [\boldsymbol{\sigma}'] = \begin{bmatrix} \dot{\gamma}\sigma'_{12} & \dot{\gamma}\sigma'_{22} & 0 \\ 0 & 0 & 0 \\ 0 & 0 & 0 \end{bmatrix} \tag{F.5-3}$$

$$[\boldsymbol{\sigma}'] \cdot [\nabla \boldsymbol{u}]^{\dagger} = \begin{bmatrix} \dot{\gamma}\sigma'_{12} & 0 & 0 \\ \dot{\gamma}\sigma'_{22} & 0 & 0 \\ 0 & 0 & 0 \end{bmatrix} \tag{F.5-4}$$

Hence:

$$\frac{\delta[\boldsymbol{\sigma}']}{\delta t} = \begin{bmatrix} -2\dot{\gamma}\sigma'_{12} & -\dot{\gamma}\sigma'_{22} & 0 \\ -\dot{\gamma}\sigma'_{22} & 0 & 0 \\ 0 & 0 & 0 \end{bmatrix} \tag{F.5-5}$$

b) The four equations for the Maxwell model are readily obtained:

$$\sigma'_{11} - 2\dot{\gamma}\theta\sigma'_{12} = 0 \tag{F.5-6}$$

$$\sigma'_{12} - \dot{\gamma}\theta\sigma'_{22} = \eta\dot{\gamma} \tag{F.5-7}$$

$$\sigma'_{22} = 0, \ \sigma'_{33} = 0 \tag{F.5-8}$$

c) In terms of material functions, these equations become:

$$\sigma'_{12} = \sigma(\dot{\gamma}) = \eta\dot{\gamma} \tag{F.5-9}$$

$$\sigma'_{11} - \sigma'_{22} = N_1(\dot{\gamma}) = 2\eta\theta\dot{\gamma}^2 \tag{F.5-10}$$

$$\sigma'_{22} - \sigma'_{33} = N_2(\dot{\gamma}) = 0 \tag{F.5-11}$$

Newtonian behaviour is predicted when the primary normal stress difference, $N_1 = \sigma'_{11} - \sigma'_{22}$, is negligible compared to the shear stress, σ (small value of the relaxation time θ).

Problem 6.1-B: Rotating disk extruder

a) The velocity field in cylindrical coordinates (r, θ, z) is for any point in the liquid:

$$\begin{cases} u = 0 \\ v = r\Omega(z) \\ w = 0 \end{cases} \tag{F.5-12}$$

and the rate-of-strain tensor is:

$$[\dot{\varepsilon}] = \begin{bmatrix} 0 & 0 & 0 \\ 0 & 0 & \frac{1}{2}r\frac{d\Omega}{dz} \\ 0 & \frac{1}{2}r\frac{d\Omega}{dz} & 0 \end{bmatrix} \tag{F.5-13}$$

The θ, z-planes are then sheared with intensity equal to

$$\dot\gamma = r\frac{d\Omega}{dz} \tag{F.5-14}$$

b) From the result of Problem 1.3-A, the stress tensor can be written in the most general form as:

$$[\boldsymbol{\sigma}] = \begin{bmatrix} \sigma_{rr} & 0 & 0 \\ 0 & \sigma_{\theta\theta} & \sigma_{\theta z} \\ 0 & \sigma_{\theta z} & \sigma_{zz} \end{bmatrix} \tag{F.5-15}$$

with the material functions defined by:

$$N_1 = \sigma_{\theta\theta} - \sigma_{zz}$$
$$N_2 = \sigma_{zz} - \sigma_{rr} \tag{F.5-16}$$
$$\sigma = \sigma_{\theta z}$$

c) The upper convected derivative for the stress tensor is:

$$\frac{\delta[\boldsymbol{\sigma}']}{\delta t} = \frac{D[\boldsymbol{\sigma}']}{Dt} - [\nabla\boldsymbol{u}]\cdot[\boldsymbol{\sigma}'] - [\boldsymbol{\sigma}']\cdot[\nabla\boldsymbol{u}]^\dagger \tag{F.5-17}$$

and the following tensors are written with the help of Appendix A:

$$[\nabla\boldsymbol{u}] = \begin{bmatrix} 0 & -\Omega & 0 \\ \Omega & 0 & r\frac{d\Omega}{dz} \\ 0 & 0 & 0 \end{bmatrix} \tag{F.5-18}$$

$$[\nabla\boldsymbol{u}]\cdot[\boldsymbol{\sigma}'] + [\boldsymbol{\sigma}']\cdot[\nabla\boldsymbol{u}]^\dagger = \begin{bmatrix} 0 & \Omega[\sigma'_{rr}-\sigma'_{\theta\theta}] & -\Omega\sigma'_{\theta z} \\ \Omega[\sigma'_{rr}-\sigma'_{\theta\theta}] & 2r\frac{d\Omega}{dz}\sigma'_{\theta z} & r\frac{d\Omega}{dz}\sigma'_{zz} \\ -\Omega\sigma'_{\theta z} & r\frac{d\Omega}{dz}\sigma'_{zz} & 0 \end{bmatrix} \tag{F.5-19}$$

On the other hand, $D\boldsymbol{\sigma}'/Dt$ may be expressed with the help of Appendix A:

$$\frac{D[\boldsymbol{\sigma}']}{Dt} = \begin{bmatrix} 0 & \Omega(\sigma'_{rr}-\sigma'_{\theta\theta}) & -\Omega\sigma'_{\theta z} \\ \Omega(\sigma'_{rr}-\sigma'_{\theta\theta}) & 0 & 0 \\ -\Omega\sigma'_{\theta z} & 0 & 0 \end{bmatrix} \tag{F.5-20}$$

As a consequence

$$\frac{\delta[\boldsymbol{\sigma}']}{\delta t} = \begin{bmatrix} 0 & 0 & 0 \\ 0 & 2r\frac{d\Omega}{dz}\sigma'_{\theta z} & r\frac{d\Omega}{dz}\sigma'_{zz} \\ 0 & r\frac{d\Omega}{dz}\sigma'_{zz} & 0 \end{bmatrix} \tag{F.5-21}$$

The use of the Maxwell model leads to the following equations

$$\sigma'_{rr} = 0 \tag{F.5-22}$$

$$\sigma'_{\theta\theta} - 2\theta r \frac{d\Omega}{dz} \sigma'_{\theta z} = 0 \tag{F.5-23}$$

$$\sigma'_{\theta z} - \theta r \frac{d\Omega}{dz} \sigma'_{zz} = \eta r \frac{d\Omega}{dz} \tag{F.5-24}$$

$$\sigma'_{zz} = 0 \tag{F.5-25}$$

As in simple shear flow, one obtains:

$$[\sigma'] = \begin{bmatrix} 0 & 0 & 0 \\ 0 & 2\eta\theta\left[r\frac{d\Omega}{dz}\right]^2 & \eta r\frac{d\Omega}{dz} \\ 0 & \eta r\frac{d\Omega}{dz} & 0 \end{bmatrix} \tag{F.5-26}$$

d) The total stress tensor is:

$$[\sigma] = \begin{bmatrix} -p' & 0 & 0 \\ 0 & -p' + 2\eta\theta\left[r\frac{d\Omega}{dz}\right]^2 & \eta r\frac{d\Omega}{dz} \\ 0 & \eta r\frac{d\Omega}{dz} & -p' \end{bmatrix} \tag{F.5-27}$$

and neglecting inertial and gravitational forces we obtain the following components for the equations of motion:

$$\frac{\partial\sigma_{rr}}{\partial r} + \frac{\sigma_{rr} - \sigma_{\theta\theta}}{r} = 0 \tag{F.5-28}$$

$$\frac{\partial\sigma_{\theta z}}{\partial z} = 0 \tag{F.5-29}$$

$$\frac{\partial\sigma_{zz}}{\partial z} = 0 \tag{F.5-30}$$

Equations (F.5-27) and (F.5-29) imply that $d\Omega/dz = $ constant, i.e.:

$$\Omega = \Omega_0 \frac{z}{z_0} \tag{F.5-31}$$

Equations (F.5-27) and (F.5-30) imply on the other hand that the arbitrary pressure p' is a unique function of r. It follows then from (F.5-27), (F.5-28) and (F.5-31) that:

$$\frac{dp'}{dr} = -2\eta\theta\left(\frac{\Omega_0}{z_0}\right)^2 r \tag{F.5-32}$$

This equation can be integrated using the following boundary condition:

$$\sigma'_{rr}(R) = -p'(R) = -p_a \tag{F.5-33}$$

where p_a is the atmospheric pressure. The solution for p' is:

$$p'(r) = p_a + \eta\theta\left(\frac{\Omega_0}{z_0}\right)^2 (R^2 - r^2) \tag{F.5-34}$$

In particular, the pressure on the axis $(r = 0)$ is:

$$p'_0 - p_a = \eta\theta\left(\frac{\Omega_0}{z_0} R\right)^2 \tag{F.5-35}$$

Remarks

In this case, p' has a specific meaning. It is equal to σ_{zz} and does represent the measurable stress at the disk wall, using for example a pressure transducer. In contrast, the thermodynamic pressure is not measurable. From the definition of p,

$$p = -\frac{1}{3} tr[\boldsymbol{\sigma}] = p' - \frac{2}{3}\eta\theta r^2\left(\frac{\Omega_0}{z_0}\right)^2 \tag{F.5-36}$$

and using (F.5-34), we obtain:

$$p = p_a + \eta\theta\left(\frac{\Omega_0}{z_0}\right)^2\left(R^2 - \frac{5}{3}r^2\right) \tag{F.5-37}$$

This result predicts for a large enough rotational speed a negative pressure at the rim of the disks. This is thermodynamically unacceptable.

e) The force that tends to push apart the disks is given by:

$$F = -\int_S \sigma_{zz}\, ds = \int_S p'(r) r\, dr\, d\theta = \frac{\pi}{2}\eta\theta\left(\frac{\Omega_0}{z_0}\right)^2 R^4 \tag{F.5-38}$$

where S is the disk surface. Combining with Equation (F.5-35), we get:

$$F = \left(p'_0 - p_a\right)\frac{\pi R^2}{2} \tag{F.5-39}$$

The torque, C, required to rotate the disk is:

$$C = \int_S \sigma_{\theta z} r\, dS = \int_S \eta r\frac{\Omega_0}{z_0} r^2\, dr\, d\theta \tag{F.5-40}$$

This can be readily integrated to obtain:

$$C = \frac{\pi}{2}\eta\frac{\Omega_0}{z_0} R^4 \tag{F.5-41}$$

We notice that the ratio F/C is equal to $\theta(\Omega_0/z_0)$. Hence the relaxation or characteristic time of a polymer can be determined using this technique.

Numerical results

$$p_0' - p_a = 2.46 \text{ MPa}$$

$$F \approx 22 \text{ kN (2.2 tons)}$$

$$C \approx 105.4 \text{ N} \cdot \text{m}$$

f) Due to the Weissenberg effect an appreciable pressure can be built up at the center part of the disk. A tubular die can be connected to the fixed disk as illustrated in Figure F.5-1 and this simple machine can extrudate polymers at high enough flow rate for moderate torque. However, two types of problems are encountered:

– The force exerted on the disks can be quite large (typically of the order of 2 tons for a 0.15 m disk). This imposed serious constraints on the design of the mechanical parts.

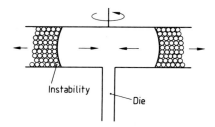

Figure F.5-1   Rotating disk extruder

– More seriously, the feeding of polymer granules is difficult and the operation is unstable.

Indeed:

• The solid polymer granules fed at the rim are subjected to centrifugal forces and forced feeding is required.
• The molten polymer flows towards the center due to the Weissenberg effect. In the plasticating zone, between the solid and the molten polymer, instabilities will develop and create perturbations on the overall operation.

This explains why the rotating disk extruder is not extensively used. Most existing machines are used as mixing devices (shearing between the disks is well controlled) in which molten polymer is fed. Feeding problems are then avoided.

Problem 6.1-C: Couette flow of a Maxwell fluid

a) When steady state regime is reached (this means that the free surface is stabilized) the kinematic hypothesis is:

$$\begin{cases} u = 0 \\ v(r) \\ w = 0 \end{cases} \quad [\nabla \boldsymbol{u}] = \begin{bmatrix} 0 & -\frac{v}{r} & 0 \\ \frac{dv}{dr} & 0 & 0 \\ 0 & 0 & 0 \end{bmatrix} \quad [\dot{\varepsilon}] = \begin{bmatrix} 0 & \frac{1}{2}\left(\frac{dv}{dr} - \frac{v}{r}\right) & 0 \\ \frac{1}{2}\left(\frac{dv}{dr} - \frac{v}{r}\right) & 0 & 0 \\ 0 & 0 & 0 \end{bmatrix} \quad \text{(F.5-42)}$$

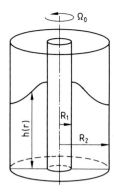

Figure F.5-2
Couette flow of a polymeric liquid.

The general form for the total stress tensor associated to $\dot{\varepsilon}$ is:

$$[\boldsymbol{\sigma}] = \begin{bmatrix} \sigma_{rr} & \sigma_{r\theta} & 0 \\ \sigma_{r\theta} & \sigma_{\theta\theta} & 0 \\ 0 & 0 & \sigma_{zz} \end{bmatrix} \quad \text{and} \quad [\boldsymbol{\sigma}'] = [\boldsymbol{\sigma}] + p'[\boldsymbol{\delta}] = \begin{bmatrix} \sigma'_{rr} & \sigma'_{r\theta} & 0 \\ \sigma'_{r\theta} & \sigma'_{\theta\theta} & 0 \\ 0 & 0 & \sigma'_{zz} \end{bmatrix} \quad \text{(F.5-43)}$$

The material derivative of the stress tensor $\boldsymbol{\sigma}'$ is (Appendix A):

$$\frac{d[\boldsymbol{\sigma}']}{dt} = \begin{bmatrix} \frac{d\sigma'_{rr}}{dt} - 2\frac{v}{r}\sigma'_{r\theta} & \frac{d\sigma'_{r\theta}}{dt} + \frac{v}{r}(\sigma'_{rr} - \sigma'_{\theta\theta}) & 0 \\ \frac{d\sigma'_{r\theta}}{dt} + \frac{v}{r}(\sigma'_{rr} - \sigma'_{\theta\theta}) & \frac{d\sigma'_{\theta\theta}}{dt} + 2\frac{v}{r}\sigma'_{r\theta} & 0 \\ 0 & 0 & \frac{d\sigma'_{zz}}{dt} \end{bmatrix}$$

$$\frac{d\sigma'_{rr}}{dt} = \underbrace{\frac{\partial\sigma'_{rr}}{\partial t}}_{\substack{=0 \\ \text{(stationary)}}} + \underbrace{u\frac{\partial\sigma'_{rr}}{\partial t}}_{\substack{=0 \\ (u=0)}} + \underbrace{\frac{v}{r}\frac{\partial\sigma'_{rr}}{\partial\theta}}_{\substack{=0 \\ \text{(axisymmetric)}}} + \underbrace{\omega\frac{\partial\sigma'_{rr}}{\partial\sigma}}_{\substack{=0 \\ (\omega=0)}}$$

The same reasoning implies $\frac{d\sigma'_{r\theta}}{dt} = \frac{d\sigma'_{\theta\theta}}{dt} = \frac{d\sigma'_{rr}}{dt} = 0$

As a consequence:

$$\frac{d[\boldsymbol{\sigma}']}{dt} = \begin{bmatrix} -2\frac{v}{r}\sigma'_{r\theta} & \frac{v}{r}(\sigma'_{rr} - \sigma'_{\theta\theta}) & 0 \\ \frac{v}{r}(\sigma'_{rr} - \sigma'_{\theta\theta}) & 2\frac{v}{r}\sigma'_{r\theta} & 0 \\ 0 & 0 & 0 \end{bmatrix} \quad \text{(F.5-44)}$$

The convective terms are:

$$-[\nabla\boldsymbol{u}]\cdot[\boldsymbol{\sigma}'] - [\boldsymbol{\sigma}']\cdot[\nabla\boldsymbol{u}]^\dagger = \begin{bmatrix} 2\frac{v}{r}\sigma'_{r\theta} & \frac{v}{r}\sigma'_{\theta\theta} - \frac{dv}{dr}\sigma_{rr} & 0 \\ \frac{v}{r}\sigma'_{\theta\theta} - \frac{dv}{dr}\sigma'_{rr} & -2\frac{dv}{dr}\sigma'_{r\theta} & 0 \end{bmatrix} \quad \text{(F.5-45)}$$

As a consequence, the convective derivative is:

$$\frac{\delta[\sigma']}{\delta t} = \frac{d[\sigma']}{dt} - [\nabla u] \cdot [\sigma'] - [\sigma'] \cdot [\nabla u]^\dagger$$

$$= \begin{bmatrix} 0 & \left(\frac{v}{r} - \frac{dv}{dr}\right)\sigma'_{rr} & 0 \\ \left(\frac{v}{r} - \frac{dv}{dr}\right)\sigma'_{rr} & 2\left(\frac{v}{r} - \frac{dv}{dr}\right)\sigma'_{r\theta} & 0 \\ 0 & 0 & 0 \end{bmatrix} \qquad (\text{F.5-46})$$

b) The Maxwell upper convected model implies, taking into account Eqs. (F.5-42) (F.5-43) and (F.5-46),

$$\sigma'_{rr} = 0 \qquad (\text{F.5-47})$$

$$\sigma'_{r\theta} + \theta\left(\frac{v}{r} - \frac{dv}{dr}\right)\sigma'_{rr} = \eta\left(\frac{dv}{dr} - \frac{v}{r}\right) \qquad (\text{F.5-48})$$

$$\sigma'_{\theta\theta} + 2\left(\frac{v}{r} - \frac{dv}{dr}\right)\sigma'_{r\theta} = 0 \qquad (\text{F.5-49})$$

$$\sigma'_{zz} = 0 \qquad (\text{F.5-50})$$

which leads to the extra stress tensor σ':

$$[\sigma'] = \begin{bmatrix} 0 & \eta\left(\frac{dv}{dr} - \frac{v}{r}\right) & 0 \\ \eta\left(\frac{dv}{dr} - \frac{v}{r}\right) & 2\eta\theta\left(\frac{dv}{dr} - \frac{v}{r}\right)^2 & 0 \\ 0 & 0 & 0 \end{bmatrix} \qquad (F.5-51)$$

and the total stress tensor σ:

$$[\sigma] = \begin{bmatrix} -p' & \eta\left(\frac{dv}{dr} - \frac{v}{r}\right) & 0 \\ \eta\left(\frac{dv}{dr} - \frac{v}{r}\right) & -p' + 2\eta\theta\left(\frac{dv}{dr} - \frac{v}{r}\right)^2 & 0 \\ 0 & 0 & -p' \end{bmatrix} \qquad (\text{F.5-52})$$

The three viscometric functions are

$$\sigma(\dot{\gamma}) = \eta\left(\frac{dv}{dr} - \frac{v}{r}\right) \qquad (\text{F.5-53})$$

$$N_1(\dot{\gamma}) = \sigma_{\theta\theta} - \sigma_{rr} = 2\eta\theta\left(\frac{dv}{dr} - \frac{v}{r}\right)^2 \qquad (\text{F.5-54})$$

$$N_2(\dot{\gamma}) = \sigma_{rr} - \sigma_{zz} = 0 \qquad (\text{F.5-55})$$

It is interesting to note that these viscometric functions are equivalent to the ones derived from Problem 6.1-A and Problem 6.1-B. The positive value of N_1 and the zero value of N_2 justify the analysis of Section 6.1-3.

c) Stress balance

Neglecting mass and inertia, the equations of motion are:

$$-\frac{\partial p'}{\partial r} - \frac{2\eta\theta}{r}\left(\frac{dv}{dr} - \frac{v}{r}\right)^2 = 0 \tag{F.5-56}$$

$$\eta\frac{d}{dr}\left(\frac{dv}{dr} - \frac{v}{r}\right) + \frac{2\eta}{r}\left(\frac{dv}{dr} - \frac{v}{r}\right) = 0 \tag{F.5-57}$$

$$-\frac{\partial p'}{\partial z} = 0 \tag{F.5-58}$$

Equation (F.5-57) can be integrated as in the Newtonian case

$$\frac{dv}{dr} - \frac{v}{r} = -2\Omega_0 \frac{R_1^2 R_2^2}{R_2^2 - R_1^2} \cdot \frac{1}{r^2} \tag{F.5-59}$$

Hence

$$v(r) = \frac{\Omega R_1^2}{R_2^2 - R_1^2}\left[\frac{R_2^2}{r} - r\right] \tag{F.5-60}$$

Equation (F.5-56) is integrated taking into account Equations (F.5-59) and (F.5-58):

$$p'(r) = C + 2\eta\theta\Omega_0^2 \frac{R_1^4 R_2^4}{(R_2^2 - R_1^2)^2} \cdot \frac{1}{r^4} \tag{F.5-61}$$

It follows that:

$$\sigma_{zz}(r) = -p' = -C - 2\eta\theta\Omega_0^2 \frac{R_1^4 R_2^4}{(R_2^2 - R_1^2)^2} \cdot \frac{1}{r^4} \tag{F.5-62}$$

d) As indicated in Section 6.1-3

$$\sigma_{zz} = -\varrho g\, h(r) \tag{6.1-30}$$

and the expression for the free surface is given by:

$$h(r) = C' + \frac{2\eta\theta\Omega_0^2}{\varrho g} \frac{R_1^4 R_2^4}{(R_2^2 - R_1^2)^2} \cdot \frac{1}{r^4} \tag{F.5-63}$$

The free surface is a decreasing function between the inner rotating cylinder and the outer cylinder. For example

$$\Delta h = h(R_1) - h(R_2) = \frac{2\eta\theta\Omega_0^2}{\varrho g} \frac{R_2^2 + R_1^2}{R_2^2 - R_1^2} \tag{F.5-64}$$

Problem 6.1-D: Drawing of a Maxwell liquid

This solution is based on the analysis of DENN and MARRUCCI (1971).

a) *Steady-state solution*

The tensor $\dot{\varepsilon}$ is constant and uniform, i.e. $D\dot{\varepsilon}/Dt = 0$ and:

$$[\dot{\varepsilon}] = [\nabla u] = [\nabla u]^{\dagger} = \begin{bmatrix} \dot{\varepsilon} & 0 & 0 \\ 0 & -\frac{\dot{\varepsilon}}{2} & 0 \\ 0 & 0 & -\frac{\dot{\varepsilon}}{2} \end{bmatrix} \tag{F.5-65}$$

and since the flow is axisymmetrical:

$$[\sigma'] = \begin{bmatrix} \sigma'_{xx} & 0 & 0 \\ 0 & \sigma'_{yy} & 0 \\ 0 & 0 & \sigma'_{zz} \end{bmatrix} \tag{F.5-66}$$

The upper convected derivative of the stress tensor becomes:

$$\frac{\delta[\sigma']}{\delta t} = \begin{bmatrix} -2\dot{\varepsilon}\sigma'_{xx} & 0 & 0 \\ 0 & \dot{\varepsilon}\sigma'_{yy} & 0 \\ 0 & 0 & \dot{\varepsilon}\sigma'_{zz} \end{bmatrix} \tag{F.5-67}$$

The Maxwell model is:

$$[\sigma'] + \theta\frac{\delta[\sigma']}{\delta t} = 2\eta[\dot{\varepsilon}] \tag{F.5-68}$$

Solving, we obtain the three following relations:

$$\sigma'_{xx} - 2\dot{\varepsilon}\theta\sigma'_{xx} = 2\eta\dot{\varepsilon} \tag{F.5-69}$$

$$\sigma'_{yy} + \dot{\varepsilon}\theta\sigma'_{yy} = -\eta\dot{\varepsilon} \tag{F.5-70}$$

$$\sigma'_{zz} + \dot{\varepsilon}\theta\sigma'_{zz} = -\eta\dot{\varepsilon} \tag{F.5-71}$$

Hence:

$$\sigma'_{xx} = \frac{2\eta\dot{\varepsilon}}{1 - 2\dot{\varepsilon}\theta} \quad \text{if} \quad \dot{\alpha}\theta \neq \frac{1}{2} \tag{F.5-72}$$

$$\sigma'_{yy} = \sigma'_{zz} = -\frac{\eta\dot{\varepsilon}}{1 + \dot{\varepsilon}\theta} \tag{F.5-73}$$

We notice that $tr\,\sigma' = \sigma'_{xx} + \sigma'_{yy} + \sigma'_{zz} \neq 0$, i.e. σ' is not a deviatoric stress tensor. The total stress tensor σ is related to σ' by:

$$[\sigma] = -p'[\delta] + [\sigma'] \tag{F.5-74}$$

Since the stress at the fiber surface is assumed to be equal to zero, we write:

$$\sigma_{yy} = -p' + \sigma'_{yy} = 0 \tag{F.5-75}$$

Therefore:

$$\sigma_{xx} = -p' + \sigma'_{xx} = \sigma'_{xx} - \sigma'_{yy} \qquad \text{(F.5-76)}$$

$$\sigma_{xx} = \frac{2\eta\dot{\varepsilon}}{1 - 2\dot{\varepsilon}\theta} + \frac{\eta\dot{\varepsilon}}{1 + \dot{\varepsilon}\theta} = \frac{3\eta\dot{\varepsilon}}{(1 - 2\dot{\varepsilon}\theta)(1 + \dot{\varepsilon}\theta)} \qquad \text{(F.5-77)}$$

and the extensional viscosity is:

$$\eta_e = \frac{\sigma_{xx}}{\dot{\varepsilon}} = \frac{3\eta}{(1 - 2\dot{\varepsilon}\theta)(1 + \dot{\varepsilon}\theta)} \qquad \text{(F.5-78)}$$

The Maxwell model predicts that the extensional viscosity increases with increasing extensional rate whereas it predicts a constant shear viscosity, η (see Chapter 6.1): It is well known that most polymers are shear thinning (in shear flow), but the rheological behaviour of molten polymers under extensional flow is not yet elucidated. In cases, the behaviour depicted in Figure F.5-3 has been observed. We also note from (F.5-78):

- if $\dot{\varepsilon} \to 1/2\theta$, $\eta_e \to \infty$
- if $\dot{\varepsilon} > 1/2\theta$, $\eta_e < 0$, this has no physical meaning.

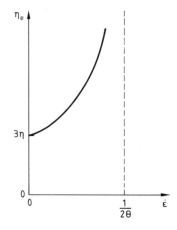

Figure F.5-3   Extensional viscosity of a Maxwell liquid

Physical interpretation

The stress σ_{xx} results from the sum of the forces acting on the macromolecules, which are oriented in the x-direction. In extensional flow there is no rotation and the macromolecules are readily oriented and stretched with increasing extensional rate. As $\dot{\varepsilon}$ reaches $1/2\theta$, σ_{xx} becomes unbounded, this would for Gaussian chains correspond to an infinite end-to-end distance (the probability of finding a macromolecule of infinite length is infinitely small). In reality, the length of polymeric chains is finite and the stress will rather reach a plateau for an extensional rate of the order of $1/2\theta$. This is shown in Figure F.5-5.

Figure F.5-4   Stretching of a macromolecule

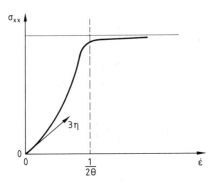

Figure F.5-5   Increase of the stress with extensional rate

b) *Transient regime*

In the drawing process illustrated in Figure F.5-6, the extensional rate $\dot{\varepsilon}$ is applied for a finite time (between the exit of the die and the entrance in the water bath where the polymer is solidified).

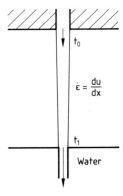

Figure F.5-6   Drawing of a polymer fiber

We will assume that at the die exit $(t = 0)$, the stresses are within an arbitrary isotropic term equal to zero, then increase with time; but we assume that the flow is steady with respect to the position. The time derivative must then be included in the Maxwell model which becomes:

$$\sigma'_{xx} + \theta\frac{\partial \sigma'_{xx}}{\partial t} - 2\dot{\varepsilon}\theta\sigma'_{xx} = \ 2\eta\dot{\varepsilon} \tag{F.5-79}$$

$$\sigma'_{yy} + \theta\frac{\partial \sigma'_{yy}}{\partial t} + \dot{\varepsilon}\theta\sigma'_{yy} = - \ \eta\dot{\varepsilon} \tag{F.5-80}$$

$$\sigma'_{zz} + \theta\frac{\partial \sigma'_{zz}}{\partial t} + \dot{\varepsilon}\theta\sigma'_{zz} = - \ \eta\dot{\varepsilon} \tag{F.5-81}$$

Obviously, $\sigma'_{yy} = \sigma'_{zz}$. Equations (F.5-79) and (F.5-80) can be solved using the initial conditions, $\sigma'_{xx}(0) = \sigma'_{yy}(0) = 0$ and assuming that $\dot{\varepsilon}$ is constant. The result is:

$$\sigma'_{xx}(t) = \frac{2\eta\dot{\varepsilon}}{1 - 2\dot{\varepsilon}\theta}\left[1 - e^{-\frac{t}{\theta}(1-2\dot{\varepsilon}\theta)}\right] \tag{F.5 - 82}$$

$$\sigma'_{yy}(t) = \frac{-\eta\dot{\varepsilon}}{1 + \dot{\varepsilon}\theta}\left[1 - e^{-\frac{t}{\theta}(1+\dot{\varepsilon}\theta)}\right] \tag{F.5 - 83}$$

As for the steady-state solution, σ'_{xx} may be unbounded. We can distinguish three cases:

b.1) $\dot{\varepsilon} < 1/2\theta$

The exponent of e is negative and σ'_{xx} remains finite and:

$$\lim_{t\to\infty} \sigma'_{xx} = \frac{2\eta\dot{\varepsilon}}{1 - 2\dot{\varepsilon}\theta} \tag{F.5-84}$$

b.2) $\dot{\varepsilon} > 1/2\theta$

The exponent of e is positive and the term tends to infinity as $t \to \infty$. For short time, however, σ'_{xx} remains finite.

b.3) $\dot{\varepsilon} = 1/2\theta$

$$\frac{1 - e^{-\frac{t}{\theta}(1-2\dot{\varepsilon}\theta)}}{1 - 2\dot{\varepsilon}\theta} \sim \frac{t}{\theta} \tag{F.5-85}$$

Hence:

$$\sigma'_{xx}(t) \sim 2\eta\dot{\varepsilon}\frac{t}{\theta} \tag{F.5-86}$$

Remark

We note that the initial slope:

$$\frac{d\sigma'_{xx}}{dt}(0) = \frac{2\eta}{\theta}\dot{\varepsilon} \tag{F.5-87}$$

increases with the extensional rate $\dot{\varepsilon}$. The three cases are summarized in Figure F.5-7.

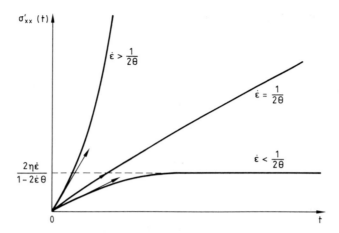

Figure F.5-7   Extensional stress variation with time in a Maxwell liquid

Extensional viscosity

As we have done in part a), we write:

$$\sigma_{xx} = \sigma'_{xx} - \sigma'_{yy} = \frac{3\eta\dot{\varepsilon}}{(1-2\dot{\varepsilon}\theta)(1+\dot{\varepsilon}\theta)} - \frac{2\eta\dot{\varepsilon}}{1-2\dot{\varepsilon}\theta} e^{-\frac{t}{\theta}(1-2\dot{\varepsilon}\theta)}$$

$$- \frac{\eta\dot{\varepsilon}}{1+\dot{\varepsilon}\theta} e^{-\frac{t}{\theta}(1+\dot{\varepsilon}\theta)} \tag{F.5-88}$$

and the transient (stress growth) extensional viscosity is expressed by:

$$\eta_e^+(t,\dot{\varepsilon}) = \frac{3\eta}{(1-2\dot{\varepsilon}\theta)(1+\dot{\varepsilon}\theta)} - \eta\left[\frac{1}{1+\dot{\varepsilon}\theta} e^{-\frac{t}{\theta}(1+\dot{\varepsilon}\theta)} + \frac{2}{1-2\dot{\varepsilon}\theta} e^{-\frac{t}{\theta}(1-2\dot{\varepsilon}\theta)}\right] \tag{F.5-89}$$

The transient extensional viscosity shows a similar behaviour as the extensional stress. It is obvious from Figure F.5-8 that steady-state conditions are attainable only for $\dot{\varepsilon} < 1/2\theta$ and $\eta_e^+(t,\dot{\varepsilon})$ remains finite for finite time.

Remarks

- $$\frac{d\eta_e^+}{dt}(\dot{\varepsilon},0) = \frac{3\eta}{\theta} \tag{F.5-90}$$

This is independent of the extensional rate, meaning that all the curves have the same slope at the origin.

- If $\dot{\varepsilon} = 1/2\theta$:

$$\eta_e^+\left(\frac{1}{2\theta}, t\right) = 2\eta\frac{t}{\theta} + \frac{2}{3}\eta\left[1 - e^{-\frac{3t}{2\theta}}\right] \tag{F.5-91}$$

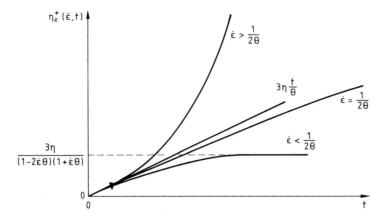

Figure F.5-8   Transient extensional viscosity for a Maxwell liquid

Notations

$\boldsymbol{a}$	tensor in Eq. (A.1-12)
a	eccentricity in an orthogonal rheometer (Figure 6.2-9)
a	length of a chain segment defined in Figure C.2-5
a	characteristic length in Eq. (1.1-1) and in Eq. (5.2-1), variable defined by Eq. (3.2-6) and Eq. (5.1-59)
a, a_0	true and apparent contact area of a granule in the feed zone of the extruder
$a_i{}^j$	transformation matrices in Eq. (B.2-10)
a^*	variable defined by Eq. (3.2-8)
a_H	variable defined by Eq. (3.2-9)
a_T	shift factor, Eq. (C.4-5)
A	dimensionless section in Eq. (5.1-35), compression factor in Eq. (4.3-16), function of z in Eq. (3.4-6)
A	parameter of the Andrade expression, Eq. (C.4-7), in the hyperbolic tangent model, Eq. (1.5-5), in the film stretching force, Eq. (5.2-46), in Eq. (3.2-27)
A_m	critical compression factor in single screw extrusion (Figure 4.3-15)
$\boldsymbol{b}$	tensorial field in Table B.2-1
b	thermal effusivity $= \sqrt{k \varrho c_p}$
b	thermal function in Eq. (5.2-1), variable defined by Eq. (5.1-59), function in Eq. (2.3-43)
b	length of coaxial cylinders in Figure 4.2-4, characteristic length in Eq. (5.2-1)
B	screw pitch
B	parameter of the hyperbolic tangent model, Eq. (1.5-5), parameter of the Andrade expression, Eq. (C.4-7)
B	coefficient in Eq. (E.2-6), parameter in Eq. (3.2-27)
B'	axial distance between two adjacent helices in a screw geometry
B_1, B_2	velocity coefficients in Eq. (D.5-6)
Bi	Biot number defined by Eq. (2.2-22)
Br	Brinkman number defined by Eq. (2.3-1)
BUR	blow up ratio defined by Eq. (5.3-2)

c	light velocity in vacuum
c	Chung's parameter defined by Eq. (4.2-2), parameter in Eq. (6.1-66)
c_p, c_{pa}, c_{pw}	heat capacity, heat capacity of air, heat capacity of water
c_{ps}, c_{pl}	heat capacity per unit mass of the solid polymer, of the liquid polymer
C	parameter in the White-Metzner equation (6.1-66), parameter in the Andrade expression, Eq. (C.4-7), parameter in Eq. (E.2-6), constant in Eq. (3.4-6), in Eq. (3.4-16)
C, C_x, C_y	torque exerted, torque components in the KEPES rheometer, (Figure 6.2-14)
C_2, C_3	parameters in Eq. (C.2-17)
$\boldsymbol{C}$	conformation tensor defined by Eq. (C.3-33)
$\boldsymbol{C}^{-1}$	Finger (non-linear) deformation tensor, Eq. (6.1-58)
$C_1{}^g$, $C_2{}^g$	WLF parameters in Eq. (C.4-5)
C_D	drag coefficient defined by Eq. (5.1-58)
C'	optical sensibility coefficient of the material, Eq. (6.3-10)
Ca	Cameron number defined by Eq. (2.3-22)
d	distance between the center of mass of two molecules, Eq. (C.2-18)
d	tube diameter
dS	surface element
dt	time interval
D	mean diameter, diameter of the tube, jet diameter
D_0, D_R	jet diameter, final diameter in fiber spinning
$\overline{D}$	mean diameter
D_1	internal diameter of the extruder barrel
D_2	diameter of the screw shaft
D_{2m}, D_{2f}	screw diameter in the metering zone, in the feed zone
D	coefficient in Eq. (C.3-11)
D/Dt	substantial derivative defined by Eq. (2.1-13) or Eq. (6.1-38)
$\mathcal{D}/\mathcal{D}t$	Jauman derivative defined by Eq. (6.1-61)
DDR	draw down ratio
DDR_1, DDR_2, DDR_c	critical draw ration
De, De_c	Deborah number defined by Eq. (6.1-54), critical Deborah number

e, e_0, e_1	film thickness
$e, \bar{e}, e_1, e_2$	flight thickness, in single screw extrusion
e	flow thickness in Eq. (6.3-10)
e_1, e_2, e_3	orthogonal coordinate system
$e(\ )$	exponential function
erf$(\ )$	error function, Eq. (2.2-7)
E	activation energy in Eq. (C.4-2), radiative heat flux
E	adimensionnal stretching force, Eq. (5.1-37)
E	parameter in Eq. (E.2-5)
$E^b, E_\lambda^b, E_\theta^b, E_{\lambda\theta}^b$	radiative heat flux for a black body
f	drag coefficient
f_0	drag coefficient on a chain segment
f, f_1, f_2	friction coefficients for polymer-barrel, polymer-screw surfaces
f_L	parameter in Eq. (4.4-34)
F, F_0	drawing force, extensional force, shear force
$\boldsymbol{F}$	gravity force
F_b	force due to the Brownian motion
$F_g, F_{g\,\text{Max}}$	force due to gravity, maximum pulling force due to gravity, Eq. (5.1-24)
F_h	hydrodynamic force
F_N	Newtonian drawing force
F_x, F_y	Force exerted in an orthogonal rheometer
g	gravitational acceleration
$\boldsymbol{g}$	metrix tensor, Eq. (B.2-13)
G	free energy of activation
G	mass flow rate
G	modulus
G', G''	storage modulus, loss modulus defined by Eq. (6.1-13)
G^*	complex modulus defined by Eq. (6.1-13)
Gr	Grashof number defined by Eq. (D.1-8)
h	characteristic length in Eq. (1.1-1), distance between two parallel plates or two parallel disks
h	Planck's constant in Eq. (6.1-35) and (D.3-3)
h	heat transfer coefficient

h, h_0	current distance, gap between the rollers of a calender, Figure 3.2-3
h_0, h_1, h_L	characteristic thickness in a flat die
h^*	characteristic distance in a variable gap geometry
h_c, h_r	heat transfer coefficient by convection, by radiation
H	enthalpy Eq. (C.2-3)
H	distance between parallel plates, bank dimension in calendering
H	screw channel depth
H_a, H_p	channel depth in the feed zone, in the metering zone
I_1, I_2, I_3	invariants defined by Eqs. (B.1-4 to 5)
J	Jacobian in Eq. (B.2-26)
J_1, J_2, J_3	invariants defined by Eqs. (B.1-7 to 9)
k	parameter in the hyperbolic tangent model, Eq. (1.5-5)
k	heat conductivity
k_a, k_{air}	heat conductivity of air
k_w	heat conductivity of water
k_l, k_s	thermal conductivity of the liquid, of the solid polymer
k_b	Boltzmann' constant
k	parameter in a flow rate, pressure drop relationship
k_f, k_p, k_c	characteristics of a die, of the metering zone, of the compression zone of an extruder
K, K_0	consistency index in the power-law expression
K	parameters in Eq. (4.2-13), Eq. (4.2-21), Eq. (4.2-28), Eq. (C.2-9), Eq. (D.4-7)
K'	parameter in Eq. (D.4-10)
l, l_0, l_1	characteristic lengths
l_x, l_y, l_z	characteristic dimensions along the axes
L, L_0	characteristic length, initial length
L	distance between die and water bath, screw length, width of a film, width of a coat hanger die
L_m, L_c, L_f	length of metering, compression, feed zone of an extruder
m	power-law parameter in Eq. (6.1-66)
$m(t, t')$	memory function

M	meridian direction, parameters in Eq. (4.2-13) and Eq. (4.2-21)
$\boldsymbol{M}, \boldsymbol{M'}$	second order tensor in Eq. (B.1-11)
M, M_w	molecular weight
M_c	critical molecular weigth
M_e	average molecular weight of the chain segments between two consecutive entanglements
M_z, M_{z+1}	second and third moment weight average molecular weight
$\boldsymbol{n}, n^p, n^{p'}$	unit normal vector
n	power law index, parameter of the hyperbolic tangent model, Eq. (1.5-5)
n	force exerted on the back flight per unit area, Eq. (4.2-9)
n_j	components of the unit normal vector
n_1	number of entangled molecules
N	concentration of molecules
N	force exerted on the back flight in the solid conveying zone, Eq. (4.2-4)
N_1, N_2	primary and secondary normal stress differences, Eqs. (F.1-24, 25)
N_{1w}	primary normal stress difference at the wall
Nu	Nusselt number defined by Eq. (2.3-24)
p, p', p_0'	hydrostatic pressure, arbitrary pressure
p	parameter in Eq. (6.1-66)
p_c, p_p	pressure at the end of the compression zone, metering zone in an extruder
p_0, p_s	entrance, exit pressure
p_I	injection pressure
p_a, P_a	atmospheric pressure
p_c	pressure necessary to deform the solid bed
P	parallel direction
Pe	Peclet number defined by Eq. (2.3-3)
Pr	Prandtl number defined by Eq. (D.1-10)
$\boldsymbol{q}$	heat flux vector, flow rate vector
q	heat flux, flow rate
q_E, q_I	flow rate
q_{12}	radiant heat flux

q_L	leakage flow between two adjacent channels
q_x, q_y	components of the flow rate in the x and y directions
Q	volumetric flow rate, heat transfer rate
Q_m	mass flow rate of the molten polymer
Q_{1p}, Q'_{1p}	shear flow rate in the metering zone
Q_{2p}, Q'_{2p}	pressure flow rate in the metering zone
Q_{1c}, Q'_{1c}	shear flow rate in the compression zone
Q_{2c}	pressure flow rate in the compression zone
Q_s	rate of energy transferred by conduction
$\boldsymbol{r}$	end to end vector
r	frequency in the change of molecular orientation, Eq. (C.2-5)
r	cylindrical or spherical coordinate
r	spread height in Eq. (3.2-5)
r_e, r_i, r^*	characteristic radii
r_0	quadratic average length of the molecule at rest
r_0	radius of the sprue in injection molding
R, R_1, R_2, R_F, R_0	characteristic radii
R, R_p, R_m	radii of curvature
R	ideal gas constant
R_F	radius of a center gated disk
Ra	Rayleigh number defined by Eq. (D.1-6)
Re	Reynolds number defined by Eq. (1.4-5)
Re{ }	real part of { }
S	entropy of activation, surface, section
S_0, S_R	initial, final section of the filament in fiber spinning
S_f	lateral surface of a flight
S_m, S_p	characteristic surfaces in film blowing
S_{ss}	steady state section
S^*, S_1	section perturbation in fiber spinning
St	Stanton number defined by Eq. (6.3-8)
$\boldsymbol{t}$	unit tangent vector
t	time, contact time, interaction time
t_F	filling time of a mold
t_1	critical residence time, time period
t_l	critical residence time
t_j	components of the unit tangential vector

tr[ ]	trace of [ ]
$\boldsymbol{T}$	force vector
T	Temperature, period of the perturbation in fiber spinning
$\overline{T}$	average temperature
T_a	ambient temperature
T_c	temperature at the center of a plaque
T_f	characteristic temperature of the fluid
T_f'	film temperature, Eq. (D.1-14)
T_g	glass transition temperature
T_i, T_{ic}	interfacial temperature
T_m	melting temperature
T_p	force per unit surface of a film, Eq. (5.2-13)
T_s	surface temperature, solid polymer temperature
T_w	wall temperature
T_0	initial, inlet, reference temperature
T_1	barrel temperature
$\boldsymbol{u}$	velocity vector
u	velocity component in the x-direction, velocity of the solid bed in single screw extrusion
$\overline{u}$	average velocity
$\boldsymbol{U}$	displacement vector
U	characteristic velocity, displacement component in the x-direction
U	internal energy
U_i, U_j, U_k	components of the displacement vector in Eq. (1.1-7) and Eq. (1.1-11)
v	velocity component in the y-direction
V	characteristic velocity, displacement component in the y-direction
V	volume
V_F	relative free volume in Eq. (C.4-8)
V_R	wind-up speed
V_S	velocity of the solid bed towards the barrel
V_{air}	air velocity in the fiber spinning process
V_0	initial velocity
V_1	linear velocity of extruder barrel, wind-up speed

V_{1x}, V_x	relative velocity component of the barrel in the direction normal to the channel axis
V_{1z}, V_z	relative velocity component of the barrel in the direction of the channel axis
w	velocity component in the z-direction
w_{ss}, w^*, w_1	steady state velocity in fiber spinning, velocity perturbation in fiber spinning
W	displacement component in the z-direction
W	work done per unit volume
$W, W_1, W_2, \overline{W}$	characteristic width, width of screw channel
W	parameter in Eq. (B.2-23), weight of a tensor defined by Eq. (B.2-23)
W, W^*	number of possible configurations for a molecule, Eq. (C.2-2)
W'	width of the liquid pocket, Eq. (4.4-32)
$\dot{W}$	rate of energy generated by viscous forces per unit volume
$\dot{W}_V$	energy dissipated by viscous forces in Volume V
We	Weissenberg number defined by Eq. (6.1-55)
x	number of segments
X	stretching length of a film
X, X'	width of the solid bed
y_p	penetration thickness
Y_c, Y_r	convective and radiant terms in the cooling of a film
z_0	distance between rotating disks
Z	helix length for one turn
Z, Z_0, Z_1	parameters of Eqs. (5.2-47) and (5.2-49)
Z_a, Z_c, Z_p	helix length of the feed zone, the compression zone, the metering zone
Z_t	length required for plastication in single screw extrusion

Greek letters

α	thermal diffusivity
α	absorbance capacity or absorptivity defined by Eq. (D.3-7)
α	cone or dihedron angle
α	angle between the fiber axis and the tangent to the surface

α	angle between the film surface and the midplane
α	cylindrical coordinate in a film blowing die
α	coat hanger angle in a flat die geometry
α	friction parameter in Eq. (4.2-2)
α	parameters in Eqs. (C.1-4), (C.4-10), (E.2-3)
α, α_f	slope of the free volume as a funtion of the temperature, Figure C.4-1
α'	compression parameter in Eq. (4.5-1)
α_m, α_p	characteristic angles in film blowing
$\boldsymbol{\beta}$	tensorial field in Eq. (B.2-32)
β	coefficient of volume expansion, defined by Eq. (D.1-5)
β	angle between the lateral surface and the stretching axis in the cast-film process
γ	coefficient of pressure dependance of the viscosity, defined in Eq. (C.4-16)
γ	strain
$\dot{\gamma}, \dot{\gamma}_0, \dot{\gamma}_c$	rate of strain, shear rate, critical shear rate
$\dot{\bar{\gamma}}$	generalized shear rate
γ_{ij}	components of the metrix tensor in curvilinear coordinates
$\dot{\gamma}_w$	shear rate at the wall
$\dot{\Gamma}_w$	Newtonian shear rate at the wall
$\dot{\Gamma}$	maximum shear rate in fiber spinning defined by Eq. (5.1-22)
$\boldsymbol{\delta}$	unit tensor
δ, δ'	film thickness, leakage thickness in a film blowing die
δ_F	clearance between the screw flight and the barrel
δ_{ij}	Kronecker delta
$\delta/\delta t$	upper convected or Oldroyd derivative defined by Eq. (6.1-44)
$\Delta(\ )$	difference between two values
Δ_{ij}	components of the strain tensor
ΔX_c	distance between two successive channels in a film blowing die
ε	surface emissivity
ε_{ij}	components of the strain tensor
$\boldsymbol{\varepsilon}$	strain tensor defined by Eq. (1.1-7)

$\dot{\boldsymbol{\varepsilon}}$	rate of strain tensor defined by Eq. (1.1-17)
$\dot{\varepsilon}$	extensional or elongational rate
$\dot{\varepsilon}_{ij}$	components of the rate of strain tensor
$\dot{\bar{\varepsilon}}$	generalized rate of deformation
$\varepsilon_{\lambda\theta}$	emission factor for the wave length λ in the θ-direction
$\dot{\varepsilon}_m, \dot{\varepsilon}_p, \dot{\varepsilon}_l$	rate of deformation in the meridian, parallel and radial directions (film blowing)
ζ	dimensionless distance in fiber-spinning
ξ^1, ξ^2, ξ^3	curvilinear coordinates in Eqs. (B.2-2 to 4)
η, η_0	viscosity, zero-shear viscosity or reference viscosity
η_a, η_w	viscosity of air, of water
η_1	viscosity of the liquid monomer
η_{be}	biaxial elongational viscosity
η_e	elongational viscosity
η_e^+	extensional viscosity in stress growth experiment
η_f	polymer viscosity evaluated at the shear rate in the flight clearance
η_g	viscosity at T_g
η^*, η', η''	complex viscosity, real and imaginary parts of the complex viscosity
η_∞	high shear-rate viscosity
θ	cylindrical or spherical coordinate
θ	fluid characteristic time or relaxation time
θ, θ_0	angle, cone angle
$\theta_1, \theta_2, \bar{\theta}$	screw flight angles
λ	characteristic time
$\lambda, \lambda_1, \lambda_2, \lambda_3$	complex eigenvalues
λ	latent heat of fusion per unit mass
λ	wave length. Lamé coefficient in Eq. (1.3-1)
λ	parameter in Eq. (1.5-4), Eq. (3.2-27), Eq. (E.2-6)
λ_1, λ_2	viscoelastic parameters in Eq. (6.1-56) and Eq. (6.1-65)
λ'	parameter in Eq. (3.2-27), Eq. (F.3-4)
μ	Clapeyron coefficient in Eq. (1.3-1)
μ_0, μ_1, μ_2	viscoelastic parameters in Eq. (6.1-65)
μ, μ', μ''	parameters in Eq. (3.2-27)

ν	Poisson's coefficient for the polymer granules
ν	kinematic viscosity defined by Eq. (1.4-3)
ν_{air}	air kinematic viscosity
ν_b	frequency of the Brownian motion
$\varrho, \varrho_a, \varrho_w$	polymer, air, water density
$\varrho_0, \varrho_\infty$	bulk density of the granules, density of the solid polymer
ϱ_g	density at T_g
ϱ_s, ϱ_l	density of solid, of liquid polymer
$\boldsymbol{\sigma}$	stress tensor
$\sigma, \sigma_0, \sigma_1, \sigma_2, \sigma_{12}$	shear stress
σ_0	yield stress of the solid polymer
σ	Stefan-Boltzmann constant, coefficient in Eq. (C. 3-11)
σ_c	critical shear stress in Eq. (6.3-11)
$\sigma_e, \sigma_m, \sigma_m^N, \sigma_p, \sigma_p^N$	stress components in film blowing
σ_n	normal stress component
$\boldsymbol{\sigma}^p, \boldsymbol{\sigma}^{p'}$	stress tensor on the surface and on the side of a stretched film
σ_{zp}, σ_{zs}	elongational stress in the main flow and in the recirculation flow (Figure 6.3-13)
σ_w	shear stress at the wall
$\boldsymbol{\sigma}', \boldsymbol{\sigma}_0'$	extra stress tensor or deviatoric stress tensor
τ	friction force per unit surface exerted by the surrounding air
φ	out of phase angle
ϕ	angle, spherical coordinate, solid conveying angle
ϕ	characteristic of the overall plastication efficiency, Eq. (4.3-6)
ϕ	dimensionless stress
ψ	function in Eq. (B.2-27)
ψ	stream function defined by Eq. (1.4-14)
ψ	characteristic angle in a film blowing die geometry
ψ	dimensionless stress
ψ, ψ_0	distribution functions
ψ_1, ψ_2	primary and secondary normal stress coefficients defined by Eqs. (6.2-7, 8)
ω	pulsation

$\omega,\ \omega'$	plastication rate
$\boldsymbol{\Omega}$	vorticity tensor defined by Eq. (1.4-16)
$\Omega,\ \Omega_0,\ \Omega_c$	angular velocity, vorticity in 2D-flows

Abreviations of polymers

ABS	acrylonitrile-butadiene-styrene
HDPE	high-density polyethylene
LDPE	low-density polyethylene
LLDPE	linear low-density polyethylene
PBT	polybutylene terephthalate
PC	polycarbonate
PET	polyethylene terephthalate
PMMA	polymethyl methacrylate
PP	polypropylene
PS	polystyrene
PVC	polyvinyl chloride
SAN	styrene-acrylonitrile

References

AGASSANT, J. F.: "Le calandrage des matières thermo-plastiques", Ph.D. Thesis, Paris VI (1980)

AGASSANT, J. F.: Rev. Gén. Therm. Fr., *279*, 221 (1985)

AGASSANT, J. F., and AVENAS, P.: J. Macromol. Sci. Phys., B*14* (3), 345 (1977)

AGASSANT, J. F., CHENOT, J. L., and NEYRET, B.: "Finite element calculations of two dimensional molten polymer flows", in: Advances in Rheology, Vol. 1, Ed.: *Mena, B., Garcia-Rejon, A., Rangel-Nafaile, C.*, UNAM, Mexico, (1984)

AGASSANT, J. F., and ESPY, M.: Polym. Eng. Sci., *25*, 118 (1985)

AGASSANT, J. F., and PHILIPPE, A.: "Interrelation between processing conditions and defects in calendered sheets", in: Interrelations between processing, structure and properties of Polymeric Materials, edited by *Seferis, J. C., Theocaris, P. S.*, Elsevier, Amsterdam (1984)

AGASSANT, J. F., SAILLARD, P., and VERGNES, B.: Rev. Gén. Therm. Fr., *272*, 477 (1984)

AGASSANT, J. F., ALLES, H., PHILIPON, S., VINCENT, M.: Polym. Eng. Sci., *28*, 460 (1988)

AGASSANT, J. F., VERGNES, B., WEY, E.: "Contribution de la modelisation à l'optimisation de l'extrusion monovis des polymères thermoplastiques", in: Les colloques de l'INRA n° 41, Ed. INRA, Paris, (1987)

AGUR, E. E., and VLACHOPOULOS, J.: 35th ANTEC, Montreal (1977)

AGUR, E. E., and VLACHOPOULOS, J.: Polym. Eng. Sci., *22*, 1084 (1982)

ALLES, H., PHILIPON, S., AGASSANT, J. F., VINCENT, M., DEHAY, G., LEREBOURS, P.: Polym. Proc. Eng., 4 *(1)*, 71 (1986)

ALSTON, W. W., and ASTILL, K. N.: J Appl. Polym. Sci., *17*, 3157 (1973)

ARDICHVILLI, G.: Kautschuk, *14*, 23 (1938)

AVENAS, P., J. Méca. Appli. *4*, n° 3, 283 (1980)

BALLENGER, T. F., CHEN, I. J., CROWDER, J. W., HAGLER, G. E., BOGUE, D. C., and WHITE, J. L.: Trans. Soc. Rheol., *15*, 195 (1971)

BALLENGER, T. F., and WHITE, J. L.: J. Appl. Polym. Sci., *15*, 1949 (1971)

BARTOS, O., and HOLOMEK, J.: Polym. Eng. Sci., *11*, 324 (1971)

BAUDIER, F. and AVENAS, P.: lecture notes in Physics, *19*, 10 (1973)

BEAUFILS, P., VERGNES, B. and AGASSANT, J. F.: Int. Polym. Proc., *4* (2), 78 (1989)

BEKIN, N. G., LITVINOV, V. V., and PETRUSHANSKII, V. YU.: Kauchuk i Rezina, *8*, 32 (1975)

BENARD, H.: "Les tourbillons cellulaires dans une nappe liquide", Revue générale des sciences pures et appliquées, *11*, 1261 (1900)

BENBOW, J. J., CHARLEY, R. B., and LAMB, P.: Nature, *192*, 223 (1961)

BENIS, A. M.: Chem. Eng. Sci., *22*, 805 (1967)

BERGEM, N.: Proc. VII Int. Cong. Rheol., Göteborg, Sweden (1976)

BERGER, J. L., and GOGOS, C. G.: Polym. Eng. Sci., *13*, 102 (1973)

BERNSTEIN, B., KEARSLEY, E. A., ZAPAS, L. J.: Trans. Soc. Rheol., *7*, 391 (1963)

BIRD, R. B., STEWART, W. E., and LIGHTFOOT, E. N.: "Transport Phenomena", Wiley and Sons, New York (1960)

BIRD, R. B., and CARREAU, P. J.: Chem. Eng. Sci., *23*, 427 (1968)

BIRD, R. B., ARMSTRONG, R. C., and HASSAGER, O.: "Dynamics of Polymeric Liquid", Vol. 1, Fluid Mechanics, Wiley (1977)

BIRD, R. B., ARMSTRONG, R. C., and HASSAGER, O.: "Dynamics of Polymeric Liquid", Vol. 1, Fluid Mechanics, 2nd edition, Wiley (1987)

BIRD, R. B., HASSAGER, O., ARMSTRONG, R. C., and CURTISS, C. F.: "Dynamics of Polymeric Liquid", Vol. 2, Kinetic Theory, Wiley (1977)

BIRD, R. B., CURTISS, C. F., ARMSTRONG, R. C., and HASSAGER, O.: "Dynamics of Polymeric Liquid", Vol. 2, Kinetic Theory, 2nd edition, Wiley (1987)

BLACK, J. R., and DENN, M. M.: J. Non Newt. Fluid Mech., *1*, 83 (1976)

BOGER, D. V., and RAMAMURTHY, A. V.: Rheol. Acta, *11*, 61 (1972)

BRAZINSKY, I., COSWAY, H. F., VALLE, C. F., CLARK JONES, R., and STORY, V.: J. App. Polym. Sci., *14*, 2771 (1970)

BRINKMAN, H. C.: Appl. Sci. Res., *A2*, 120 (1951)

BROYER, E., GUTFINGER, C., and TADMOR, Z.: Trans. Soc. Rheol., *19*, 423 (1975)

BROYER, E., and TADMOR, Z.: Polym. Eng. Sci., *12*, 12 (1972)

BUECHE, F.: J. Chem. Phys., *20*, 1959 (1952)

BUECHE, F.: J. Chem. Phys., *48*, 4781 (1968)

CAMERON, A.: "Principles of Lubrication", Lougmans Green, London (1966)

CARREAU, P. J.: Trans. Soc. Rheol., *16*, 99 (1972)

CARREAU, P. J., DE KEE, D., and DAROUX, M.: Can. J. Chem. Eng., *57*, 135 (1979)

CARSLAW, H. S., and JAEGER, J. C.: "Conduction of Heat in Solids", Oxford University Press (1959)

CASWELL, B., and TANNER, R. I.: Polym. Eng. Sci., *18*, 416 (1978)

CHONG, J. S.: J. Appl. Polym. Sci., *12*, 191 (1968)

CHOO, K. P., NEELAKENTAN, N. R., and PITTMAN, J. F. T.: Polym. Eng. Sci., *20*, 349 (1980)

CHUNG, C. I.: Mod. Plast., *45*, 178 (1968)

CHUNG, C. I.: SPE Journal, *26*, 32 (1970)

CHUNG, C. I.: Polym. Eng. Sci., *15*, 29 (1975)

CHUNG, C. I.: Plast. Eng., *6*, 48 (1976)

COHEN, M., and TURNBULL, D.: J. Chem. Phys., *31*, 1164 (1959)

COLEMAN, C. J.: J. Non-Newt. Fluid Mech., *8*, 261 (1981)

COTTO, D.: "Etude de la fabrication de films de polypropylène par extrusion et refroidissement sur rouleau thermostaté", Ph.D. Thesis, Ecole des Mines de Paris (1984)

COTTO, D., SAILLARD, P., AGASSANT, J. F., and HAUDIN, J. M.: "Influence of processing conditions on two step biaxial stretching of polypropylene films", in: "Interrelations Between Processing, Structure and Properties of Polymeric Materials", edited by *Siferis, J. C.*, and *Theocaris, P. S.*, Elsevier, Amsterdam (1984)

COX, A. P. D., and FENNER, R. T.: Polym. Eng. Sci., *20*, 562 (1980)

COX, H. W., and MACOSKO, C. W.: AIchE J., *20*, 785 (1974)

CROCHET, M. J., DUPONT, S., and MARCHAL, J. M.: "The Numerical Simulation of the Flow of Viscoelastic Fluids of the Differential and the Integral Types: a Comparison", Proc. *IX* Int. Cong. on Rheology, Acapulco, Mexico (1984)

DARNELL, W. H., and MOL, E. A. J.: Soc. Plast. Engs. J., *12*, 20 (1956)

DEBBAUT, R., MARCHAL, J. M. and CROCHET, M. J.: J. Non Newt. Fluid Mech., *29*, 119–146 (1988)

DEMAY, Y., and AGASSANT, J. F.: J. de Méc. Théorique et Appliquée, *1*, 763 (1982)

DEMAY, Y.: "Instabilité d'étirage et bifurcation de Hopf", Ph.D. Thesis, Nice (1983)

DENN, M. M., and MARRUCCI, G.: AIChE J., *17*, 101 (1971)

DENN, M. M., PETRIE, C. J. S., and AVENAS, P.: AIChE J., *21*, 791 (1975)

DE WITT, T. W.: J. Appl. Phys., *26*, 889 (1955)

DINH, S. M., and ARMSTRONG, R. C.: AIChE J., *28*, 294 (1982)

DOBROTH, T. and ERWIN, L.: Polym. Eng. Sci., *26*, 462 (1986)

DONOVAN, R. C.: Polym. Eng. Sci., *11*, 247 (1971)

DOOLITTLE, A. K.: J. Appl. Phys., *22*, 1031 (1951)

DUONG DANG, V. I.: J. Appl. Polym. Sci., *23*, 3077 (1979)

EDMONDSON, I. R., and FENNER, R. T.: Polymer, *16*, 48 (1975)

ELBIRLI, B., LINDT, J. T., GOTTGETREN, S. R., and BOBO, S. M.: Polym. Eng. Sci., *23*, 86 (1983)

EL KISSI, N.: "Ecoulements de Polymères fondus dans les contractions brusques. Etude des écoulements secondaires et des défauts d'extrusion. Mise en évidence et modelisation du glissement à la paroi", Ph. D. Thesis, Grenoble (1989)

FENNER, R. T.: "Extruder Screw Design", Iliffe Books, London (1970)

FENNER, R. T.: Plast. and Polym., *6*, 114 (1974)

FENNER, R. T., and NADIRI, F.: Polym. Eng. Sci., *19*, 203 (1979)

FENNER, R. T., and WILLIAMS, J. G.: Trans. Plast. Inst., *35*, 701 (1967)

FERRY, J. D.: "Viscoelastic Properties of Polymers", 2nd Edition, John Wiley and Sons, New York (1970)

FISHER, R. J., and DENN, M. M.: Chem. Eng. Sci., *30*, 1129 (1975)

FISHER, R. J., and DENN, M. M.: AIChE J., *22*, 236 (1976)

FISHER, R. J., and DENN, M. M.: AIChE J., *23*, 23 (1977)

FINSTON, M.: J. Appl. Mech. (Trans. ASME), *18*, 12 (1951)

FOREST, G., and WILKINSON, W. L.: Trans. Chem. Eng., *51*, 331 (1973)

FRANZKOCH, B., and MENGES, G.. 36th ANTEC (1978)

FREDRICKSON, A. G., and BIRD, R. B.: Ind. Eng. Chem., *50*, 347 (1958)

FUKASE, H., KUNIO, T., SHINYA, S., and NOMURA, A.: Polym. Eng. Sci., *22*, 578 (1982)

GAO, H. W., RAMACHANDRAN, S., and CHRISTIANSEN, A. B.: J. Rheol., *25*, 213 (1981)

GASKELL, R. E.: J. Appl. Mech. (Trans. ASME), *17*, 334 (1950)

GAGON, D. K.: "An Iterative Computer Simulation of Polymer Melt Spinning", M. Sc. Thesis, University of Delaware (1980)

GAVIS, J., and LAURENCE, R. L.: Ind. Eng. Chem. Fund., *7*, 232 (1968)

GEE, R. E., and LYON, J. B.: Ind. Eng. Chem., *49*, 956 (1957)

GERMAIN, P.: "Cours de mécanique des milieux continus", Masson, Paris (1973)

GLASSTONE, S., LAIDLER, K., and EYRING, H.: "The theory of rate processes", McGraw Hill, New York (1941)

GODDARD, J. D., and MILLER, C.: Rheol. Acta, *5*, 177 (1966)

GOULD, J., and SMITH, F. S.: J. Text. Inst., *1*, 38 (1980)

GRAESSLEY, W. W.: J. Chem. Phys., *43*, 2696 (1965)

GRAESSLEY, W. W.: J. Chem. Phys., *47*, 1942 (1967)

GRAETZ, L.: Am. Phys., *25*, 337 (1889)

GRMELA, M., and CARREAU, P. J.: J. Non-Newt. Fluid Mech., *23*, 271 (1987)

GUPTA, R. K., METZNER, A. B., and WISSBURN, K. F.: Polym. Eng. Sci., *22*, 172 (1982)

GUPTA, S. K., HAMILTON, G. M., and HIRST, W.: Proc. Int. Engrs., *184*, 148 (1969–1970)

D'HALEWYN, S., AGASSANT, J. F. and DEMAY, Y.: Polym. Eng. Sci., *30*, 335 (1990)

HALMOS, A. L., PEARSON, J. R. A. and TROTTNOW, R.: Polymer, *19*, 1199 (1978)

HAMI, M. L., and PITTMAN, J. F. T.: Polym. Eng. Sci., *20*, 339 (1980)

HAMMOND, L. R.: Wire and Wire Products, 725 (June 1960)

HAN, C. D., and PARK, J. Y.: J. Appl. Polym. Sci., *19*, 3257 (1975)

HAN, C. D., and PARK, J. Y.: J. Appl. Polym. Sci., *19*, 3277 (1975)

HAN, C. D.: "Rheology in Polymer Processing", Academic Press, New York (1976)

HELMY, H. A. A., and WORTH, R. A.: In: Proceedings of the VIII International Congress on Rheology, Naples (1980)

HULATT, M., and WILKINSON, W. L.: Polym. Eng. Sci., *18*, 1148 (1978)

HUSEBY, T. W.: Trans. Soc. Rheol., *10*, 181 (1966)

HUTTON, J. F.: Nature, *200*, 646 (1963)

INGEN HOUSZ, J. F., and MEIJER, H. E. H.: Polym. Eng. Sci., *21*, 352 (1981)

IOOSS, G., and JOSEPH, D. D.: "Elementary Stability and Bifurcation Theory", Springer Verlag, New York (1980)

ISHIHARA, H., and KASE, S.: J. Appl. Polym. Sci., *19*, 557 (1975)

ITO, K.: Japan Plastics, *4*, 27 (1970)

JANESCHITZ-KRIEGL, H.: Advan. Polym. Soc., *6*, 170 (1969)

KACIR, L., and TADMOR, Z.: Polym. Eng. Sci., *12*, 307 (1972)

KALONI, P. N.: J. Phys. Soc. Japan, *20*, 132 (1965)

KAMAL, M. R., and KENIG, S.: Polym. Eng. Sci., *12*, 294 (1972)

KASE, S., and MATSUO, T.: J. Polym. Sci., *A3*, 2541 (1965)

KERTSCHER, E.: 22nd Int. Wire and Cable Symp. (1973)

KEUNINGS, R. and CROCHET, M. J.: J. Non-Newt. Fluid Mech., *14*, 279–299 (1984)

KEUNINGS, R.: "Simulation of viscoelastic fluid flow", in "Fundamentals of computer modeling for polymer processing", edited by Tucker, C. L., *III*, Hanser Publishers, (1989)

KIPARISSIDES, C., and VLACHOPOULOS, J.: Polym. Eng. Sci., *16*, 712 (1976)

KIPARISSIDES, C., and VLACHOPOULOS, J.: Polym. Eng. Sci., *18*, 210 (1978)

KLEIN, I.: "Computer Applications to Processing", in: Plastics Polymer Science and Technology, John Wiley and Sons, New York (1982)

KLEIN, I., and KLEIN, R.: SPE. J., *29*, 33 (1973)

LAMONTE, R. R., and HAN, C. D.: J. Appl. Polym. Sci., *16*, 3285 (1972)

LeRoy, P., and Pierrard, J. M.: Rheol. Acta, *12*, 449 (1973)

Lin, S. H., and Hsu, W. K.: Heat Trans. J., *102*, 382 (1980)

Lindt, J. T.: Polym. Eng. Sci., *16*, 284 (1976)

Lindt, J. T.: Polym. Eng. Sci., *21*, 1162 (1981)

Lodge, A. S.: "Elastic Liquids", Academic Press, London (1960)

Losson, J. M.: 32th ANTEC, 231 (1974)

Losson, J. M.: Japan Plastics, *1*, 22 (1974)

Lovegrove, J. G. A., and Williams, J. G.: Polym. Eng. Sci., *14*, 589 (1974)

Luo, X. L. and Tanner, R. I.: J. Non-Newt. Fluid Mech., *22*, 61 (1986)

Luo, X. L. and Tanner, R. I.: J. Non-Newt. Fluid Mech., *31*, 143 (1989)

McAdams, W. H.: "Heat transmission", Mc Graw Hill, New York (1954)

McClelland, D. E., and Chung, C. I.: Polym. Eng. Sci., *23*, 100 (1983)

McIntire, L. V.: J. Appl. Polym. Sci., *16*, 2901 (1972)

McKelvey, J. M.: "Polymer Processing", John Wiley and Sons, New York (1962)

Maddock, B. H.: Soc. Plast. Engs. J., *15*, 383 (1959)

Maders, H.: Une modelisation par elements finis de l'écoulement d'un fluide viscoélastique dans une filière bidimensionnelle, Ph. D. Thesis, Ecole des Mines de Paris (1990)

Maillefer, C.: Swiss Patent no. 363149 (1959)

Malone, M. F., and Middleman, S.: "A Finite Element Scheme for Two-Dimensional Viscoelastic Fluid Flow", AIChE meeting (1979)

Mandel, J.: "Introduction à la mécanique des milieux continus déformables", Editions scientifiques de Pologne, Varsovie (1974)

Mandel. J.: "Cours de Mécanique des milieux continus", Annexe XXI, Gauthiers-Villard, Paris (1966)

Marchal, J. M. and Crochet, M. J.: J. Non-Newt. Fluid Mech., *26*, 77 (1987)

Matovich, M. A., and Pearson, J. R. A.. Ind. Eng. Chem. Fund., *8*, 512 (1969)

Matsubara, Y.: Polym. Eng. Sci., *19*, 169 (1979)

Matsui, M.: Trans. Soc. Rheol., *20:3*, 465 (1976)

Maxwell, B., and Chartoff, R. P.: Trans. Soc. Rheol., *9-1*, 41 (1965)

Meister, B. J.: Trans. Soc. Rheol., *15*, 63 (1971)

Menges, G., and Klenk, P.: Kunststoffe, *57*, 598 (1967)

Mennig, G.: Plast. and Polym., *12*, 330 (1972)

Middleman, S.: "Fundamentals of Polymer Processing", McGraw Hill (1977)

Mitsoulis, E., Vlachopoulos, J., and Mirza, F. A.: Polym. Eng. Sci., *25*, 6 (1985)

Morette, R. A., and Gogos, C. G.: Polym. Eng. Sci., *8*, 272 (1968)

Mount, E. M., and Chung, C. I.: Polym. Eng. Sci., *18*, 711 (1978)

Mount, E. M., Watson, J. G., and Chung, C. I.: Polym. Eng. Sci., *22*, 729 (1982)

Nakajima, N., and Collins, E. A.: Polym. Eng. Sci., *14*, 137 (1974)

Nebrensky, J., Pittman, J. F. T., and Smith, J. M.: Polym. Eng. Sci., *13*, 209 (1973)

Neyret, B.: "Calcul par la méthode des éléments finis des écoulements de polymères fondus dans les géométries convergentes", Ph.D. Thesis, Ecole des Mines de Paris (1985)

Nunn, R. E., and Fenner, R. T.: Polym. Eng. Sci., *17*, 811 (1977)

Oldroyd, J. G.: Proc. Roy. Soc. London, A *200*, 523 (1950)

OLDROYD, J. G.: Proc. Roy. Soc. London, A 345, 278 (1958)

DEN OTTER, J. L.: Rheol. Acta, 10, 200 (1971)

DEN OTTER, J. L.: Plast. Polym., 38, 155 (1970)

PEARSON, J. R. A.: "Mechanical Principles of Polymer Melt Processing", Pergamon Press, Oxford (1966)

PEARSON, J. R. A.: Polym. Eng. Sci., 18, 222 (1978)

PEARSON, J. R. A. and MATOVICH, M. A.: Ind. Eng. Chem. Fund., 8, 606 (1969)

PEARSON, J. R. A., and PETRIE, C. J. S.: Plast. Polym., 38, 85 (1970)

PEARSON, J. R. A., and PETRIE, C. J. S.: J. Fluid. Mech., 40, 1 (1970)

PEIFFER. H.: Ph.D. Thesis, Technische Hochschule, Aachen (1981)

PERERA, M. G. N., and STRAUSS, K.: "Direct Numerical Solution of the Equations for Viscoelastic Fluid Flow", IUTAM Symposium on Non-Newtonian Fluid Mechanics, Louvain-la-Neuve, Belgium (1978)

PETRIE, C. J. S.: Rheol. Acta, 12, 92 (1973)

PETRIE, C. J. S.: AIChE J., 21, 275 (1975)

PHAN-THIEN, N.: Trans. Soc. Rheol., 22, 259 (1978)

PHILIPPE, A.: "Contribution à l'étude de écoulements viscoélastiques dans les convergents", Ph.D. Thesis, Ecole des Mines de Paris (1981)

PHILIPON, S., AGASSANT, J. F., and VINCENT, M.: Rev. Gén. Therm. Fr., 279, 291 (1985)

PIANA, A.: "Etudes des relations entre mise en forme, orientation et retraction dans les films de polyéthylène basse densité réalisés par soufflage de gaine", Ph.D. Thesis, Ecole des Mines de Paris (1984)

PINTO, G., and TADMOR, Z.: Polym. Eng. Sci., 10, 279 (1970)

POWELL, R. E., ROSEVEARE, W. E., and EYRING, H.: Ind. Eng. Chem., 33, 430 (1941)

PRASTARO, A., and PARRINI, P.: Textile Research J., 118 (1975)

PROCTER, B.: S.P.E. J., 28, 34 (1972)

RANZ, W. E., and MARSHALL, W. R.: J. Chem. Eng. Proc., 48, 141 (1952)

RAUTENBACH, R., and PEIFFER, H.: Kunststoffe, 72, 137 (1982)

RAUWENDAAL, C., and FERNANDEZ, F.: 42th ANTEC, New Orleans (1984)

LORD RAYLEIGH: Phil. Mag., 32, 529 (1916)

RIVLIN, R. S., and ERICKSEN, J. L.: J. Rat. Mech. Anal., 4, 323 (1955)

RIVLIN, R. S.: "Some Recent Results on the Flow of Non-Newtonian Fluids", IUTAM Symposium on Non-Newtonian Fluid Mechanics, Louvain-la-Neuve, Belgium (1978)

ROTHENBERGER, R., McCOY, H., and DENN, M. M.: Trans. Soc. Rheol., $17{:}2$, 259 (1973)

ROUSE, P. E.: J. Chem. Phys. 21, 7, 1272 (1953)

RUDIN, A.: "The Elements of Polymer Science and Engineering – An Introduction Text for Engineers and Chemists", Academic Press, New York (1982)

SAILLARD, P.: "Les phénomènes thermiques dans les outillages d'extrusion des matières plastiques", Ph. D. Thesis, Ecole des Mines de Paris (1982)

SAILLARD, P., and AGASSANT, J. F.: Polym. Proc. Eng., 2, 37 (1984)

SAILLARD, P., VERGNES, B. and AGASSANT, J. F.: Polym. Proc. Eng., 2, 53 (1984)

SCHLICHTING, H.: "Boundary-Layer Theory", McGraw Hill, New York (1955); Sixth edition (1968)

SCHOTT, H.: J. Appl. Polym. Sci., *6*, S29 (1962)

SERGENT, J. PH.: "Etude de deux procédés de fabrication de films, le soufflage de gaine et l'extrusion de film à plat", Ph.D. Thesis, Université Louis Pasteur, Strasbourg (1977)

SHAH, Y. T., and PEARSON, J. R. A.: Ind. Eng. Chem. Fund., *11*, 145 (1972)

SHAPIRO, J., HALMOS, A. L., and PEARSON, J. R. A.: Polymer, *17*, 905 (1976)

SILVERSTON, P. L.: Forsch. Ing. Wes., *24*, 29 (1958)

SORNBERGER, G., QUANTIN, J. C., FAJOLLE, R., VERGNES, B. and AGASSANT, J. F.: J. Non-Newt. Fluid Mech., *23*, 123 (1987)

SOSKEY, P. R., and WINTER, H. H.: 40th ANTEC, San Fransicso (1982)

SPRIGGS, T. W.: Private Communication (1965), in: "Dynamics of polymeric liquids", *Bird, R. B., Armstrong, R. C., et Hassager, O.*, Wiley and Sons, New York (1977)

STREET, L. F.: Int. Plast. Engs., *1*, 283 (1961)

STREETER, V. L.: "Handbook of Fluid Dynamics", Mc Graw-Hill, New York (1961)

TADMOR, Z., and BROYER, E.: Polym. Eng. Sci., *12*, 378 (1972)

TADMOR, Z.: J. Appl. Polym. Sci., *18*, 1753 (1974)

TADMOR, Z., BROYER, E., and GUTFINGER, C.: Polym. Eng. Sci., *14*, 660 (1974)

TADMOR, Z., and KLEIN, I.: "Engineering Principles of Plasticating Extrusion", Van Nostrand Reinhold Company, New York (1970)

TANNER, R. I.: J. Polym. Sci., *18*, 2067 (1970)

TANNER, R. I.: "Engineering Rheology", Clarendon Press, Oxford (1985)

TASKERMAN-KROZER, R., SCHENKEL, G., and EHRMANN, G.: Rheol. Acta, *14*, 1066 (1975)

TAYLOR, G. I.: Phil. Trans., *223 A*, 289 (1923)

TORDELLA, J. P.: J. Appl. Phys. *27*, 454 (1956)

TORDELLA, J. P.: J. Appl. Polym. Sci., *7*, 215 (1963)

TORNER, R. V.: "Grundprozesse der Verarbeitung von Polymeren", V.E.B. Deutscher Verlag für Grundstoffindustrie, Leipzig (1974)

TORRE-GROSSA TOMAS, J. M.: "Défauts d'extrusion du polyéthylène de haute densité: nouveaux aspects théoriques et expérimentaux". Ph.D. Thesis, Université Louis Pasteur, Strasbourg (1984)

TROUTON, F. T.: Proc. Roy. Soc., London, A 77 (1906)

TUNG, T. T., and LAURENCE, R. L.: Polym. Eng. Sci., *15*, 401 (1975)

TURNBULL, D., and COHEN, M.: J. Chem. Phys., *34*, 120 (1961)

UHLAND, E.: Rheol. Acta, *15*, 30 (1976)

UNKRUER, W.: Kunststoffe, *62*, 7 (1972)

VAN WIJNGAARDEN, H., DIJKSMAN, J. F., and WESSELING, P.: J. Non-Newt. Fluid Mech., *11*, 175 (1982)

VERGNES, B.: "Modélisation de l'écoulement d'un polymère dans une filière de cablerie", Thesis, Ecole des Mines de Paris (1979)

VERGNES, B.: "Calcul des écoulements de polymères fondus dans les filières d'extrusion", Ph. D. Thesis, Université de Nice (1985)

VERGNES, B.: Matériaux et Techniques, 187 (June–July 1981)

VERGNES, B., SAILLARD, P., and AGASSANT, J. F.: IX International Congress on Rheology, Acapulco, Mexico (1984)

VERGNES, B., SAILLARD, P., and AGASSANT, J. F.: Polym. Eng. Sci., *24*, 980 (1984)

VERGNES, B., SAILLARD, P., and PLANTAMURA. B.: Kunststoffe, *11*, 752 (1980)

VERGNES, B., WEY, E., and AGASSANT, J. F.: Rev. Gal. des Caout. et Plast., *633*, 81 (1983)

VERGNES, B., and AGASSANT, J. F.: Advances in Polymer Technology, *6*, 441 (1986)

VINCENT, M.: "Etude de l'orientation de fibres de verre courtes lors de la mise en oeuvre de thermoplastiques chargés", Ph.D. Thesis, Ecole des Mines de Paris (1984)

VINOGRADOV, G. V., and MANIN, V. N.: Kolloid $z - z$. Polym., *201*, 93 (1965)

VINOGRADOV, G. V., INSAROVA, N. I., BOIKO, B. B., and BORISENKOVA, K. E.: Polym. Eng. Sci., *12*, 232 (1972)

VIRIYAYUTHAKORN, M., and KASSAHUN, B.: 42th ANTEC (1984)

VIRIYAYUTHAKORN, M. and CASWELL, B.: J. Non-Newt. Fluid Mech., *6*, 245 (1980)

VLACHOPOULOS, J., and ALAM, M.: Polym. Eng. Sci., *12*, 184 (1972)

VLACHOPOULOS, J., and KEUNG, C. K. I.: AIChE J., *18*, 1272 (1972)

WAGNER, M. H.: Rheol. Acta, *15*, 136 (1976)

WALES, J. L. S.: Technical Report TNO no. P 748 B, "A Collaborative Study of Capillary Flow of a Highly Lubricated Unplasticized Polyvinylchloride" (1975)

WALTERS, K.: "Rheometry", Chapman and Hall, London (1975)

WALTERS, K.: "Developments in Non-Newtonian Fluid Mechanics – A Personal View". IUTAM Symposium on Non-Newtonian Fluid Mechanics, Louvain-la-Neuve, Belgium (1978)

WEILL, A.: "Oscillation de relaxation du polyéthylène de haute densité et défauts d'extrusion", Ph.D. Thesis, Université Louis Pasteur, Strasbourg (1978)

WEILL, A.: Rheol. Acta, *19*, 623 (1980)

WEISSENBERG, K.: Proc. 1st Int. Rheology Cong. (1948)

WEY, E.: "Etude de la plastification des polymères thermoplastiques en injection", Ph.D. Thesis, Ecole des Mines de Paris (1984)

WHITE, J. L., and METZNER, A. B.: J. Appl. Polym. Sci., *7*, 1867 (1963)

WHITE, J. L.: J. Appl. Polym. Sci., *8*, 2339 (1964)

WHITE, J. L.: Polym. Eng. Sci., *15*, 44 (1975)

WILLIAMS, G., and LORD, H. A.: Polym. Eng. Sci., *15*, 553 (1975)

WINTER, H. H.: Polym. Eng. Sci., *15*, 84 and 460 (1975)

WINTER, H. H.: Adv. Heat Transf., *13*, 205 (1977)

WINTER, H. H. and FISHER, E.: Polym. Eng. Sci., *21*, 366 (1981)

WU. P. C., HUANG, C. F., and GOGOS, C. G.: Polym. Eng. Sci., *14*, 223 (1974)

ZAVADSKY, E., and KARNIS, J.: IXth Int. Congr. Rheology, Acapulco, Mexico (1984)

ZEICHNER, G. R.: "Spinnability of Viscoelastic Fluids", M.Sc. Thesis, University of Delaware (1973)

ZIABICKI, A., and KEDZIERSKA, K.: Kolloid Z., *171*, 111 (1961)

ZIMM, B.: J. Chem. Phys., *24*, 2, 269 (1956)

Subject Index